Petroleum Reservoir Management

Petroleum Reservoir Management

Considerations and Practices

Ashok K. Pathak

CRC Press is an imprint of the
Taylor & Francis Group, an **informa** business

First edition published 2022
by CRC Press
6000 Broken Sound Parkway NW, Suite 300, Boca Raton, FL 33487-2742

and by CRC Press
2 Park Square, Milton Park, Abingdon, Oxon, OX14 4RN

CRC Press is an imprint of Taylor & Francis Group, LLC

ISBN: 978-0-367-51922-3 (hbk)
ISBN: 978-0-367-51927-8 (pbk)
ISBN: 978-1-003-05570-9 (ebk)

Typeset in Palatino
by MPS Limited, Dehradun

This book is dedicated to

- *My alma mater – the Indian School of Mines for awakening me to the world of petroleum*
- *My home country India, the very basis of who I am*
- *My second home Kuwait and its people for treating me as one of their own*
- *My family and friends for having accepted me with my shortcomings*

Contents

Foreword

प्रो. राजीव शेखर
निदेशक

Prof. Rajiv Shekhar
Director

भारतीय प्रौद्योगिकी संस्थान (भारतीय खनि विद्यापीठ), धनबाद
धनबाद-826004, झारखण्ड, भारत

Indian Institute of Technology (Indian School of Mines), Dhanbad
Dhanbad-826004, Jharkhand, India

Foreword

Energy is as important for human beings as food and water. Proper management and planning of energy resources are crucial for sustaining the national growth and development. In this context, I am pleased to see a firsthand account of the issues and challenges faced by the upstream segment of the petroleum industry. It unfolds with an eye-catching visual of *Kalpa Vriksha* on the cover, symbolizing the significance of petroleum for the modern world. This picture is worth a thousand words and presents a holistic perspective on the development of petroleum resources.

In his book "Petroleum Reservoir Management Considerations and Practices," the author combines his analytical skills with creative ability and experience to address complex technological, financial, and management issues. In the same inimitable style, he establishes a rapport with his readers to generate a lasting interest in the intricate E&P industry matters.

Coming from an alumnus of the Indian School of Mines who spent over four decades in the upstream sector of the petroleum industry, I hope his book will serve as a bridge between academia and industry.

I congratulate him on this spectacular effort.

(Prof. Rajiv Shekhar)

1926 से राष्ट्र की सेवा में | In the service of Nation since 1926

Phones : +91-326-2296601 • Telefax : +91-326-2296603 • e-mail : director@iitism.ac.in • Web: www.iitism.ac.in

Foreword

N.M. Borah
(Former) Chairman & Managing Director,
Oil India Limited
(Former) Technical Member (Petroleum and Natural Gas),
Appellate Tribunal for Electricity

FOREWORD

The science and engineering of petroleum are complex and full of glorious uncertainties. The management of oil assets is challenging, inter alia, due to their technical complexity and remote location below the earth's surface. Still, many other factors such as geopolitical, social, economic, environmental, and regulatory issues add to the difficulties but are seldom discussed in depth in textbooks. Here, I see a book that takes the reader through an oil reservoir's arduous life cycle journey, cutting across the whole range of inter-related issues and considerations.

'Petroleum Reservoir Management Consideration and Practices' is a book highlighting the need for holistic integration in application of geosciences to optimize recovery and production rate from oilfields. The individual domain specialists in multi-disciplinary teams make their contributions to maximize objective functions as and when required. Being a music lover, I am tempted to compare the entire reservoir management effort with a musical performance where each member of the band knows when to pitch in on what scale and when to fade, for others to take over. It is this team performance that, if executed perfectly, makes the listening experience a lifetime pleasure both for singers and audiences. The crucial importance of the Team effort, which is true for the perfect blending of music from individual players in an orchestra, is equally valid for reservoir management. It demands seamless integration of diverse perspectives of the geologists, geophysicists, petrophysicists, geochemists, petroleum, and reservoir engineers and the contributions from petroleum economists.

I have had the pleasure of knowing the author as a dear friend and an esteemed batchmate of Petroleum Engineering. After leaving the Indian School of Mines, recently rechristened as IIT (ISM), Mr. Ashok Pathak and I followed different career paths with separate organizations. Still, spanning across geography and time, we maintained close professional and social links. Knowing him the way I do, the decision to share his wide experience, both in-country and overseas, with the fraternity via this book does not come as any surprise.

The author has deliberately kept the subject matter simple while maintaining an admirable distance from complex mathematical equations, which can be intimidating for some. My humble assessment is that for all you worthies, including students of petroleum engineering, fresh entrants to the profession, experienced industry personnel, and senior management alike, this book is a must in your reading bucket list.

(NM Borah)

Foreword

शशि शंकर
Shashi Shanker
अध्यक्ष एवं प्रबन्ध निदेशक
Chairman & Managing Director
ऑयल एण्ड नेचुरल गैस कॉरपोरेशन लि.
Oil and Natural Gas Corporation Ltd.

FOREWORD

Petroleum reservoir management is a dynamic process that aims to recognize the uncertainties in reservoir performance and seeks to mitigate the effects of these uncertainties through systematic application of multidisciplinary approach. It is essentially a strategy for applying multiple technologies in an optimal way to achieve maximum recovery. Requiring close interaction between technical, operating, and management groups, the modern reservoir management process involves goal setting, planning, implementation, monitoring, evaluation, and revising plans, taking reservoir operation and control as an integrated system, rather than as a set of disconnected functions.

This book ***"Petroleum Reservoir Management Considerations and Practices"*** by Shri Ashok K. Pathak, an alumnus of my Alma mater IIT (ISM), Dhanbad discusses various aspects of reservoir management and presents a wide spectrum of options and opportunities for efficient exploitation of the same. Starting from fundamental premises, the book covers the pertinent issues like, the reservoir lifecycles, the operating arrangement and reservoir development objectives, sound reservoir management etc. including policy frameworks for reservoir.

In my view, the depth of the issues discussed in the book shall empower young as well as seasoned geoscientists, engineers, and financial forecasters as well to systematically resolve uncertainties and reduce risks in the development of hydrocarbon resources.

The author's long association with various national and international oil companies puts him in a unique position to review and comment on different approaches and strategies for the exploitation of oil and gas reserves. He has incisively analyzed pitfalls with a positive frame of mind and suggested innovative changes along with mid-course corrections which could be implemented gainfully.

This book reflects the huge work experience of Shri Pathak and could be a valuable resource for all practicing oil industry professionals.

I wish him good luck and success in his effort.

(Shashi Shanker)

दीनदयाल ऊर्जा भवन, 5, नेल्सन मंडेला मार्ग, वसंत कुंज, नई दिल्ली -110 070 (भारत)
Deendayal Urja Bhawan, 5, Nelson Mandela Marg, Vasant Kunj, New Delhi -110070 (India)
दूरभाष (Tel) + 91 11 26129001 / 2612 9011, फैक्स (Fax): + 91 11 26129021, 2612 9041, ई-मेल (e-mail): cmd@ongc.co.in
वेबसाइट (Website) : ongcindia.com, CIN : L74899DL1993GO1054155

Foreword

FOREWORD

E&P operations have come a long way since the drilling of the first 70 feet deep well by a retired railway conductor, Colonel Edwin L. Drake, in Pennsylvania in 1859. Today's oilfield operations are much more sophisticated and capable of extracting oil and gas from the depths several kilometers below the surface and beneath the sea-floor.

It will not be an overstatement to say that Petroleum, referred to as 'black gold,' has been the focus of the modern world's attention for many decades and will continue to occupy the same position for many years in the future. However, continually diminishing reserves and the escalating cost of production underline the ever-increasing importance of proper Reservoir Management. The range and scale of operations further add to the complexity of the mission. Today Petroleum Reservoir Management is not only about managing the capital, technology, operations, and processes; it must also deal with the expectations of the people, investors, and the government.

Ashok Pathak is a Reservoir Management Expert who I met when I was Manager - Fields Development for North Kuwait. His experience has been a source of inspiration and knowledge for us. His book 'Petroleum Reservoir Management Considerations and Practices' demonstrates his thought, leadership in science and art of managing oil & gas reservoirs. It provides a holistic understanding of the technical, business and regulatory dimensions needed to extract oil and gas in compliance with prevailing industry norms and ecosystems. For me, it is a signature of a veteran who has lived through the professional challenges experiencing the resultant agonies and ecstasies. I am sure, his book can prove a handy reference not only for geoscience and engineering professionals, but also for the leaders who care to instill a sense of excellence in the upstream sector of the industry.

My sincere greetings and congratulations on the publication of the book!

Bader Ahmed Al Munaifi
Dy. Chief Executive Officer – South & East Kuwait

On the Cover

Petroleum Kalpa Vriksha

Petroleum is the new *Kalpa Vriksha* of Indian mythology for the modern world. It has the magical powers to fulfill every wish of the present civilization. However, it is a non-renewable source of energy, and its occurrence is sporadic. Mother Nature has granted this valuable endowment to a few selected countries. Countries lucky to have this resource must be careful in its consumption because it will not last forever. The countries that do not have it or do not have enough must pay a fortune to buy it and its products.

Ownership and control of petroleum resources are central to the philosophy of their exploitation. Reservoir management is the interconnecting link between the resource and the resulting benefits. It not only provides a strong anchor to the large and heavy frame of the tree, but it also ensures the supply of nutrients to the lush, leafy, and fruity tops symbolized by rate, recovery, and returns.

With the advent of NOCs, the role of reservoir management has increased in scope, scale, and responsibility. Operating companies now have to go beyond maximizing their cash flow. They must now ensure that an oil and gas deposit is exploited optimally within the legal, financial, social, and environmental framework using industry best practices. It is a complex mission that requires deep insights into the reservoir management considerations and practices to reap the benefits of the petroleum *Kalpa Vriksha*.

Preface

In today's energy-hungry world, petroleum is a prime source of energy essential for a nation's survival and growth. It is a strategic commodity that powers the global economy and modern civilization. Its use is no more limited to illuminating the households of the masses and firing up their cooking stoves; it fulfills a common necessity of traveling at high speed on the ground and in the air. Its products and byproducts define our life and lifestyles.

The strategic significance of petroleum is evident from its position on the center stage in the middle of sovereign nations' energy security, international relations, and economic policies. Mankind's constantly growing need for energy and its dependence on petroleum to meet this requirement on a sustained basis causes energy security concerns among oil and gas importing nations. Even a notional risk to supplies of petroleum and its products has the potential to spark political tensions and conflicts.

A resource of this importance requires proper management across the entire supply chain from the upstream to the downstream end. Reservoir management is not a core subject of petroleum engineering curricula; it is a value system based on the operating philosophy and business practices of the E&P companies. It shapes its contours based on the considerations of ownership and style of control and management.

The objective of reservoir management is to maximize three Rs: Rate, Recovery, and Returns. Maximizing all three Rs at the same time is an exercise in conflict management. Therefore, operating companies focus on maximizing an "R" that is the primary focus of their mandate. The success of any business enterprise is measured by the benefits of wealth and value creation. Each enterprise must decide how much emphasis it wants to place on profit and how much on value. Some organizations can choose to take advantage of the oil rate, while others can capitalize on oil recovery or overall returns in terms of monetary or other benefits. This approach defines the nature and character of E&P operating companies and is a differentiating factor between IOC, NOC, and POC.[1]

NOCs are a corporate vehicle for the state's participation in petroleum E&P ventures. However, under the influence of their political leadership, they can be conveniently used to pursue the national and international agenda besides their core commercial activities. Today's oil companies, both NOCs and IOCs, cannot limit their undertakings to oil/gas production, reserve growth, and cost reduction. They have to ensure that their operations comply with local and international regulations; their products also get the market price they deserve.

Large oil and gas projects are not a "one-man show." Their success depends on the mutually beneficial and well-rounded relationships among various companies, functional teams, and service providers of diverse experience and skill-sets. Such relationships built on trust and shared understanding of common goals must promote collaboration to enhance performance and project value. In practice, it is easier said than done. Commercial relationships are fragile and can be adversely affected by a lack of communication and the slightest mistrust. Contractors are often criticized for over-promise and under-delivery and the clients for harsh contract terms, unilateral decision-making, and delayed payments. There is also a tendency to shift the project risks and delays towards each other. Good man-management is the key to a successful E&P project. In a climate of risks and uncertainties, appreciation of genuine limitations and constraints can create an environment where people are willing to consider company interests above their own.

Petroleum literature is full of excellent textbooks, monographs, and technical papers, providing significant reservoir management material. However, most of this material is either focused on a specific module or follows a standard template of contents dealing with Decline Rate, Material Balance Equation, waterflooding, etc. The result of this is that the core narrative of a subject as crucial as reservoir management gets blindfolded to its national and international dimensions, limiting its scope and significance. It does not present an exciting picture to the young and ambitious industry professionals looking for higher goals and motivations in their careers. Without understanding the connection and impact of global factors on domestic and foreign oil and gas production, they cannot comprehend its complete influence on the business.

Unfortunately, there is no single book available today, which provides a 360° view of petroleum reservoir management issues and considerations. Lack of a full and seamless perspective limits the understanding of issues and affects decision-making ability. In the absence of the basic reference material, undergraduate students, industry interns, and staff members cannot connect the dots and draw a "big picture." The situation is further complicated because there are no set and simple answers to most reservoir management questions such as reservoir health, well quality, project completion schedule, oil/gas recovery, and so forth.

National and international oil and gas magazines are doing a tremendous service in this respect because they aim to convey the message without going after semantics. They build a storyline relating the details of the technical, logistic, social, environmental, and geopolitical challenges involved in the pursuit of oil and gas in the business context and highlight the strategies used to overcome barriers. These success stories capture the fancy of younger generations who are keen to write a new script for tomorrow's corporations.

This book's outline is shaped based on my discussions with young friends and senior colleagues in various formal and informal settings. Persistent queries of fresh graduates, working geologists, engineers, management members, and most of all, my self-questioning have strengthened my resolve to present a fuller landscape of reservoir management aspects in a simple storytelling format. It finds its genesis in an imaginary father walking with his teenage kid sharing his life experiences. This setting allows a simple discussion of issues, considerations, and practices. It has no room for complex mathematical equations and intricate details that are all over the petroleum literature. Professionals and executives can pick a subject or an issue from here and dive deep to find answers to specific questions if necessary.

It is important to mention that this is not a core petroleum engineering book; this book is about reservoir management. It uses the knowledge and concepts of petroleum engineering to set and achieve reservoir management objectives. Therefore, its scope extends beyond the typical subsurface/surface issues to include equally important subjects of crude oil markets, financial institutions, and oil pricing mechanisms, criteria, and strategies. A well-orchestrated business plan around these considerations is necessary to realize the full value of the investments made into the E&P operations.

The book discusses the methods and strategies to bridge the gaps in planning and execution, which play a decisive role in a project's economics and value creation potential. It puts together a broad reservoir management framework and its governance model to ensure that company policies and guidelines are followed by all concerned. Operating companies and host governments can tweak the framework suitably to fit their requirements.

The industry has always been aware of the risks involved in the oil and gas E&P activities and worked hard to minimize them by streamlining and standardizing its goods, products, and services. It has constantly focused on using technology to extend its reach to inaccessible reserves and promote operational efficiency, safety, and environmental protection. Standards of materials and goods developed through various codes of practice are already an integral part of operations in the oil and gas industry. Embracing recommended/best practices and the ISO-family of standards has significantly improved overall asset integrity and safety compliance. Ongoing efforts to improve people's quality, processes, and decision-making will ultimately shape the corporate work culture to overcome the challenges of redundant inventory, operational downtime, and cost overruns.

Performance measurement is an essential management tool to evaluate change resulting from an action's outcome. Most companies use qualitative or quantitative measures to assess individuals, teams, and the company's performance. Qualitative assessment is based on impressions, attitudes, and behaviors. It is, therefore, prone to bias and dispute. On the other hand, quantitative measures are more scientific because they use tests, evidence, and benchmarks and are supported by numbers and statistics. This approach provides a solid rationale for decision-making since it leaves no room for personal latitude.

In many oil companies, quality of well completion, drilling efficiency, reservoir health assessment, major projects, training & development, employee output, and so forth have not been properly nailed down by quantitative measures. Devising standards/performance measures for such activities is an important and ingenious task of the reservoir management execution program.

I admit that the book *Petroleum Reservoir Management Considerations and Practices* should ideally include natural gas in its fold. However, in my opinion, crude oil and natural gas, despite being from the same family, are two vastly different species that need separate space and time for discussion. The same argument applies to unconventional petroleum resources. I wish and hope that many other authors, more knowledgeable and experienced in those subjects, will come forward to fill in those spaces.

Ashok K. Pathak

Note

1 IOC: International Oil Company, NOC: National Oil Company, POC: Private Oil Company

Acknowledgments

This book is a compilation of knowledge that I acquired in my years of working with the Oil and Natural Gas Corporation Ltd. (ONGC) and Kuwait Oil Company (KOC). I want to thank these organizations for providing me with the best exposure and opportunities to learn and share knowledge without any barriers.

The idea of writing this book took root in the vacuum caused by my retirement. It was encouraged by two industry experts, J.M. Joshi of ONGC and Rupert R. Weber of Shell. God's ways are strange. Unfortunately, Mr. Joshi left for his heavenly abode the day I sat down to write this acknowledgment.

This book was completed in times of unprecedented pain and suffering caused by Covid-19. It was difficult for sensitive people to maintain their composure when electronic and print media poured stories of human devastation caused by a deadly virus. This unwelcome situation, however, provided the solitude I needed to focus on and complete the task.

I feel for my family members and friends scattered in distant countries who feigned bravado as they talked to us on the phone. The increased frequency of calls was a telltale sign of their nervousness and worry for our welfare. Ironically the pandemic has brought people closer despite the new social order of maintaining a physical distance. While on this subject, I take this opportunity to pay tributes to those who risked their lives to save others. Also, my heartfelt sympathies to those who lost their dear ones or could not be with them in their last moment.

Finally, I am personally satisfied that about 2 years of hard work has come to an end. However, it would be much more gratifying if the year of completion of my book 2020 were without a reference to the global contagion.

Author

Ashok K. Pathak is a petroleum engineer by qualification with more than four decades of professional experience as a reservoir engineer in various positions in Oil and Natural Gas Corporation Ltd. (ONGC) and Kuwait Oil Company (KOC). Reservoir development and management are his core areas of expertise where he built a reputation for himself to provide simple and practical solutions. In his long association with national and international oil companies/consultants, he provided technical leadership and successfully navigated various complex reservoir studies and projects for value creation. Several new path-breaking projects were executed under his stewardship at times when old mindsets were proving a barrier to enhance oil production, reserve growth, and cost optimization.

Building and developing young people on the way has been one of Ashok Pathak's passions. Transiting through his roles of engineer, manager, specialist, and adviser, he played a cameo to coach and mentor his colleagues with mantras of self-discipline, commitment, and excellence. He believes that assurance is an important ingredient of performance, and success comes not just by chance. It is the result of a structured and organized process intertwined with hard work. He endorses systems, processes, and procedures as long as they promote transparency, efficiency, quality, and cost savings.

He has a master's and a bachelor's degree in petroleum engineering from the Indian School of Mines.

1

Petroleum Reservoir Lifecycle

1.1 Introduction

Metaphorically, oilfields are comparable to humans. They are born when discovered and die when abandoned due to depletion or economic reasons. They may have the same physiology, yet they are so much different from each other in their physics and chemistry! Analogies work only to a limited extent.

Petroleum is a Latin word for "rock oil" formed from the remains of plants, animals, and ancient marine organisms. It is a fossil fuel like coal that can naturally occur in colorless to very dark brown/black color in gaseous, liquid, and semi-solid phases such as natural gas, condensate, crude oil, and tar. Primarily composed of an organic compound of two elements, hydrogen, and carbon, petroleum is also known as a hydrocarbon.

It is important to highlight the difference between a "Reservoir" and a "Field" before making extensive use of these terms in our discussion. A petroleum reservoir is an individual, separate, dynamically pressure-connected formation system characterized by three essential properties: porosity, permeability, and caprock. Porosity allows the rock formation to hold the oil and natural gas accumulation, and permeability allows its movement through the rock when pressure equilibrium is disturbed. The caprock is a tight or impermeable rock that constitutes the sealing mechanism to prevent the escape of hydrocarbons from the top under normal, natural conditions. In simple words, a field may have single or multiple reservoirs overlying each other, which may or not be in communication with each other.

1.2 Life Cycle

Like all living beings, an oil reservoir goes through the life cycle of exploration (embryonic), discovery (birth), delineation (childhood), growth (youth), decline (aging), and abandonment (termination). This progressive life cycle, as seen in Figure 1.1, will apply to each reservoir that an oilfield

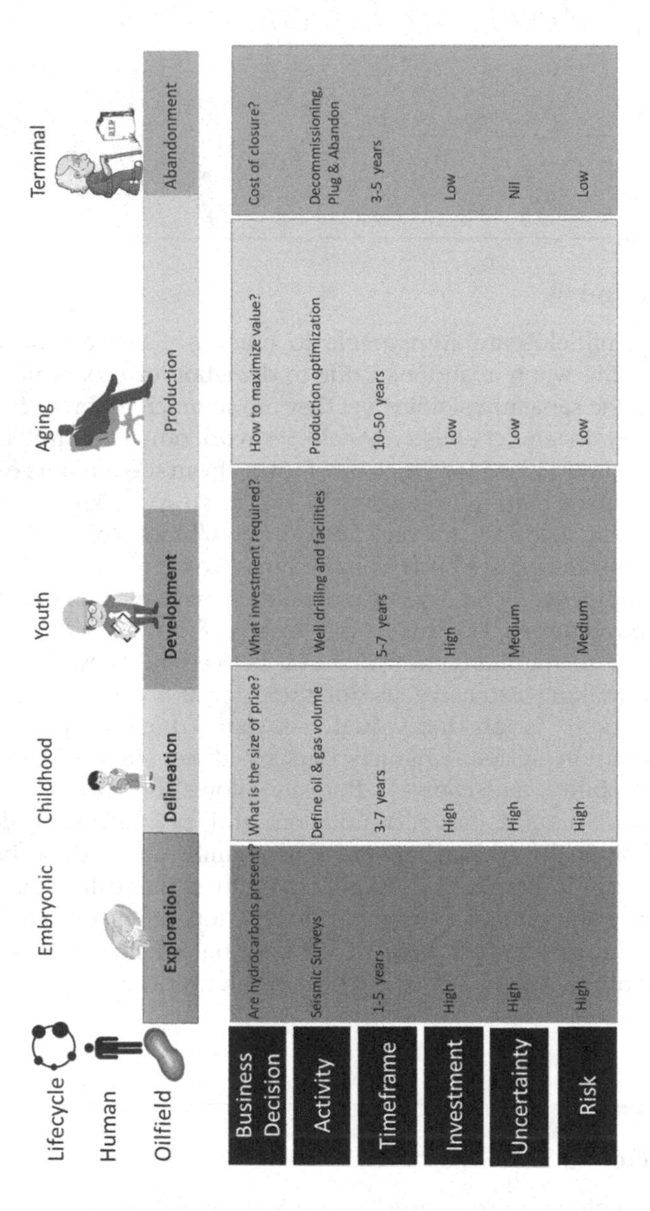

FIGURE 1.1
Symbolic Equivalence between Human and Oilfield Lifecycle.

might contain. Depending on the size, potential, commercial viability, and business interest, an operating company must devise a strategy to manage the entire reservoir lifecycle to maximize the desired benefits. Good reservoir management strategy is fundamental for higher productivity, economic gains, and extension of the field life.

Each of the oilfield phases, that is, exploration, appraisal, development, production, and abandonment, has its own set of unique challenges and characteristics. Most of the discussions presented in this book will cover the reservoir management aspects of the development and production phases, which form a large part of the reservoir lifecycle and generate most of the cash flow to keep operating companies interested in the business.

An oilfield over its lifecycle may be operated by a National Oil Company (NOC), International Oil Company (IOC), a Private Oil Company (POC), or a combination. The oilfield ownership and operating arrangements may change from time to time, calling for changes in the reservoir management strategy affecting the investment and the cash flow profiles based on the projected production rates and recoveries.

1.2.1 Exploration Phase

The first goal of the exploration phase is to examine the prospectivity of a petroleum province and rank geological plays and prospects to decide their priority. The most fundamental question for a company to pursue exploration work on a geological prospect is to ascertain the possibility of hydrocarbons. Exploration is a risky and costly business. E&P companies usually work for several years before spudding their first exploratory well. Their work involves applying geological and geophysical methods to establish the most favorable location for oil entrapment. Today, with the help of advanced computing systems, geophysicists can interpret the vast and complex 3-D volumes of seismic data relatively quickly and generate an image of the subsurface for locating the possible hydrocarbon accumulation. The integration and evaluation of geological and geophysical information by the subject matter experts prepare the ground for high-grading a prospect for drilling the first exploratory well.

An exploratory well is drilled to confirm the presence of oil/gas on a structure. This process proves or disproves the geological hypothesis that is used to drill exploratory wells. Coring, well logging, and well testing methods permit collecting additional information about the hydrocarbon-bearing formation. Coring is a direct method to obtain a sample of rock from the wellbore. Several plugs can be extracted from this rock to perform Routine Core Analysis (RCA) and Special Core Analysis (SCAL). Well-logs provide quantitative estimates of pay thickness and fluid saturations for the geological formations. A well-test is conducted to measure the production rate of oil, gas, and water. Multiple choke test is an excellent method to determine the production potential of a well and optimize its flow rates. The success of the exploratory oil/gas well kicks off the field lifecycle as hypothetical resources are upgraded to proven estimates of hydrocarbons-in-place and reserves.

The exploration phase is usually characterized by high geological uncertainties and capital investments in seismic surveys and exploratory drilling. Exploratory areas have high uncertainty about the discovery and expose investments to many risks, including the risk of dry holes or uncommercial ventures. Drilling operations in some exploratory areas may be unsuccessful due to unexpected drilling conditions such as abnormal pressure and temperature, causing equipment failures and blowouts. The exploration risks cannot be eliminated but can be reduced significantly by adopting appropriate workflows and technological innovations.

1.2.2 Delineation Phase

The next most important questions for an E&P company, having discovered oil or gas, are about the discovery size and its commercial viability. Seismic data is undoubtedly beneficial to map the size of the subsurface structure. Still, the only way to prove whether this structure or what part contains oil is to drill delineation appraisal wells.

The delineation/appraisal phase aims to address these questions by drilling wells that help define the reservoir limit and understand the variation in well-productivity across the reservoir. Extensive analysis of data gathered from drilling, logging, coring, sampling, and testing reduces uncertainty in the size of the oil or gas reservoir, its properties, and production rates.

The drilling and completion program of delineation wells is carefully planned to minimize operational risks and maximize gains. Continuous improvement and optimization of this program aids in building a learning curve that is useful for assessing the drilling time, cost, and challenges associated with the development drilling program. Delineation wells are usually extensively tested to ascertain the areal/vertical extent, connectivity, and variations in fluid quality and production rates. Geological cross-sections and well-correlations supported by extended production test results and well-surveillance data provide vital insights into reservoir behavior, rate sustainability, and operating drive mechanisms. Proper analysis and integration of the static (initial or virgin reservoir conditions) and dynamic (after the start of production) data enable the Field Development team to build a reservoir model representing the reservoir architecture and flow behavior. The model is further validated and tested against the information acquired from new well drilling and by matching the production history at the well and reservoir level. It serves as an essential tool to make production forecast and evaluate possible development options.

This phase is characterized by high uncertainty and capital investment, significantly reduced by the successful drilling of delineation wells, data acquisition, and testing. By the end of this phase, the Field Development team is comfortable with a reasonable assessment of hydrocarbons' volume, production potential, and commerciality. The new reservoir model helps design a conceptual development plan (CDP) that includes an in-depth analysis of various development options. The most appropriate option can then be chosen based on the feasibility of application and estimates of rate, recovery, and revenue.

1.2.3 Development Phase

The development phase starts after the appraisal phase is over and before the onset of the production phase. This phase provides impetus to development activity wherein low-risk wells, according to the CDP, are high-graded for drilling. The new wells generate enormous subsurface and production data that validate the reservoir model configuration and provide assurance for performance forecasting. The development team is now more confident of its in-place estimates and reserves calculations. It can now design an evidence-based Field Development Plan (FDP) with long-term production forecasts to support the business strategy and goals. The FDP documents the following:

1. Objectives of the FDP
2. Review of the CDP
    - Status of existing locations
    - Reservoir performance analysis
    - Need for modification in the CDP or its extension

3. New data, its quality, relevance, and impact on the development planning
4. Possible development options and their evaluation based on the business objectives
5. The recommended option for reservoir development
    - New performance forecast, recovery factor, and reserves
    - Type, location, number, and scheduling of wells
    - Long-term oil, gas, and water production forecasts

6. Production handling and injection facility
    - Project goals and background relevant to understanding the design
    - Underlying assumptions about the need and schedule, design criteria, and considerations (such as capacity, energy consumption,

Health, Safety and Environment (HSE) standards/regulations, sustainability, etc.)
- Location of the facility/facilities
- Limiting conditions for system operations
- Schedule and budget limitations

7. Need and timing for artificial lift methods
8. Effluent Water Disposal (EWD) and management plans
9. Cost of the total plan
10. Data gaps, uncertainties, and risks
11. HSE issues and concerns
12. Human resources requirement

Depending on the field's size and resources at the company's disposal, the development phase can last 5 to 7 years, or even longer. Sometimes, the need for a pressure maintenance program may not be imminent immediately. It can be dealt with by a revised FDP at the appropriate time.

This phase is characterized by high capital investment in drilling development wells and the simultaneous construction of surface facilities. Operating companies usually start production from the onshore fields using a temporary production facility and *ad hoc* transportation arrangement. Ideally speaking, this phase aims at building a subsurface and surface production capacity that can deliver the plateau rate of oil/gas production without interruption as the field transits from the development to the production phase. There is a significant reduction in uncertainty by this time and, therefore, a lower risk to investments.

1.2.4 Production Phase

The production phase marks the first commercial production of oil and gas from the subsurface via production wells drilled and connected to the production facility for processing. It also marks a turning point in an oil/gas field's lifecycle when the investment curve shifts from expenditure to earnings.

Depending on the field's size, it can last for 10 to 50 years and end up yielding significant profits after the recovery of all the investments made over the reservoir lifecycle. The production rate from a reservoir is the sum of production rates of individual wells. It usually manifests three distinct periods identified as: buildup, plateau, and decline. During the buildup period, oil/gas wells come online, and production builds up progressively. The production buildup depends on the availability and allocation of resources such as drilling rigs and manpower; the higher the resources shorter the buildup period, and vice versa. The plateau period is characterized by a constant rate of production from the reservoir. Production rate is controlled by well-productivity, well-life, infill drilling, well-stimulation, and well-intervention/maintenance activities. A continuous decline in oil rate after the plateau period indicates either the loss of reservoir energy or the waning production potential of oil/gas wells (Figure 1.2). Monitoring the performance of individual wells is, therefore, the key to maintain production from a reservoir. It calls for concerted efforts to measure oil, gas, water production rates of individual wells, bottom-hole pressures, gas oil ratio (GOR), and water cuts to back up well-diagnostics.

At times, the flowline network and facilities can also act as a choke in the production system. Connecting several wells of varying production capacity and reservoir pressures through a single shared flowline to the production facility can result in a net loss of production. Similarly, facilities may have conditions that limit well performance. A good well, flowline, and facility management is the key to excellent reservoir management.

Well-workovers, stimulation, artificial lift, new well-drilling, and so forth, are carried out based on the cost–benefit analysis and help extend the production phase and field's overall life.

Assuming that the model can correctly predict the reservoir performance, extraction of oil/gas is maintained in line with the plan to achieve the projected recovery and profits. Low capital expenditure (CAPEX), high operating expenditure (OPEX), and high cash flow are the characteristic features of the production phase, as most revenue is generated in this phase. New infill wells must be drilled to compensate for the production loss caused by wells that go out of production because either mobile oil is flushed out or moved away from the production wells. Surveillance-based operational activities to support well diagnostics and interventions constitute a large portion of the OPEX budget.

1.2.5 Abandonment or Decommissioning Phase

Abandonment is the terminal phase in the lifetime of an oilfield characterized by an economic limit. However, the term "abandonment" is equally applicable in the context of production, injection, and disposal wells. Decommissioning is a broader and more inclusive term than abandonment. It includes the plugging and abandonment (P&A) activities of wells besides dismantling of surface facilities to restore the oilfield area to its original state.

The economic limit is described by the low production rate when it is no longer economical to continue production because operating costs are higher than the net revenue. It implies that no "commercial quantity" of recoverable oil is left behind in an oilfield at the time of abandonment. The term commercial quantity is determined based on the economic limit, whereby OPEX exceeds the earnings from a barrel of oil. The terms, economic limit, and commercial quantity referred to here are specific to a time, field, oil price, and company (due to accounting practices) used to calculate OPEX. Under these

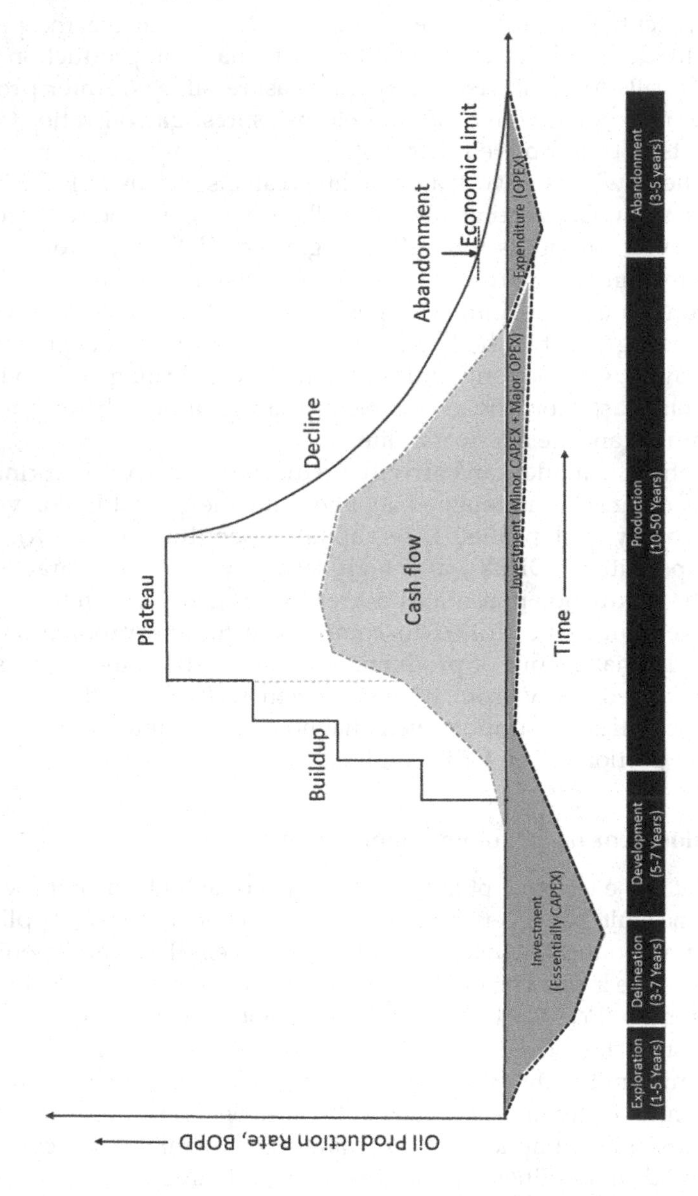

FIGURE 1.2
Significance of Production in the Lifecycle of an Oil and Gas Reservoir.

conditions, the operator may decide to close down the operations and surrender the acreage after plugging and abandoning the wells.

Modern abandonment operations are quite sophisticated and elaborate. Any spills, blowouts, leakage, seepage, or interference of hydrocarbons with land and subsurface aquifers can cause severe damage to air, soil, and the aquatic environment in the neighborhood. Therefore, operators take no chances and temporarily abandon a well for various reasons in compliance with regulatory provisions.

- Successful exploratory wells, both onshore or offshore, maybe temporarily abandoned by placing downhole plugs until production and transportation facilities are constructed.
- Production wells with remaining reserves may also be temporarily plugged.
- Low oil prices may force the postponement of secondary recovery or EOR operations for several years.
- Sub-economic producers with no potential for reserves may have alternate use later, such as water injection or EOR.

The temporary abandonment procedure is simple and allows the wells to be brought back online quickly when required. The basic requirements for permanent abandonment of offshore and onshore wells are fundamentally the same but more rigorous. P&A operations ensure that all freshwater and near freshwater zones apart from future possible commercial zones are adequately isolated and protected. The process requires removing the tubing from the well and plugging of hydrocarbon-bearing sections of the well with cement to prevent any potential leak/communication with the aquifers. All surface equipment is removed, and the casing pipe is cut below the surface before relinquishing the field. A sturdy steel cap is then welded on top of the well.

Decommissioning onshore and offshore facilities involve dismantling and recycling/reusing the equipment and material in other locations after or without refurbishment. The onshore decommissioning process is relatively straightforward and cheaper than offshore. In the US and Canada, the state governments regulate onshore well decommissioning.[1] They ensure that operating companies have enough funds to fulfill this obligation when they surrender the lease.

The decommissioning of offshore fields includes topsides and substructure. Topside is the entire structure of the platform that is visible above the waterline. The substructure comprises the parts and equipment that are below the water. It may be wholly or partially removed as obligatory under the law/contract. Both, the topsides and substructure, are removed and shipped back to shore for recycling after refurbishment, reuse elsewhere, or disposal as waste. Subsea pipelines or power cables can remain in place unless they interfere with marine or fishing operations.

Out of more than 65,000 known oil and gas fields, about 94% of oil is concentrated in fewer than 1500 fields,[2] most of which are still at some stage of exploitation. Extensive web browsing revealed that there is hardly any entirely abandoned oil/gas field in the world except for Mesa within the city limits of Santa Barbara, California, in the US.

In contrast, more than 27,000 wells have been abandoned in the world so far. The development and application of new technologies have significantly improved oil recovery and reduced the cost of production. The growth in the strategic importance of petroleum, coupled with its high prices (Figure 1.3), has contributed to many mature fields' extended economic life.

1.3 Phasing of Development

Petroleum reservoir management objectives require a good understanding of the recovery mechanisms all along their lifecycle. Recovery mechanisms constitute a sound basis for phasing the reservoir development and management of oil and gas reservoirs. Each recovery mechanism has its own characteristics (Figure 1.4) that provide useful insights for planning.

Recovery Discriminator (RD) is a term coined to explain the fundamental difference between primary, secondary, and tertiary recovery mechanisms. It is a key factor responsible for incremental oil recovery from a specific recovery mechanism.

The primary recovery mechanism takes advantage of the energy initially available in the system. RD for a typical depletion drive mechanism is ΔP or the pressure drop across the reservoir. Wells flow naturally to the surface or by artificial lift methods under the pressure drop applied by surface chokes or pumps. Production uncertainty and costs are low.

The secondary recovery mechanism utilizes the supplementary energy that is provided to the reservoir by water or gasflood. The RD for such a recovery process is the mobility ratio (M) because it controls the sweep and is a critical success factor for the secondary recovery mechanism. The secondary recovery factor has a higher uncertainty than the primary recovery factor due to the constant migration of in-situ fluids under the influence of water or gasfloods. The production cost from secondary recovery methods is also higher due to the requirement of additional wells and facilities for water/gas injection.

The tertiary/enhanced oil recovery (EOR) processes are facilitated by thermal, chemical, or gas injection methods. They are associated with an increase in the capillary number (N_C) described as the RD for EOR

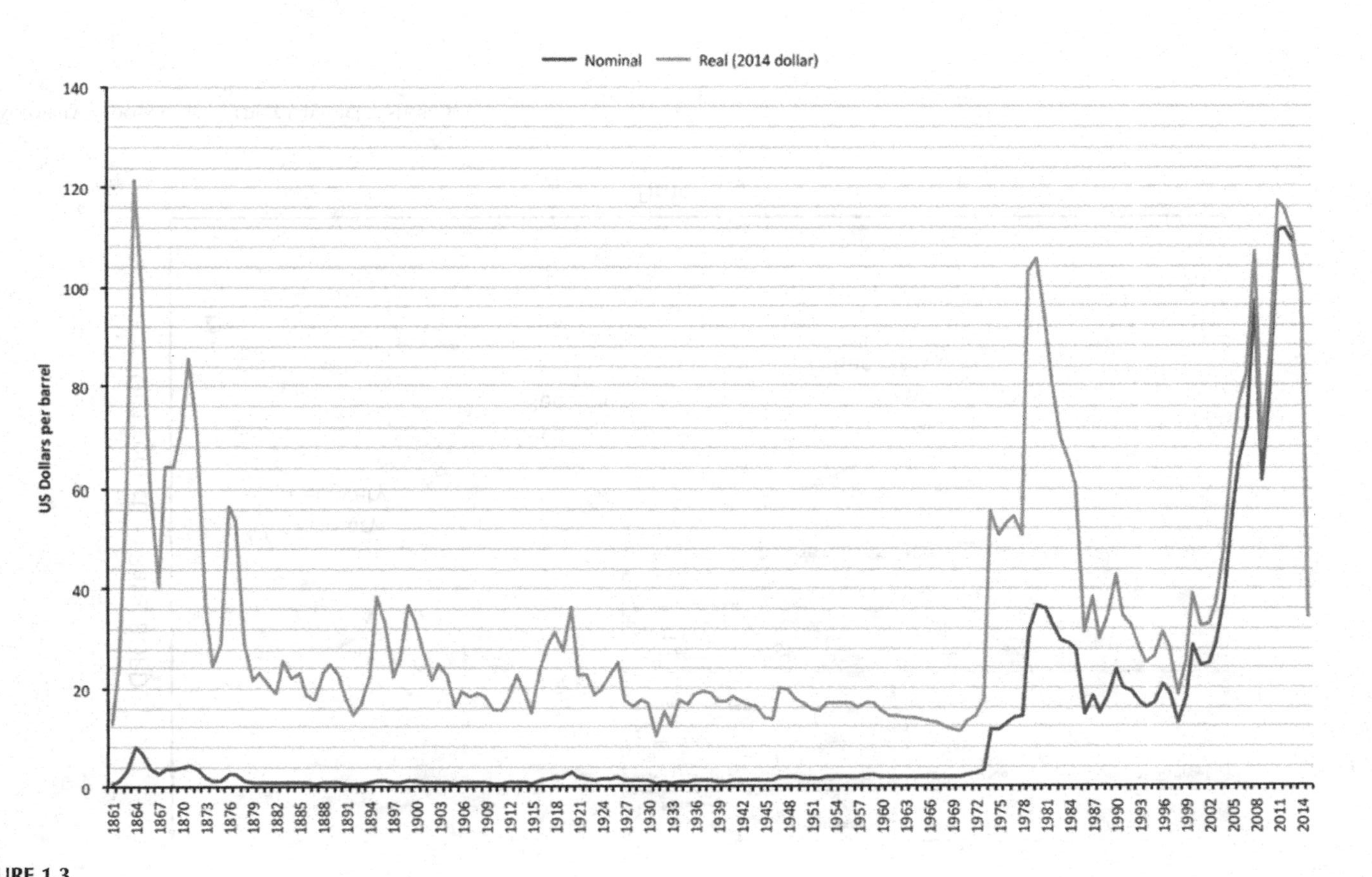

FIGURE 1.3
Crude Oil Prices Since 1861[3].

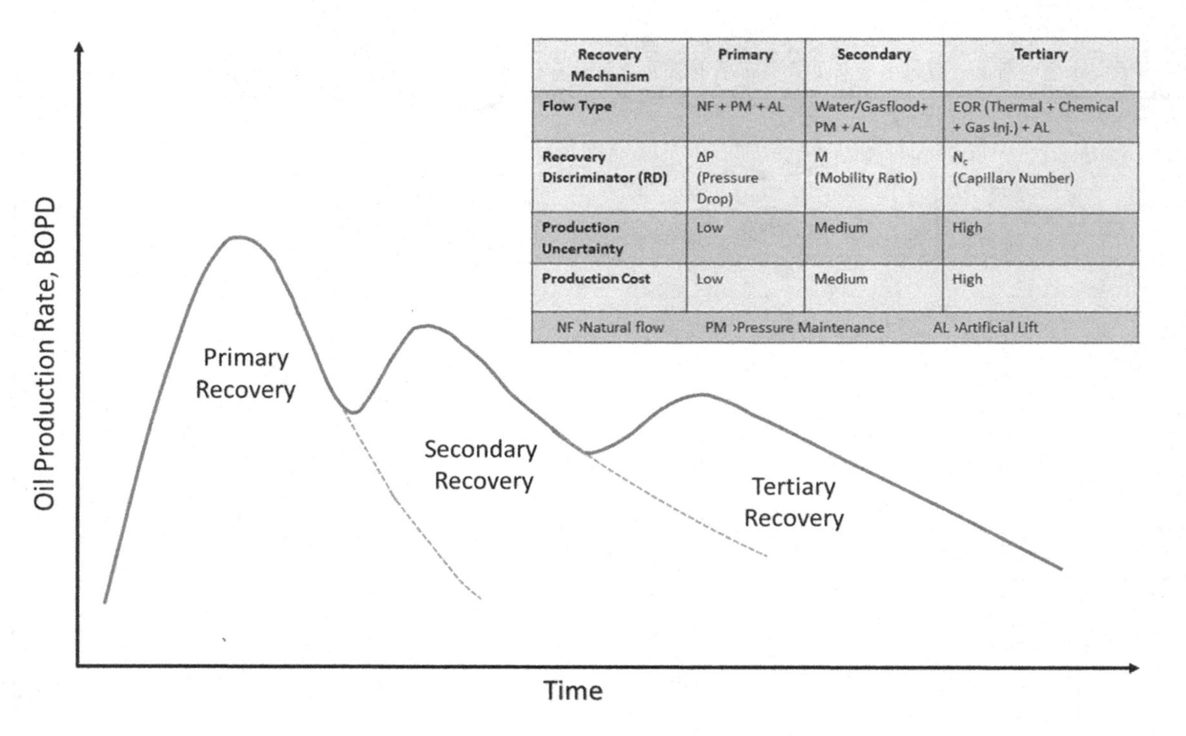

Recovery Mechanism	Primary	Secondary	Tertiary
Flow Type	NF + PM + AL	Water/Gasflood+ PM + AL	EOR (Thermal + Chemical + Gas Inj.) + AL
Recovery Discriminator (RD)	ΔP (Pressure Drop)	M (Mobility Ratio)	N_c (Capillary Number)
Production Uncertainty	Low	Medium	High
Production Cost	Low	Medium	High
NF ›Natural flow	PM ›Pressure Maintenance		AL ›Artificial Lift

FIGURE 1.4
Oil Recovery Processes and Their Characteristics.

processes. EOR methods are knowledge and capital intensive, need unique completion material, equipment, and facilities for handling injection and production. Production costs and uncertainties are high for these processes.

The operating company needs to deploy significant resources towards data acquisition, analytics, feasibility studies, pilot tests, development planning, well drilling, and facility creation for executing the development plan.

1.3.1 Primary Recovery

Oil reservoirs need the energy to drive oil out to the surface. This energy comes from the natural gas dissolved in oil or from the associated gas cap overlying the oil zone or from an aquifer (underground source of water) that may underlie the oil column or on its edge.

Reservoir rock, fluids, and aquifer are highly compressed at reservoir conditions due to massive overburden and gravitational forces acting over them. Gas being highly compressible compared to the formation and liquids, it is the most impactful contributor to reservoir energy whether as dissolved gas or free gas phase in the form of gas cap. A summary of the compressibility values of formation rock and fluids is presented in Table 1.1.

Gravitational forces play an important role in oil recovery. These forces are dominant when the reservoir withdrawal rate is low and vice versa. Oil and gas production from the reservoir causes a drop in reservoir pressure inducing water movement from the aquifer or gas from the gas cap if there is one. Depending on the size of the aquifer and gas cap, this phenomenon may partially or fully supplement the reservoir energy depleted by oil and gas production.

Primary oil recovery is the first phase of oil and gas production, which utilizes the inherent reservoir energy to bring the underground resource to the surface by successfully drilling and completing a set of wells. It may deploy artificial lift methods such as Gas Lift and Electrical Submersible

TABLE 1.1

Reservoir Rock and Fluid Compressibility Values[4]

Reservoir Rock and Fluid	Compressibility, psi^{-1}	
	Range	Typical Value
Consolidated Rock	2×10^{-6} to 7×10^{-6}	3×10^{-6}
Oil	5×10^{-6} to 100×10^{-6}	10×10^{-6}
Water	2×10^{-6} to 4×10^{-6}	3×10^{-6}
Gas	At 1000 psi	$1{,}000 \times 10^{-6}$
	At 5000 psi	100×10^{-6}

Pumps (ESPs). The primary recovery phase continues until the reservoir pressure is inadequate to produce a commercial quantity of oil.

Reservoirs with large gas caps and large aquifers may not require additional pressure support to displace oil from the pore spaces. However, in some cases, the natural drive mechanisms may not be strong enough to support high withdrawal rates imposed on the reservoirs manifested by a sharp decline in reservoir pressure and production rate. It is common to inject water or gas into such reservoirs to maintain reservoir pressure above the bubble point. It arrests the declining reservoir pressure, conserves the reservoir energy, sustains the production rate, and increases oil recovery. Pressure maintenance methods by water or gas injection are considered an extension of the primary oil recovery process.

The efficiency of the primary recovery process can vary significantly depending on the reservoir as follows:

- rock type (sandstone, carbonate)
- rock quality (Ø, k, NTG, heterogeneity, faulting/compartmentalization)
- fluid saturations (S_o, S_g, S_w) and
- fluid quality defined by pressure, volume, and temperature (PVT) properties

Under natural depletion, the production from subsurface to the surface continues as long as the pressure drawdown applied on the wells is higher than the sum of viscous forces in the formation, hydrostatic head of fluids, and frictional forces in the tubing. The use of artificial lift methods might be incumbent if the pressure drop applied on a naturally flowing well is not sufficient to deliver a reasonable oil rate. The duration of primary recovery under depletion drive may be relatively short and limited by natural reservoir energy.

1.3.2 Secondary Recovery

An expanding aquifer or a gas cap initially present with the petroleum reservoir assists in pressure maintenance and supports the primary recovery. The secondary recovery methods rely on injecting extraneous water or gas to drive out oil trapped in the reservoir due to a lack of energy. As a thumb rule, these methods can double the oil recovery achieved by primary recovery methods. It is a practice to employ pressure maintenance methods to arrest the continual decline in production rate and reservoir pressure by injecting water or gas at or above bubble point. It is common to inject water to maintain reservoir pressure due to its ready availability, relative simplicity, and effectiveness in most oil reservoirs.

Petroleum engineering literature makes rather unselective use of the two terms—pressure maintenance and flooding—due to their design and operation similarities. It strengthens the general belief that there is no difference between the two. However, as highlighted by Morris Muskat, there is a subtle difference between these two terms. Pressure maintenance by water injection is resorted to when the reservoir pressure nears the bubble point and is not necessarily fully depleted. In other words, the pressure maintenance method is used during the primary recovery phase and not at the end of it. As mentioned in the previous section, it helps preserve the reservoir energy to improve production and oil recovery.

Injection at the original bubble point requires higher injection pressure with upfront expenditure and preparations. Waterflooding is a secondary recovery method employed at the end of primary recovery and is not necessarily a pressure maintenance method. It offers the advantages of injecting water at much lower reservoir pressure and free gas phase, improving waterflood efficiency. By this time, the operator is more confident about the project's technical and economic success based on the additional reservoir information available to execute the plan at lower risk and more cash available in hand. However, this method suffers from the disadvantage of loss of energy due to excessive gas production which may be counterproductive to the objectives of ultimate oil recovery.

Water for pressure maintenance or waterflood is injected into some identified wells called "injectors" and produced from other wells called "producers." Incremental oil recovery from this process is directly proportional to the areal sweep efficiency defined as the percent volume of oil contacted by injection water. The injection water pumped through injectors contacts the oil in the reservoir and sweeps it areally towards producers. The sweep efficiency increases steadily before the water breakthrough in the production wells; after that, it increases at a slower rate. For developed patterns, it may reach 100% with continued water injection after breakthrough.[5]

Areal sweep needs to be combined with vertical sweep efficiency to account for heterogeneities in all three dimensions. Changes in permeability cause an irregular and uneven waterflood front (Figure 1.5), resulting in a poor vertical volumetric sweep efficiency. Water moves faster through high permeability layers than in the lower permeability layers. Some water layers in the reservoir may act as a direct conduit or "thief zone" between the injector and producer to cause a quick breakthrough in the production well. Other layers may be too tight and may not accept any water at all. The volumetric sweep efficiency accounts for the three-dimensional (3-D) effect of reservoir heterogeneities and is a product of the areal and vertical sweep efficiencies.

Volumetric sweep efficiency is a function of pattern-type (injection-production well arrangement) and mobility ratio that affects areal sweep. Other factors such as formation dip, permeability variations, presence of fractures, and off-pattern wells, areal and vertical heterogeneity, the position of gas-oil and oil/water contacts, reservoir thickness, the density

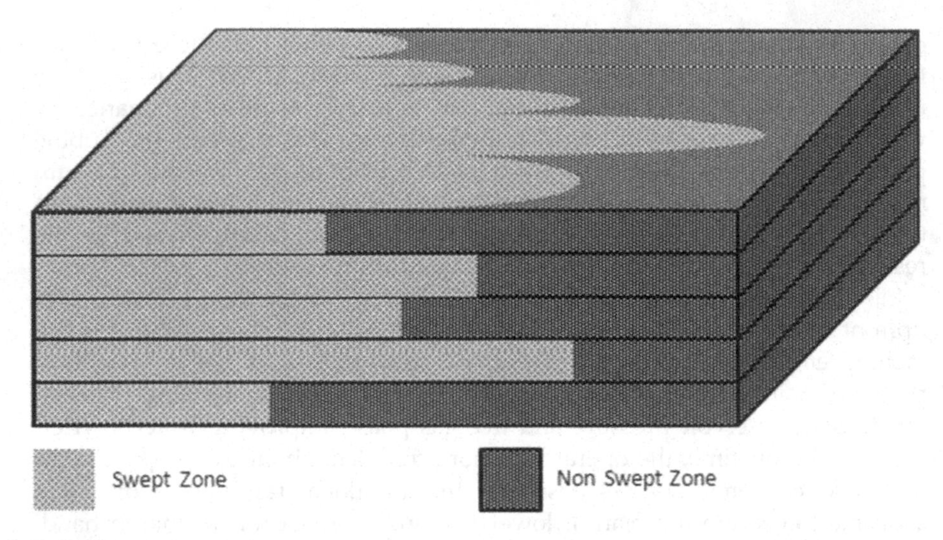

FIGURE 1.5
Volumetric Sweep Efficiency[6].

difference between the displacing and the displaced fluid, and flow rate also influence the volumetric sweep and, therefore, the waterflood recovery.

The mobility of a reservoir fluid is defined by the ratio of a rock's effective permeability to a fluid and the fluid viscosity (k/μ). In the case of a waterflood, the displacing phase is the water, and the displaced phase is oil. Accordingly, the Mobility Ratio (M) is the ratio of water to oil mobilities:

$$M = \frac{(k/\mu)_w}{(k/\mu)_o} \quad (1.1)$$

The values of mobility ratios less than one are considered favorable, and more than one adverse for the waterflood performance.

There are divergent views on whether oil-wet reservoirs should be waterflooded. However, many oil-wet reservoirs are reported to have been successfully waterflooded and continue to make massive profits.[7]

1.3.3 Tertiary Recovery

There is a fundamental difference in the concept that is behind the tertiary and secondary recovery mechanisms. Tertiary recovery primarily consists of EOR Methods that may follow secondary recovery processes if economically justified. Primary and secondary recovery methods can recover a possible 50% of oil-in-place. It still leaves behind much oil that is waiting to be recovered. On the contrary, Tertiary/EOR processes focus on oil recovery by changing the physical and chemical properties of the displacing and displaced fluids and the reservoir rock. Tertiary/EOR methods are known to accomplish higher oil recovery by altering the reservoir's rock or fluid properties, such as wettability, density, interfacial tension (IFT), and viscosity. Secondary recovery methods of waterflood or gasflood comprise pushing the oil by injected water or injected gas towards the production well.

The mechanism of oil trapping and the fundamental principle of its release from the pore spaces by EOR methods can be explained by the Capillary Number (N_c) equation given as follows:

$$N_c = \frac{\mu v}{\sigma \cos \theta} \tag{1.2}$$

Here, μ is the viscosity of the displacing phase, v is the velocity, σ is the interfacial tension, and Θ is the contact angle between the two fluid phases. The capillary number, therefore, is the ratio of viscous forces and capillary forces.

With continued water injection, the reservoir's oil phase is reduced to isolated patches of residual oil and is trapped in pore spaces. The efficient and practical way to recover this oil is to increase the capillary number by reducing the IFT (σ) and improving the viscosity (μ) of displacing fluids. From a practical standpoint of EOR, an increase in capillary number by three orders of magnitude can halve the residual oil saturation.[8]

Waterflood works best for oils that have a lower viscosity than water. It implies that the mobility ratio will be favorable, and sweep efficiency for such oils will be higher. However, the abundant availability of seawater or groundwater makes it a natural choice for waterfloods. Considerations of volume requirements and the price will limit the selection of any other displacing fluid.

From the discussion of the mobility ratio above, it is clear that water is the ideal displacing fluid for crude oils that have viscosities lower than water. Any oil with higher oil viscosity than water will have an unfavorable mobility ratio and, therefore, poor sweep. Adverse mobility ratios between displacing and displaced fluids, reservoir heterogeneities, and capillary forces will be the key factors responsible for poor waterflood performance.

Mature waterfloods that can no longer deliver commercial oil production qualify for the application of EOR methods. Oil recovery by EOR methods is accomplished by thermal and nonthermal means. Thermal methods are essentially heat based and applied to low American Petroleum Institute (API) gravity, high viscosity oils that are not mobile under reservoir conditions. Nonthermal methods that include gas-based and chemical processes are used in reservoirs containing light crudes.

1.4 Thermal Recovery

Thermal methods aim to provide heat to heavy/viscous crude oils (10°–20° API, reservoir viscosity 100–10,000 cp plus) that cannot flow on their own to the surface. Temperature decreases oil viscosity and makes the mobility ratio favorable between crude oil and the displacing fluids. Steam cycling, steam flooding, and in-situ combustion are known thermal methods. Steam flooding is an effective method that reduces oil viscosity and induces flow. Steam is injected into the injector well, which heats the oil and forces it up the producer well to the surface. Cyclic steam stimulation (CSS), also known as "huff-and-puff" or "steam soak," uses a single well for steam injection. The injection can last over months, with cycle soak times of weeks. With CSS, it is condensation, not heat, that leads to oil flow. Steam-assisted gravity drainage (SAGD) is one of the most favored variants of thermal recovery methods. It requires drilling a pair of horizontal wells on top of each other. The upper well is used for continuous steam injection. Heat mobilizes the extra-heavy oil or bitumen above it and lets it flow under gravity through the lower well.

In situ combustion techniques involve initiating a combustion front and propagating it through the reservoir with an air supply. The toe-to-heel air injection (THAI) method softens the heavy oil for extraction from the reservoir.

1.5 Nonthermal Methods

Nonthermal methods consist of chemical floods and gas-injection EOR methods. Chemical floods improve waterfloods and enhance oil recovery by altering the displacing fluid's viscosity and reducing the IFT between the displacing and displaced fluids. Some of the conventional chemical methods are briefly touched on here.

1.5.1 Chemical Floods

Polymer applications have been successfully used to achieve mobility control in highly heterogeneous formations. Polymers such as hydrolyzed polyacrylamides are long-chain molecules with high molecular weight. By adding to the injection water, polymers increase the viscosity and control mobility of the displacing phase. It promotes incremental oil production due to reduced viscous fingering and improved sweep efficiency. Injection of polymers into fractured reservoirs can cause channeling and waste of chemicals.

Surfactants are highly surface-active agents which can cleanse the oil out of the reservoir like the detergents clean the used cookware. Surfactant floods, also known as micellar (micro-emulsion) floods, use low concentration slugs that are pushed by chase water. Surfactant molecules come in contact with the pores and oil-water interfaces as the slug moves through the reservoir. The contact reduces the capillary pressure responsible for trapping the oil in the pores and the IFT between oil-water, creating favorable oil displacement conditions.

Alkaline flooding is also referred to as caustic flooding. It employs sodium hydroxide, sodium orthosilicate, or sodium carbonate as pre-flushes to micellar/polymer flooding projects or waterflooding operations. Alkalis react with some oils to form surfactants and reduce IFT between oil and water to enhance oil recovery.

One of the limitations of Alkaline and Surfactant flooding is that they have low viscosities and, therefore, adverse mobility ratio and poor reservoir sweep efficiency. However, adding polymer viscosifiers to these floods improves the process performance significantly due to improved mobility control.

1.5.2 Gas Injection Methods

Gas injection EOR methods are based on the learnings of the secondary gas floods. The lower residual oil saturations for gas displacement (S_{org}) than those for waterflooding (S_{orw}) achieved in the laboratory corefloods justify the application and viability of gas-based EOR methods. This difference in residual oil for gasfloods and waterfloods could be much higher if the injected gas is miscible with oil.

Gas injection involves injecting natural gas, nitrogen, or carbon dioxide into the reservoir. These gases can improve crude oil recovery by expanding and pushing hydrocarbon through the reservoir or dissolving in the oil to reduce oil viscosity and enhance oil production.

The availability of nitrogen is not an issue because it can be extracted from the air at a low cost. It is inert and highly compressible, forms a miscible slug at approximately 15,000 psi. The slug lowers the interfacial tension between oil and water and sweeps oil and gas from isolated sections of the reservoir. Being an inert gas, it has no corrosive effect on pipelines.

Early carbon dioxide EOR projects used naturally occurring carbon dioxide deposits. Later applications utilized CO2 generated as a byproduct of large-scale industrial emissions. Like nitrogen, CO_2 is also a crucial yet small constituent of air. CO_2 is a colorless, acidic gas that is highly soluble in water. It can be extracted from the CO_2-rich flue gas stream of large-scale chemical production, refinery, or power plant. CO_2 flooding offers multiple benefits in re-pressurization of the reservoir, swelling, and viscosity reduction, and possible lower Minimum Miscibility Pressure (MMP) than competing alternatives such as N_2. MMP is a principal parameter required for the optimization of gas-based EOR projects. It is the minimum pressure at which injected CO_2 achieves miscibility with the native crude.

A miscible gas process can use various gases for injection into the reservoir, such as carbon dioxide, nitrogen, hydrocarbon gas, and LPG. Their suitability and economic viability are decided based on laboratory experiments.

1.6 Hydrocarbon Resource and Reserves

The presence and size of a hydrocarbon resource are indicators of the wealth of a country. Only big private, international, and national oil companies venture into the E&P undertakings of these resources because they have the commercial strength and know-how to support these projects. Their capabilities and track record are a source of assurance for all stakeholders against any risks. Governments may join hands with the operating company to plan and initiate projects that can charge the domestic economy, create jobs, and improve people's living standards.

Resource and reserve estimation is an area that requires in-depth knowledge and experience. Despite the process being quite sophisticated, there is an element of subjectivity involved in the estimates. It requires applying a scientific approach for unbiased integration of multidisciplinary data, expert interpretations, and best-practice methodologies. A reliable, realistic, and consistent estimate of hydrocarbon resources is in the interest of investors, regulators, governments, and consumers. These estimates can be used to map out regional hydrocarbon-dependent energy supply and demand over different time windows, creating potential business opportunities. In the absence of a comprehensive and globally accepted code, individual countries and companies use their reserve evaluation systems that make global comparisons difficult. SPE and the WPC have combined to prepare detailed guidelines for estimating and categorizing petroleum resources/reserves espoused by many oil companies.

The term "resource" describes the total quantity of discovered and undiscovered hydrocarbons in a particular area. It represents reserves that may or not be technologically or economically recoverable from the known oil and gas deposits, plus all other undiscovered hydrocarbons and potentially recoverable reserves[9] in the future.

The difference between prospective and contingent resources needs to be clearly explained before moving further. A prospective resource is potentially recoverable from undiscovered hydrocarbon accumulations by the application of development projects. Contingent resources are

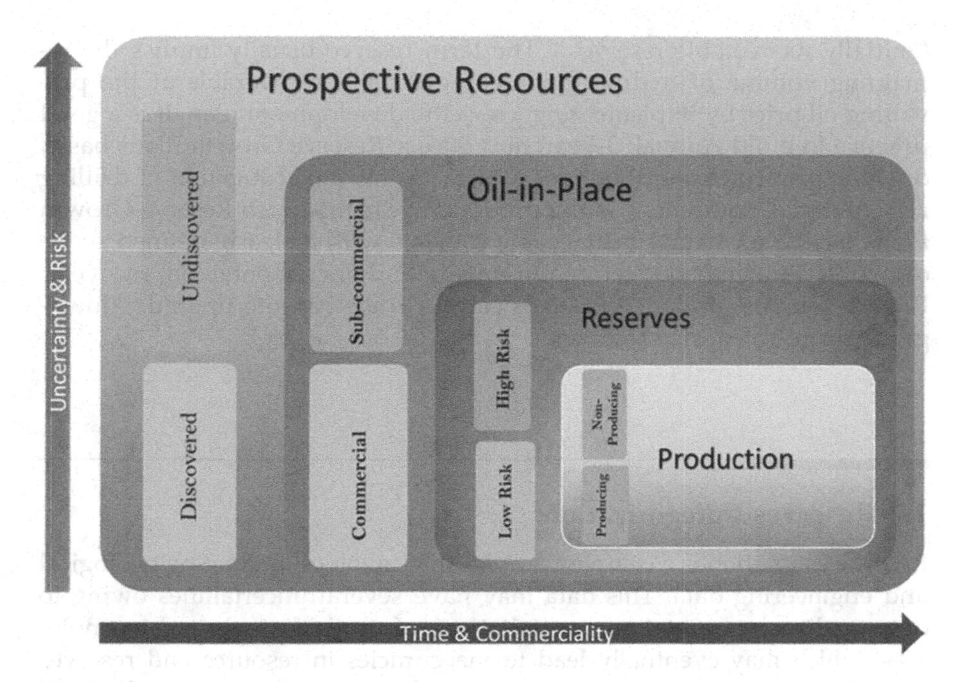

FIGURE 1.6
Hydrocarbon Resource to Production.

described as estimated quantities of hydrocarbons that are potentially recoverable from known accumulations. But unlike prospective resources, the projects/plans are not yet considered mature enough for commercial development due to one or more contingencies such as social or geopolitical conditions.

It is not enough for a company to discover a petroleum resource. It must transfer the "resource" into the "reserves" by following a standard process that ensures lower risks in its extraction (Figure 1.6). A clear understanding of uncertainties and risks associated with the project is a prerequisite for successfully executing such a process. Removing uncertainties may sometimes require very little effort, such as a well-log reinterpretation, collecting pressure data, or an oil sample from specific horizons in one or a few pre-drilled oil/gas wells. On other occasions, it may call for testing a horizon by a rig workover, drilling a new or few locations, or a pilot test for detailed evaluation. It is a good practice to start developing an area that is understood better than the other. A live data acquisition plan for each project must continue to reduce uncertainties and risks to incremental development projects.

Reserves are a function of time, recovery process, development plan, and prevailing oil price. The reserves envelope shrinks in size with time due to withdrawals. The estimated volume of hydrocarbons that can be commercially extracted before the start of any production is called "initially recoverable reserves." The term reserve usually implies the remaining volume of hydrocarbons commercially recoverable at the prevailing oil price by implementing a specific development plan. It is a good practice to build Annual, 3-Year, and 5-Year Reserve Growth Plans based on an approved financial budget to provide for the acquisition of drilling rigs, material, equipment, and a production facility. Such Reserve Growth Plans facilitate creating manageable projects with a clearly defined scope of work, timeline, budget requirement, and incremental oil recovery. Projects' success, failure, or delays can form the basis to upgrade, downgrade, or re-categorize reserves.

1.7 Reserves Categorization

Reserve estimation is a complex process that involves integrating geological and engineering data. This data may have several uncertainties owing to data quality, interpretation, or limitations of available tools and technologies, which may eventually lead to inaccuracies in resource and reserves estimates. Inconsistent application of guidelines and interpretation-bias can also influence the quality of estimates significantly.

Resource and reserve estimates are categorized in Proved, Probable, and Possible categories, usually associated with P90, P50, and P10 estimates. This methodology offers the following benefits:

- It incorporates the entire range of petroleum reserves Proved, Probable, and Possible, clearly indicating what portion of reserves is more or less risk-prone for development. Professional estimators are always at work to devise plans for upgrading the lower category of reserves into a higher grade. More often than not, reserve estimation and categorization might follow a routine workflow. Still, exceptional innovation and new technologies, coupled with the estimator's general mindset and confidence level, can greatly impact the estimates.
- Reserves categorization helps formulate short-, medium-, and long-term Reserve Growth Plans in line with the business needs and strategy. These plans must be converted into well-defined projects with clear objectives and deliverables.
- Such categorization allows the development of reservoirs in phases (drilling in segments, infill drilling, waterflood, etc.) to facilitate capital availability for projects with a clear scope of work, timeframe, estimated expenditure, and production profile, which can be used to generate revenue forecast.

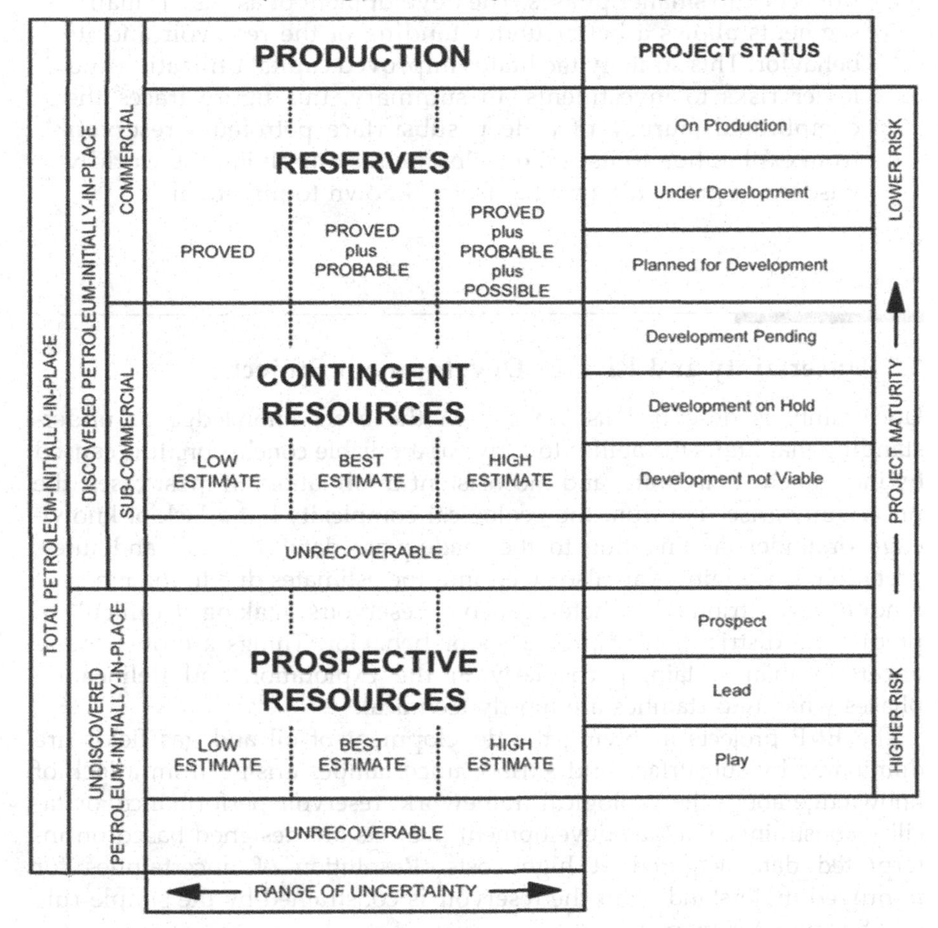

Note: For illustrative purposes only. Not to scale.

FIGURE 1.7
Hydrocarbon/Petroleum Resource Classification System[10].

- These Reserve Growth Plans can be periodically reviewed and adjusted to reflect reality in terms of production/revenue growth. Factors such as organizational capability (capital, workforce, experience, technology, project cycle time, etc.) must be considered while formulating projects that help to achieve reserves growth and production into tanks.
- Figure 1.7 illustrates the progression of the hydrocarbon resource base from high uncertainty phases (exploration and appraisal) to low uncertainty phases (development and production). It

underscores the need for breaking the massive development projects into smaller phases. The development of assets in smaller segments allows a better understanding of the reservoir and its behavior. This strategy facilitates improved capital utilization and lower risks to investments. In summary, this figure traces the complicated journey of a deep subsurface petroleum reservoir from exploration to its full development by applying the worldly wisdom of gradually moving from "known to unknown."

1.8 Uncertainty and Risk in Development Projects

Uncertainty is described as the gap in the exact knowledge or understanding that limits the ability to arrive at a reliable conclusion. It is caused by incomplete, inaccurate, and inconsistent information. In most cases, the uncertainty arises not from the geological complexity but a lack of knowledge or understanding due to the inadequate data coverage and interpretation. Uncertainty can also creep into the estimates due to the inherent randomness attributed to heterogeneous reservoirs, making it difficult to predict the distribution of properties or behavior. Things are often more uncertain than certain, particularly in the exploration and delineation phases when uncertainties are mostly technical.

The E&P projects involving the development of oil and gas fields are dominated by subsurface and surface uncertainties arising from a lack of knowledge about the geological framework, reservoir performance, or facility constraints. Oil/gas development projects are designed based on interpreted data acquired at high costs. Resolution of uncertainties for improved understanding of the reservoir is constrained by the simple rule of "the value for money."

Some of the typical examples of uncertainties associated with reservoir development and management projects are as follows:

- Fault system geometry and fault-seal capacities
- Definition of fluid contacts
- Areal or vertical extension of reservoirs
- Variation in areal.and vertical
- Interpretation of well/field and laboratory data
- Aquifer size and strength
- Reservoir performance and behavior
- The selling price of oil in the revenue forecasts etc.

In reservoir management, the risk is the consequence of an action or inaction, resulting in loss of oil rate, recovery, or returns on the investments made on a project. Uncertainty forces consideration of a range of options based on multiple possibilities of their occurrence. However, the outcome of none of the occurrences is certain. Traditional approaches for planning and decision-making do not work well under uncertainty. Uncertainty tends to blur the vision and dampen the "sixth sense" of the decision makers. It can also push the risk-averse managers into a defensive mindset preventing the company from taking advantage of positives that it can often offer. It can cause a situation of policy paralysis inhibiting development and growth.

A common approach to deal with uncertainty is aptly echoed by the age-old adage "hope for the best, prepare for the worst." This philosophy incorporates designing a full range of scenarios with associated uncertainties and the probability of their occurrence. Planning for uncertainties usually implies higher project costs, justified by reduced risks to project success and lower financial losses.

At present, the following approaches are used to account for uncertainties and deal with risks in the development of oil and gas resources and reserves-

- Deterministic or risk-based (also called incremental)
- Probabilistic or uncertainty-based approach (also called cumulative)

"Risk is defined as the probability of a discrete event to occur or not to occur, whereas uncertainty is described as the range of possible outcomes in an estimate."[11] The risk-based approach considers the proved, probable, and possible categories of reserves estimated deterministically without assigning any degree of uncertainty to probable or possible reserves. The risk for probable and possible case estimates not being present or recovered is carried by an expression like "more likely than not to be recoverable," which is qualitative and, to some extent, subjective. Proved resources will most likely be present than not. These volumes represent a downside potential. Probable resources are more likely to be present than not, with a risk of less than 50% to be present. These estimates usually represent "Most Likely" volumes. Possible resources/reserves are less likely to be present than probable reserves. Accordingly, there is a higher risk of "possible" than "probable" resources for not being present. This category refers to the upside potential in a reservoir where commercial development is subject to some doubt. The same rationale applies to reserves.

Uncertainties with a significant impact of reservoir heterogeneities or behavior can only be dealt with by probabilistic analysis. In the case of significant heterogeneities in reservoir description, no particular description is labeled as P10, P50, or P90 because these numbers represent the probabilistic results, not the description. A probabilistic distribution of the

TABLE 1.2

Probability of Exceedance and Non-Exceedance

		Probability of Exceedance			
#	**Data**	**Data**	**Frequency**	**Probability (%)**	**Probability of Exceedance (%)**
1	50	50	2	13	100
2	60	60	3	20	87
3	65	65	5	33	67
4	70	70	3	20	33
5	75	75	2	13	13
6	65	**Total**	**15**	**100**	
7	60				
8	75	**Probability of Non-Exceedance**			
9	65	**Data**	**Frequency**	**Probability (%)**	**Probability of Non-Exceedance (%)**
10	70	50	2	13	13
11	65	60	3	20	33
12	70	65	5	33	67
13	65	70	3	20	87
14	50	75	2	13	100
15	60	**Total**	**15**	**100**	

answer to any specific modeling question is generated by running it through applicable scenarios wherein various combinations of representative reservoir uncertainties are built.

The use of both, Exceedance and Non-Exceedance probabilities,[12] is common and often confusing without qualified statements. The difference between these two probabilities can be explained with the help of a sample dataset consisting of 15 values which could represent the number of any objects or estimates (Table 1.2).

Careful consideration of the data presented in Table 1.2 indicates that the probability of exceedance is obtained by cumulating the probabilities of data from larger to smaller data (75, 70, 65, 60, and 50) and the probability of non-exceedance is achieved by adding the probabilities of the data in ascending order (50, 60, 65, 70, and 75).

In the case of exceedance probability, a P90 value will be exceeded in 90% of the realizations/cases. Accordingly, the Exceedance probability of P90 will represent a lower downside value and higher confidence in the estimate. Similarly, a P10 value will be exceeded by only 10% of the cases and represent a higher upside estimate with lower confidence.

On the other hand, non-exceedance or cumulative probability is exactly the opposite. A P90 value in case of non-exceedance will be less or equal in 90% of the cases implying a higher upside figure and lower confidence.

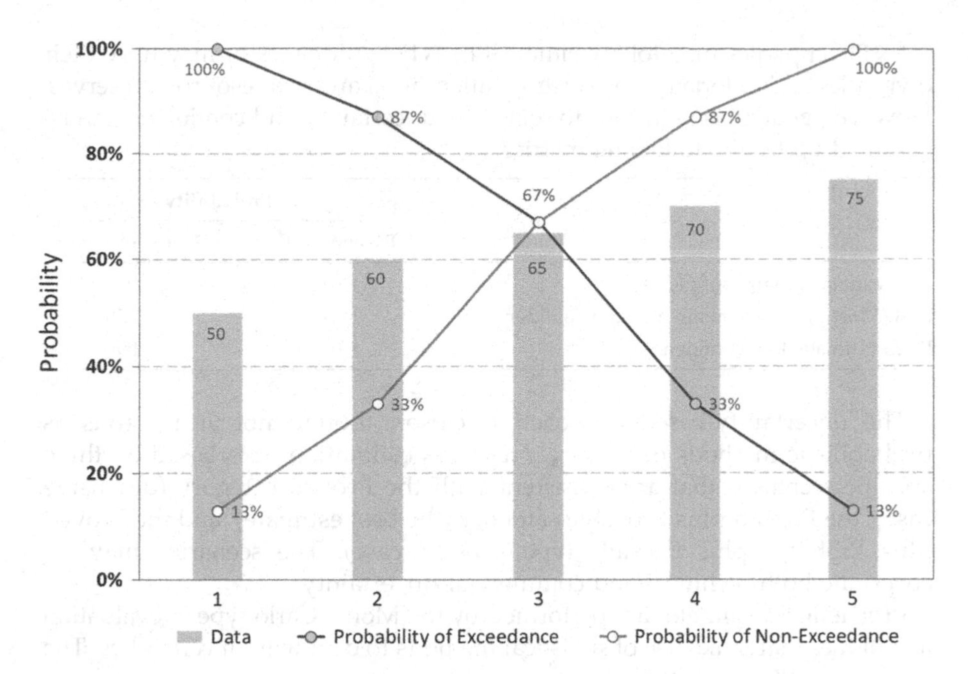

FIGURE 1.8
Probability of Exceedance and Non-Exceedance for the Sample Dataset of Table 1.2.

Likewise, a P10 value for this probability will be less or equal in only 10% of the cases, implying a lower (downside) figure and higher confidence. Due to the apparent contradiction in the P10 and P90 values in exceedance and non-exceedance probabilities, it is important to specify the type of probability linked with the probabilistic estimate (Figure 1.8).

Initial estimates of hydrocarbon resources and reserves are usually kept in the lower (low certainty) category due to a large number and magnitude of uncertainties in input parameters. With continued development drilling, new data acquisition, and improved understanding of the reservoir as the project matures, resources/reserves are upgraded from lower to higher categories.

The degree of uncertainty can relate to the magnitude of the risk—the higher the uncertainty in data, maps, and reservoir parameters, the higher the possibility of risk in field development and production. Understanding this terminology, methodology, and approaches, along with their limitations, is useful for decision makers.

Proved reserves represent a downside potential and will most likely be recoverable than not. Probable reserves represent "Most Likely" volumes and are more likely to be recoverable than not, with a risk of less than 50% to be recovered. Possible reserves refer to the upside recovery potential in a reservoir where commercial development is subject to doubt. These reserves are less likely to be recovered than probable reserves, and there is a higher risk of possible than probable reserves for not being recovered.

E&P companies may follow either SPE/WPC guidelines or may have their own rules and criteria for the categorization of oil and gas resources/reserves. However, general conclusions to relate the uncertainty and confidence can be summed up by the following metrics.

	Probability	
	Exceedance	**Non-Exceedance**
Low estimate, high confidence	P90	P10
Best/Most Likely estimate, medium confidence	P50	P50
High estimate, low confidence	P10	P90

The uncertainty-based approach is closer, though not as rigorous as probabilistic methods of resource/reserves estimation. It is based on three specific scenarios that are consistent with the Proved category (downside case), the Proved plus Probable category (the best estimate), and the Proved plus Probable plus Possible type (upside case). The scenarios may incorporate both technical and commercial uncertainty.

Probabilistic estimates are performed by the Monte Carlo type of evaluation techniques that make use of statistical methods to deal with uncertainties. The use of P90, P50, and P10 estimates of hydrocarbon resources or reserves is a common industry best practice. The boundary conditions or the scenarios represent the uncertainty in deterministic estimates of resources/reserves. The probabilistic methodology integrates the degree of uncertainty with a measure of risk. The level of probability is represented by distribution and the risk in estimates by a low (P90), medium (P50), and high (P10).

By definition:

- P90 represents at least a 90% probability of occurrence of hydrocarbon quantities/potentially recoverable quantity from a technical standpoint by the development of the reservoir. This probability is high, which implies that the estimated volume of hydrocarbon resources/reserves has lower risk and higher confidence as compared to P50 or P10 estimates.
- P50 represents at least a 50% probability of occurrence of hydrocarbon quantities/ potentially recoverable quantity from a technical standpoint by the development of the reservoir. This probability is lower than 90%. It implies that the estimated volume of hydrocarbon resources/reserves have a higher risk and lower confidence than those for P90 estimates. P50 estimates are considered to be an excellent middle estimate.
- P10 represents at least a 10% probability of occurrence of hydrocarbon quantities/potentially recoverable quantity from a technical standpoint by the development of the reservoir. This probability is even

lower than 50%, and it implies that the estimated volume of hydrocarbon resources/reserves would be higher than P50 estimates. These estimates are considered high-end estimates of resources and reserves, representing the lowest confidence and the highest risks.

1.9 Summary of Reserve Estimation Methods

One or more than one of the following methods are generally selected to estimate reserves depending on the amount and quality of data available -

- Analogy
- Volumetric
- Use of classical reservoir engineering methods
  - Material balance equation
  - Decline curve analysis (DCA)
- Reservoir modeling

A summary of these methods is presented in Table 1.3.

1.10 Reservoir Management and Development Planning

Development Planning Team must integrate the inputs of the members from the management, finance, subsurface, surface, and service providers for the best results.

Management Input:

- The expectation of oil production rate, including plateau rate based on reserve estimates
- Available resources (workforce, budget, equipment, material, services, expertise, drilling and workover rigs, and technology)
- Reservoir Management Policy
  - Safety and security of oilfield and its infrastructure
  - Environmental considerations for oil/gas field operations
  - Health and safety of the workforce
  - Maximization of profit or maximization of recovery

TABLE 1.3

Summary of Methods Used to Estimate Hydrocarbon Reserves

Method Based On	Works Best When	Application	Accuracy	Comments
Analogy	• Same lithology • Similar petrophysical properties • Similar Drive Mechanism • Comparable quality of rock and fluids	• Useful in early exploration, delineation, or initial development phase to assess- ○ Initial Hydrocarbon in place ○ Recovery factor	• Highly dependent on the similarity of reservoir characteristics • Reserve estimates are often very general	• Estimation not accurate
Volumetric	• There is low uncertainty about the following: ○ Compartmentalization ○ Petrophysical properties ○ Bulk rock volume ○ Fluid contacts ○ Recovery Factor	• Always used to determine ○ Initial Hydrocarbon in Place ○ Recovery factor • Useful in all phases of the reservoir life cycle • Estimate improves with improved characterization as development proceeds	• Highly dependent on the quality of reservoir description	• Reserves estimates are on the high side because this method does not consider the problem of reservoir heterogeneity
DCA	• The team has a good idea of compartmentalization • Operating conditions in a given compartment are consistent, not impacting decline rate (such as no change of horizon or recovery mechanism)	• Very commonly used method to estimate Recovery Factor and Initial Hydrocarbon in Place • Best applicable to post-plateau rate period • Decline Rates reveal the state of reservoir health over a given period	• The accuracy of the DCA method is affected by the change of horizons, new well drilling, switching from primary recovery to secondary recovery programs, under-investments, geopolitical	• DCA is virtually independent of the size and shape of the reservoir or the actual drive mechanism; decline rates depend on the size of reservoirs • Decline rates are known to be

(*Continued*)

TABLE 1.3 (Continued)

Summary of Methods Used to Estimate Hydrocarbon Reserves

Method Based On	Works Best When	Application	Accuracy	Comments
			compulsions and sabotages, etc.	steeper for small reservoirs in comparison to large reservoirs • The only data required for DCA and extrapolation are production data, which are relatively easy to obtain for field
Material Balance Equation	The team has - • A good idea of the compartmentalization • Adequate reservoir pressure-production history • A clear and precise drop in reservoir pressure is observed • A good sense of aquifer size and strength if applicable • Good knowledge of gas cap size if applicable • Reliable records of produced and injected water and gas available • Good handle on PVT properties	• Very commonly used to determine Initial Hydrocarbon In Place and Recoverable Reserves • Best used in a mature field with abundant geological, petrophysical, and engineering data	• Depends on the quality of the reservoir description and the amount of production data available • Reserve estimates variable	• This method is sensitive to inaccuracies in reservoir pressure measurements. It breaks down when no substantial drop occurs in reservoir pressure.

(*Continued*)

TABLE 1.3 (Continued)

Summary of Methods Used to Estimate Hydrocarbon Reserves

Method Based On	Works Best When	Application	Accuracy	Comments
Reservoir Modeling	Good understanding of the following- • The size of the reservoir • Fault seals and their conductivity • Reservoir layering and barriers • Petrophysical properties • Reservoir pressure behavior • Fluid properties • Well productivity • Fluid contacts • Drive mechanism ○ Aquifer size and strength ○ Gas cap size	• Most versatile and globally acceptable method to address development strategies and other vital questions	• Depends on the amount of production history available. Reserve estimates tend to be realistic.	• "Fit for purpose" decision-based well, sector, or 3-D models support reservoir development and management programs • Once accurately history-matched can provide reasonable estimates of Initial Hydrocarbons in Place and Recoverable Reserves.

    - The acceptable rate of return on invested capital
    - Reservoir health norms
    - Marketing and sale arrangements for oil and gas
    - Guidelines on Reserve Replacement Ratio
- Establish processes and procedures for approval of high-cost or high-impact activities such as production and reserves growth target setting, reservoir development plan, surface facilities design and the timing of their commissioning, large-scale horizontal or multi-lateral drilling, pressure maintenance plans, associated gas flaring, disposal of produced water, reporting and handling of HSSE (Health, Safety, Security and Environmental) issues, and so forth
- Be aware of the risks and take actions to update/mitigate those listed in the Risk Register

Finance Input:

- Representative/realistic cost estimates of various operations and services
- Budget allocation and timely release of funds
- Perform financial forecasting, reporting, and operational metrics tracking
- Analyze financial data—and create economic models for decision support
- Market research, data mining, business intelligence, and valuation comparisons

Subsurface Input:

- Devise development plans to achieve company/management objectives
- Divide the reservoir into manageable segments for phased development
- Identify significant uncertainties and prepare a plan to overcome the risks due to those uncertainties
- Review all possible development options/scenarios and rank them based on techno-economic criteria
- Determine the timing and injection volumes for pressure maintenance requirement
- Decide the timing and method of artificial lift
- Plan for the rig and rigless work over programs for bringing sick wells on production/injection

- Well logging, coring, sampling, testing, and other well-related capabilities
- Core and PVT laboratory study capabilities
- Acquire well and reservoir performance data to improve reservoir characterization and for continuous optimization as well as mid-course corrections to the development plan
- Improve understanding of reservoirs and their performance through various kind of studies and modeling
- Other subsurface activities

Surface Input:

- Planning construction and maintenance of facilities for production (oil, gas, and water), Gas Compression Station/Booster Compressor Station
- Planning construction and maintenance of facilities for injection (water, gas)
- Planning, construction, and maintenance of trunk pipeline
- Prepare and execute the master plan to handle-
    - Immediate hook up of wells as they get ready to deliver oil to the surface
    - Produced fluids, that is, oil, water, and gas
    - Disposal of produced water
    - Well Hook Up plan to production/injection facilities
    - Inspection and maintenance of wellheads and flowline networks
    - Other surface operations management and so forth

Drillers Input:

- Number and type of drilling and workover rig requirements
- Drilling and workover rig capabilities and workload
- The capability of rig crews to carry out drilling/workover jobs safely and efficiently
- Well logging, coring, sampling, testing, and other well-related capabilities
- Planning, building, and maintaining production and injection facilities
- Production and injection well hook up plans and arrangements
- Well stimulation, maintenance, and servicing, and so forth

Proper reservoir management and field development require seamless integration of all the teams and cohesive teamwork. It requires a clear

TABLE 1.4

Key Roles and Responsibility for Reservoir Management and Development (RACI Model)

Role ➡ / Team ⬇	Reservoir Management Policy Including HSSE	Available Resources	Budgeting	Design and Implementation of Development Plans to Achieve Corporate Objectives	Identify and Mitigate Uncertainties and Risks in the Development Plan	Surface Master Planning	Design, Construction, Operation, and Maintenance of Facilities	Acquire and Manage Necessary Rigs, Crews, Materials, Equipment, and Technologies
Management	A, R	A, R	A, R, I	A, I	A, I	A, I	A, I	A, I
Finance	I	I	R, A	I	I	I	I	I
Subsurface	R	R	C, I	R, A	R, A	C	C	C, I
Surface	R	R	C, I	C	C	R, A	R, A	I, C
Service Providers	I	R	C, I	I	I	I	I	R, A

where

R: Responsible: Those who recommend or do the work to complete the task.

A: Accountable: The final approving authority is ultimately answerable for the correct and thorough completion of the deliverable or task.

C: Consulted: Those whose opinions are sought when necessary, like Consultants or Counsel, typically specialists and subject matter experts. Two-way communication with them is essential.

I: Informed: Those who must receive updates about the progress of the task or deliverable; and with whom there is just one-way communication.

The above matrix is useful in assigning roles to individuals and maintain the status of progress until completion.

understanding of who is responsible for what. The following RACI matrix (Table 1.4) helps understand various departments' roles, responsibilities, and interactions.

Notes

1 Oil and gas production decommissioning; https://www.ektinteractive.com/upstream-oil-and-gas-production-decommissioning/
2 From Wikipedia The Free Encyclopedia- List of oilfields; https://en.wikipedia.org/wiki/List_of_oil_fields
3 Wikipedia The Free Encyclopedia; https://en.wikipedia.org/wiki/File:Oil_Prices_1861_2007.svg
4 Standard Handbook of Petroleum and Natural Gas Engineering (Third Edition) by William C. Lyons, Ph.D., P.E. Gary J. Plisga, B.S. Michael D. Lorenz, B.S.
5 Reservoir Engineering aspects of waterflooding; Forrest F. Craig, Jr.; SPE of AIME; Monograph Volume 3, Henry L. Doherty series; p 53
6 Fundamentals of Fluid Flow in Porous Media; Chapter 4 Immiscible Displacement Vertical and Volumetric Sweep Efficiencies; PERM Inc. TIPM Laboratory; https://perminc.com/resources/
7 Waterflooding of preferential oil-wet carbonates: Oil recovery related to reservoir temperature and brine composition; by Eli Jens Hognesen, Tor Austad; SPE-94166-MS; SPE Europec/EAGE Annual Conference, 13-16 June, 2005
8 S. Thomas, "Enhanced Oil Recovery - An Overview," Oil & Gas Sci. Technol. - Rev. IFP 63, 9 (2008)
9 Reserves estimation; https://wiki.aapg.org/Reserves_estimation
10 Guidelines for the Evaluation of Petroleum Reserves and Resources; a supplement to the SPE/WPC Petroleum Reserves Definitions and the SPE/WPC/AAPG Petroleum Resources Definitions
11 Guidelines for the Evaluation of Petroleum Reserves and Resources; a supplement to the SPE/WPC Petroleum Reserves Definitions and the SPE/WPC/AAPG Petroleum Resources Definition
12 P10, P50 and P90- What are these parameters? Why are they so important? https://blogs.dnvgl.com/software/2016/12/p10-p50-and-p90/)

2

Operating Arrangements and Reservoir Development Objectives

2.1 Introduction

The development of petroleum reservoirs needs to be accelerated after their discovery and appraisal to generate revenue. Detailed planning for construction and commissioning of crude oil and gas processing, storage, and transportation facilities is required before starting full-fledged development drilling. Accurate reservoir characterization and information gathering are critical for a successful drilling campaign.

A petroleum development project, being energy-critical, raises an air of excitement and expectancy within and outside the company. A company's reservoir management vision of a petroleum asset originates from the considerations of whether a company has a local or foreign origin, whether it is in the private or public sector, or whether it has a short-term or long-term business engagement. These are the primary antecedents that set the tone for business operations.

The issue of a company being of a local or foreign origin is fundamental to the business. Local companies are a natural fit because they have the advantage of knowing the local language, laws, culture, traditions, and climate. However, many local oil companies may not have the necessary experience and technology to extract oil and gas. They may also not have sufficient capital and commercial space for oil/gas E&P operations. On the other hand, foreign companies may find the challenges of a remote location, language, culture, climate, and relocating the business overwhelming. A company incorporated in one country with its business operations in the other has to deal with the complexities of law-and-order, logistics, compliance issues, and tax codes.

In private-sector businesses, the owners who make monetary investments share the entire profit and multiply their capital. As a result, wealth is concentrated in the hands of a few stakeholders. In contrast, the public-sector goal is to disseminate the economic benefits amongst a large number of shareholders. It has other wide-ranging objectives such as expanding employment opportunities, meeting far-reaching national and international aspirations such as attainment of self-reliance, and removal of regional imbalances.

The issue of a short-term and long-term engagement of a company is also vital. Short commitment will not motivate a company to invest enough in the long-term objectives. It will also be interested in recovering the most capital out of the project quickly, notwithstanding the long-term consequences of its decisions. The opposite of this is equally valid. The long-term engagement of a company may not necessarily stimulate it to achieve aggressive short-term objectives.

An organization's performance reflects the style of its CEO. Jack Welch, the retired CEO of General Electric, once said, "Anybody can manage short. Anybody can manage long. Balancing those two things is what management is." Striking a balance between short- and long-term objectives is often difficult due to their imminent conflict. CEOs of public limited companies have a limited tenure and short performance evaluation intervals to show results to their shareholders. This environment constrains them to focus on short-term objectives that will produce tangible outcomes during their tenure. They do not have enough time to focus on the organization's long-term goals, such as attaining self-reliance, developing small and ancillary industries, and achieving operational excellence. The need for periodic investment control measures to improve its annual balance sheet also affects a company's long-term objectives. In essence, "short-termism often results in earnings management rather than building the long-term value of the company."[1]

Crude oil and natural gas are primary energy supply sources for the world's social and economic development. The presence or absence of the hydrocarbons, their quality, quantity, and extent are determined by nature based on geological factors. However, ownership, control, and management of the resources/reserves are decided by geographical and political boundaries.

2.2 Oil and Gas Reserves Ownership and Control

Laws of oil and gas ownership vary globally. Oil and gas interests in the US and a few other countries are generally owned by privately or publicly owned enterprises.[2] In the majority of other countries, oil and gas interests are held by the national governments.

The US and some other countries adopted the land and property laws for granting the ownership rights of minerals, including oil and gas. These laws advocated that everything under the land surface belonged to the landowner, and he/she had full control over the mineral/resource above it or underneath. The laws were clear that the owner had the full right to earn income from the land and resource contained by renting, leasing, transferring, or selling. Oil and gas reserves ownership rights were traditionally vested in the owner of the land.

However, the subject of ownership and control of natural resources based on this law generated a great deal of controversy, conflicts, and lawsuits. A brief discussion below will put various oil and gas ownership theories in perspective.

2.2.1 Ownership In-Place Theory

This theory entitles the owner to access the resource under or over the land surface owned by him/her through a negotiated lease arrangement subject to the government's regulatory laws. This theory also grants the landowner the right to separate the surface rights from the mineral rights.

2.2.2 Non-Ownership Theory

This theory recognizes the migratory nature of oil and gas in the subsurface and "is not capable of being owned until it is produced and reduced to possession." It was based on the rationale that oil and gas were "fugacious" like air, water, sunshine, and wild animals who would not become the landowner's property if they pass over a piece of land. Therefore, fugacious resources such as oil and gas can only be subject to ownership when captured.

2.2.3 Qualified Ownership Theory

This theory was propagated to resolve separate pieces of land on the surface with shared underground oil and gas reservoirs. The conclusion was that each property owner was entitled to exploit oil and gas from a common reservoir by drilling wells in his/her part of the property. However, the law enjoined the owners not to damage or waste the resource, reservoir, and environment during extraction.

2.2.4 Ownership of Strata Theory

Strata theory granted ownership of the sedimentary layer containing oil and gas to the landowner "within the limits of the vertical planes representing the boundaries of his tract."

2.2.5 Servitude Theory (Profit a' Prendre)

Profit a' Prendre is a French term that means "right of taking" something off another person's land. Servitude theory applies to non-fugacious minerals and allows the "right to enter the land of another person and to take some profit of the soil, or a portion of the soil itself" to facilitate the exploitation of the underground mineral resources and has no bearing whatsoever on the ownership of the land surface.

These theories formed the stepping stone and provided a framework of oil and gas ownership rights in some countries. Today, in most countries except the US, either the subsoil is essentially state-owned, or the state retains a veto on its use.

A detailed discussion of these theories can be found in a well-documented paper, "Ownership and Control of Oil, Gas, and Mineral Resources in Nigeria: Between Legality and Legitimacy" by Lanre Aladeitan.[3]

2.3 State Ownership of Oil and Gas

In countries where natural resources such as metals, minerals, and oil and gas are considered national resources, the above theories or laws of land and property rights do not apply. The resources located on, above, or below the land/property owned by an individual, the state has the first right of access after suitable recompense.

The creation of the NOCs followed the rationale that natural resources were the gift of nature to humanity, like sunlight and air. The potential of this game-changing resource must be used for energy security and the economic upliftment of the masses. The countries with oil and gas resources decided to use the opportunity to go beyond the mere production of oil and gas and create lasting value for the nation via NOCs with the following goals:[4]

- Fostering the transfer of technology
- Creating employment opportunities
- Enhancing local ownership and control
- Promoting economic growth and diversification
- Contributing to energy self-sufficiency and security of supplies

The first NOC in the oil sector was established in 1908 in Austria-Hungary when the government built a topside plant to process the excess supply of crude. As oil grew in strategic importance, national governments took a deep interest in the petroleum industry's upstream and downstream operations. In 1914, the second NOC came into being when the UK government acquired a 51% stake in the ownership of the Anglo-Persian Oil Company (APOC), active in Iran and the Middle East. A list of selected NOCs is presented in Table 2.1.[5,6,7,8,9,10,11]

TABLE 2.1

Selected NOCs with Founding Dates

No.	Country	NOC Name	Year
1	United Kingdom	British Petroleum (BP)[11]	1914
2	Argentina	Yacimientos Petrolíferos Fiscales (YPF)[11]	1922
3	Bahrain	Bahrain Petroleum Company (BAPSCO)[7]	1929
4	Saudi Arabia	Saudi Aramco[7]	1933
5	Kuwait	Kuwait Oil Company (KOC)[7]	1934
6	Oman	Petroleum Development Oman (PDO)[7]	1937
7	Mexico	Petróleos Mexicanos (Pemex)[11]	1938
8	Iran	National Iranian Oil Company (NIOC)[11]	1951
9	Brazil	Petróleo Brasileiro S.A. (Petrobras)[11]	1953
10	India	Oil & Natural gas Corporation Ltd. (ONGC)[11]	1956
11	Algeria	Sonatrach[11]	1965
12	Iraq	Iraq National Oil Company (INOC)[11]	1967
13	Libya	National Oil Corporation (NOC)[7]	1970
14	Indonesia	Pertamina[11]	1971
15	Norway	Statoil[11]	1972
16	Qatar	QGPC[11]	1974
17	Malaysia	Petronas[11]	1974
18	Venezuela	Petróleos de Venezuela, S.A.(PDVSA)[11]	1975
19	Vietnam	Petrovietnam[11]	1975
20	Canada	Petro-Canada[11]	1975
21	Angola	Sonangol[11]	1976
22	Nigeria	Nigerian National Petroleum Corporation (NNPC)[11]	1977
23	China	China National Offshore Oil Corporation (CNOOC)[7]	1982
24	Uzbekistan	Uzbekneftegaz[5]	1992
25	Azerbaijan	State Oil Company of Azerbaijan Republic (SOCAR)[6]	1992
26	Russia	Rosneft[7]	1993
27	Sudan	Sudapet [11]	1997
28	Kazakhstan	KazMunayGas (KMG)[8]	2002
29	Serbia	Naftna industrija Srbije (NIS)[9]	2005
30	Turkmenistan	Turkmennebit[10]	2016

State-controlled NOCs dominate today's oil and gas landscape across the globe. Modern business methods allow separation between the right of ownership and control. Their direct participation through ownership is regarded as key to achieving the aforementioned goals. The creation of NOCs has enabled the national governments to use petroleum's strategic power to their advantage in peace and conflict times.

IOCs have always been a dominant player in the upstream and downstream petroleum sectors. Until the 1970s, they controlled most oil and gas reserves, nearly all crude oil production, and the bulk of transportation infrastructure.[11]

However, before the end of the century, they had lost this advantage. The story of their decline coincided with the emergence of NOCs, which took control of their oil and gas reserves, production, and transportation. Today, a significant share of the world's oil and gas reserves is owned by the NOCs. Most of this share has increased because much of the world's best, easy-oil acreage is out of IOCs' bounds. Today, IOCs may not own as much oil and gas reserves; they still have the capital and know how to find and extract oil/gas from difficult locations. NOCs need technological support, risk capital, and project management experience to discover the "difficult" oil, reduce the production cost, and improve secondary and tertiary recovery factors. From this perspective, these two entities, NOCs and IOCs, have to come together for their shared economic interests and energy security.

Even today, the IOCs are heavily engaged in E&P operations in most parts of the world, although their monopoly in the E&P sector is significantly eroded. Now they work as independents in areas that are outside NOCs jurisdiction or play a complementary role to provide technical support and knowhow to NOCs. NOC being the representative of the national government is the resource owner. Yet, the control and management of operations, depending on the commercial agreement, could very well be the joint responsibility of the NOC and IOC. This model is an excellent example to explain the difference between ownership and control. Ownership means having legal title to a property or resource, whereas control means having the ability to determine its use.[12] Ownership is about equity share and risk-bearing functions. Control is about corporate governance and decision-making functions.

NOC is a public limited company with many owners in the form of shareholders who do not directly control its policies, programs, activities, or finances. State ownership has increased NOCs' monopoly over the oil and gas sector. As a government's instrument, NOCs maintain control over oil and gas production and reserves and manage to keep the revenue from oil and gas in the country. National governments use these funds to pursue their policy agendas by providing cheap food, education, and hospitals. They may subsidize fuel, transport, and fertilizer to promote agriculture and make political gains from a feel-good factor.

NOCs, in return, receive government protection for their business. Protectionism triggers inefficiencies and a lack of competition. All this translates into mediocrity's growth over excellence and an aversion to risk-taking, ultimately leading to no discoveries or poor success ratio of exploratory wells in terms of exploration. Countries that are overly protective of their oil and gas resources may potentially miss out on lucrative risk capital from IOCs. NOCs usually pursue the low-risk-low-cost strategy of exploration and development. It is expected that almost all of them, including the NOCs of the oil-rich gulf-states, will need technological support from IOCs as they venture into high-risk or high-cost petroleum basins.

Inadequate resource allocation, coupled with poor reservoir and production management, can lead to a failure to deliver objectives and an increased production cost. Oil and gas operations are costly, complicated, and risky. Conscientious employees can reduce wasteful expenditure in a broad range of drilling and production operations/activities.

The NOCs story is full of mixed experiences. Almost all NOCs have remarkably supported their governments to realize the oil sector's benefits for broader national development. Yet, there is a general perception that they are inefficient managers of national resources and natural obstacle to private investment. There is an exaggerated tendency to mingle the organization's commercial and social roles at the government and company levels.

In many oil-producing countries that are a net importer of oil, production has stagnated. Many governments have enunciated new policy regimes for exploration licensing to stimulate exploratory efforts with foreign investments. Some governments have introduced new policy initiatives such as Open Acreage Licensing Policy and concessional royalty regimes to attract IOCs and other interested countries. Under the Open Acreage Licensing Policy, a bidder may seek exploration of any blocks not already covered by exploration. The concessional royalty regime applies to oil and gas production from deepwater and ultra-deepwater areas where production costs are higher than in shallow water areas.

2.4 Oil and Gas Ownership Rights in the Seawaters

Ownership of petroleum resources present in the sea is governed by the world oceans' maritime use for business and other activities. It defines nations' rights, responsibilities, and obligations concerning the exploration and exploitation of minerals and hydrocarbons. Oil and gas discovery in offshore waters is subject to the Law of the Sea (Figure 2.1), which controls the public order at sea. This law came into effect after the third international United Nations Convention on the Law of the Sea (UNCLOS) in 1982.

General provisions of this law for oil and gas activities in the sea are controlled by Article 2 of the Third United Nations Convention on the Law of the Sea under Part II, which states:

> *Legal status of the territorial sea, of the air space over the territorial sea and of its bed and subsoil.*
>
> *The sovereignty of a coastal State extends beyond its land territory and internal waters and, in the case of an archipelagic State, its archipelagic waters, to an adjacent belt of sea, described as the territorial sea.*
>
> *This sovereignty extends to the air space over the territorial sea as well as to its bed and subsoil.*

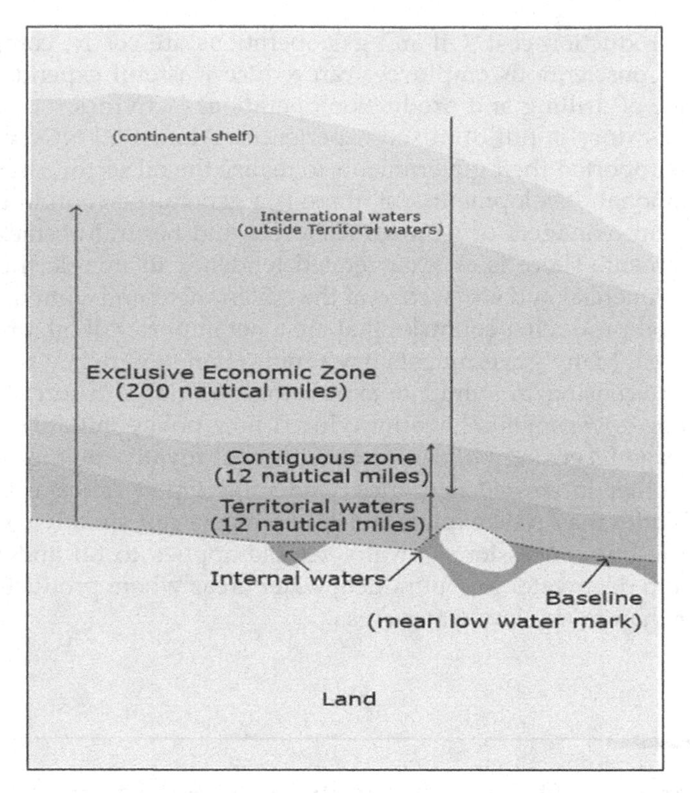

FIGURE 2.1
Definition of Seawaters for Oil and Gas and Mineral Rights (Plan View)[13].

> *The sovereignty over the territorial sea is exercised subject to this Convention and other rules of international law.*

Earlier, coastal states' territorial jurisdiction extended to a meager three nautical miles (5.6 km) from their coastline. The US was the first nation to expand its territorial waters beyond this limit to include its continental shelf's natural resources in 1945. Much later, in 1982, after prolonged discussions, UNCLOS extended the territorial water limits of all sovereign coastal nations to a maximum of 12 nautical miles (22 km) with further allowance to establish an exclusive economic zone (EEZ).[14] The EEZ can stretch up to 200 nautical miles (370 km) or the outer edge of the continental margin (Figure 2.2), whichever is farther subject to an overall limit of 350 nautical miles (650 km) from the coast where a coastal country has its sole jurisdiction over natural resources.[15] However, foreign nations invariably enjoy the "high seas freedom," a term representing freedom of navigation, overflight, and laying submarine pipes and cables with the coastal states' consent. The far side of the EEZ constitutes the *de facto* seaward boundary of a coastal nation to explore and recover natural resources from the seas.

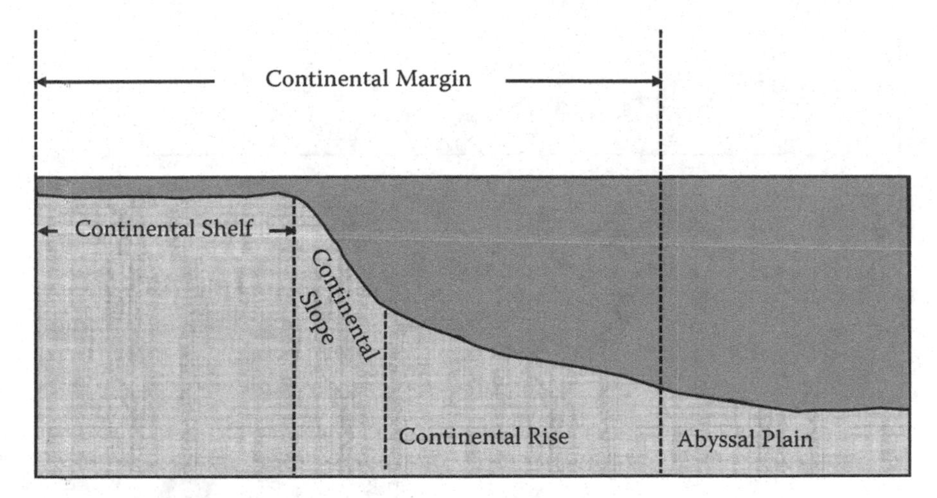

FIGURE 2.2
Continental Margin[16].

Waters beyond this boundary are treated as international waters and can be accessed by any country to extract oil and gas.

The continental shelf is considered a natural extension of land territory into the sea. It can extend to the far edge of the continental margin or a distance of 200 nautical miles (370 km) to a maximum of 350 nautical miles (650 km) from the coastal state's baseline, whichever is greater. The continental margin is a part of the ocean floor and is described as the shallow water area (water depth < 200 m) adjacent to continents. It comprises the continental shelf, continental slope, and continental rise.

2.5 Value of Vision and Mission Statements of E&P Companies

The vision is essential for a successful business. It helps the leadership pursue its dreams and set a strategy to achieve goals, but it also sets the working environment's tone. Experience has shown that success is not a fluke; it can be realized through a structured process (Figure 2.3) coupled with human qualities and efforts. A clear statement of the vision and mission of an organization is a part of this process.

Vision, mission, and value statements are like the "soul" of a corporate "body." The vision statement is about the ultimate destination, a long-term goal, where the organization wants to be in the future. Vision is the big picture or the "dream" that a company sets out to chase.

The mission statement defines a short-term goal, an action-oriented description of its purpose, guiding strategic planning, and roadmaps.

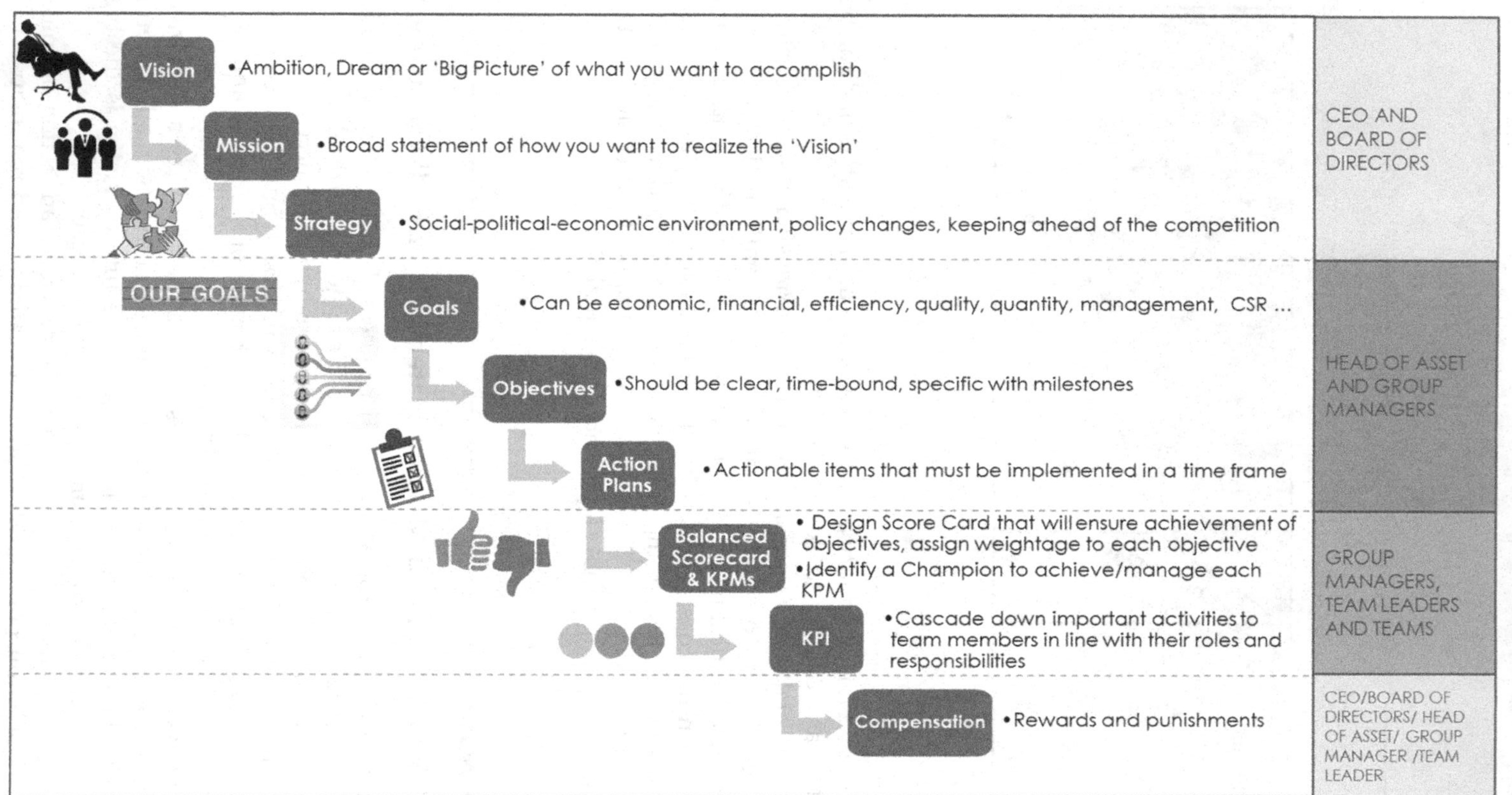

FIGURE 2.3
Stepdown Process for Managing Organizational Performance.

Values are a set of ground rules, statements of ethical practices, or a list of clear-cut do's and don'ts to deliver a service or product. Value statements represent the organizational beliefs and behavior in dealing with employees, clients, competitors, and society. An understanding of core values helps everyone in the organization to decide what is right and what is wrong. Sharing this value system with all stakeholders within and outside the organization, including clients and government, enhances its credibility.

The strategy aims to accomplish the mission under a given set of conditions. An organization that proactively accounts for uncertainties and adapts to change management programs can steer its way to success. Anticipation, flexibility, swiftness, and decisive actions need to be packaged together to benefit from unplanned social-economic-political windfall opportunities.

The vision, mission, value, and strategy statements provide a collective framework for corporate governance. By publishing these statements, the organization sets a level-playing field where each player can perform according to his/her talent, potential, and capability.

Preparing a roadmap and action plans based on organizational capability is the next most critical task. This job is best done by the asset and group managers who have personal knowledge of their team's abilities. Group managers, team leaders, and SMEs work together to schedule the tasks/projects based on priority. It sets up the stage to prepare the Balanced Score Card (BSC) and Key Performance Measures (KPMs). An organization's BSC is like a dashboard of an automobile displaying its key performance indicators (KPIs) in a visual graphic or tabular format. KPMs can be broken down into KPIs for assigning tasks to various assets, teams, and individuals based on their specialization, functional roles, and responsibilities that they must accomplish within a timeline.

Open and objective communication between the leadership and teams is essential for distributing work and sharing responsibilities in their specialization fields. True professionals like to take on challenges. They have a tremendous appetite for work and loyalty to their organization. They are looking for deeper motivations in life than a mere paycheck. A good leader can capitalize on their profound sense of duty and commitment to achieve his mission's goals.

This process integrates both the management and the employees with the long and short-term goals of the organization. It uses a structured and incremental approach to success. However, neither the process nor its elements are cast in stone. Innovations and improvisations are essential elements for the delivery of results. When combined with periodic performance management reviews, it is easy for everyone to see how far or near the organization is to achieve its set goal.

2.6 Private and Public Sector Oil Companies

Oil-rich countries have managed to ensure energy security and improve their people's lifestyles in a short time. That is why many governments want to have their control over oil and gas resources and management. The companies that are owned wholly or in the majority by their home country's government are called the NOCs. They leverage from the government's ownership of reserves and autonomy to access scarce national or international resources such as capital, technology, infrastructure, and equipment for operations. As a government representative, they benefit from its political clout and network to negotiate export and sales deals. NOCs also benefit from government protection when it comes to policy, pricing, and planning.

IOCs are the private sector companies that operate globally and have been in the oil business for decades. They have a distinct advantage over NOCs based on the expertise gained from their global international operations supported by advanced technological developments, let alone their money power for financing huge oil and gas projects. Based on their varied experience IOCs have developed a reputation for efficient, cost-conscious, result-oriented organizations that combine delivery with lean and robust business processes.

Private Sector and Public Sector and Private Limited and Public Limited companies are radically different. The term private sector refers to companies or organizations not owned by the government or where providing goods/services is not the government's responsibility. Public Sector companies are essentially government-owned companies or organizations. These are mostly the companies where the host government holds the full or majority share in the equity and is responsible for providing goods or services (Table 2.2).

A private sector company can be either Private Limited or Public Limited. A company is called Private Limited if:

- a group of promoters owns it
- its shares are in private hands
- it is not listed on the Stock Exchange and
- gains or losses made by the company are shared by its promoters/owners.

On the other hand, a public limited company:

- is owned by individuals (not by a government)
- is a corporation listed on a recognized stock exchange and traded publicly
- shareholders share profit or loss made by the company

TABLE 2.2
Public Sector versus Private Sector

Differentiating Feature	Public Sector	Private Sector
1. Ownership, Control, and Management	Full or majority share in the ownership, control, and management by the national government	Owned, controlled, and managed by private individuals or entities
2. Primary Objective	Value creation for the host country	Wealth creation for stakeholders
3. Capital Funding	By national government	Little/rare financial support by the national government
4. Listing on the Stock Exchange	Publicly traded on stock exchanges	Publicly traded on stock exchanges
5. Overseas Operations	Relatively easy	Require government support
6. Work Culture	Relaxed, high job security, lower emoluments	Aggressive, low job security, compensation based on merit and performance

Public limited companies are regulated by law and are required to publish their complete and correct financial position so that investors can determine the true worth of their stock. By the above definition, IOCs are a part of the Private Sector, and NOCs are part of the Public Sector.

The petroleum sector's need for investments and technology is everlasting. The world needs technology as well as the finances of both the private and public sectors. The companies from these two sectors can operate independently or collaborate in harvesting the energy from petroleum for growth and development.

2.7 Examples of Vision and Mission Statements of Oil Companies

The basic grounding in company structure is essential to understand the psyche and mode of operation of companies engaged in oil and gas sector activities. Private sector companies' focus is on creating wealth for a few owners if it is a private limited company or their shareholders if it is a public limited company. But the focus of public sector companies is not only wealth creation; its objective is to provide service/products to the masses and create lasting value for the national economy. That is why governments prefer the NOC instrument despite the possibility of a lower rate of return on investments in oil and gas projects.

In this background, some well-known NOCs and IOCs' vision and mission statements are reviewed to study the motivation of their businesses.

a. **Three Public Sector companies (all NOCs)**
   Saudi Aramco, China National Petroleum Corporation (CNPC), and Petróleos de Venezuela S.A. (PDVSA)
b. **Three large-size Private Sector, Public Limited IOCs**
   Exxon Mobil, Royal Dutch Shell, British Petroleum (BP)
c. **Three relatively small-size Private Sector, Public Limited companies**

   Anadarko, Apache Petroleum, Japan Exploration Petroleum Company (JAPEX)

Sampling for company selection is based on the type of sector, public or private, global geographical location, and annual revenue size. Most of the information presented below about the vision, mission, and core values of companies have been sourced from company/internet websites. However, the annual revenue data is extracted from Wikipedia.

2.8 Public Sector NOCs

Saudi Aramco, China National Petroleum Corporation (CNPC), and Petróleos de Venezuela S.A. (PDVSA)

2.8.1 Saudi Aramco[17]

Company Name	Company Logo	Status	Headquarters	Revenue Billion US $
Saudi Aramco	ارامكو السعودية Saudi Aramco	NOC	Dhahran, Saudi Arabia	356

Vision

Saudi Aramco is the world's leading integrated energy and chemicals company, focused on maximizing income, facilitating the sustainable and diversified expansion of the Kingdom's economy, and enabling a globally competitive and vibrant Saudi energy sector.

Mission

Saudi Aramco delivers on its core mission of reliably supplying energy to the Kingdom of Saudi Arabia and the world. It aims to excel as the world's leading integrated energy and chemicals enterprise, a top refiner, and a creator of energy technologies.

Values

Its core business values are safety, accountability, excellence, and integrity.

2.8.2 China National Petroleum Corporation[18]

Company Name	Company Logo	Status	Headquarters	Revenue Billion US $
CNPC		NOC	Beijing, China	429

Vision

CNPC's vision is to be a leader in the hydrocarbon industry and ensuring the gradual increase of profitability in its operations. It plans to achieve these objectives through strategic development, technological innovation, and accelerated change, emphasizing quality, and efficiency.

Mission

CNPC's commitment to efficient and sustainable energy use is reflected in its motto, "Caring for energy, Caring for you." The company's goal is to accomplish a harmonious relationship between operations and safety, energy and environment, corporate and community interests, and employers and employees. It is committed to fulfilling its social responsibility, promoting local growth and economic development.

Values

CNPC's core values include dynamism, loyalty, honesty, commitment, and excellence through innovation and integrity.

2.8.3 Petróleos de Venezuela S.A.[19]

Company Name	Company Logo	Status	He	
PDVSA	PDVSA	NOC	Maracaibo, Venezuela	48

Vision

PDVSA's vision is to become the energy firm of the world reference for excellence.

Mission

The company aims to satisfy the energy needs of society and create the maximum value for Venezuela.

Values

Not available.

2.9 Private Sector, Large Size Public Limited IOCs

Exxon Mobil, Royal Dutch Shell, BP

2.9.1 Exxon Mobil[19]

Company Name	Company Logo	Status	Headquarters	Revenue Billion US $
Exxon Mobil	ExxonMobil	Public IOC	Irving, Texas, USA	269

Vision

Exxon Mobil Corporation is committed to being the world's premier petroleum and petrochemical company. To that end, it must continuously achieve superior financial and operating results while adhering to the highest standards of business conduct. These unwavering expectations provide the foundation for the company's commitments to those with whom it interacts.

Mission

Energy is fundamental to the world's economies. Improving living standards around the globe requires affordable, reliable energy. Providing this energy is an enormous challenge that must be met practically, safely, and in an environmentally and socially responsible manner.

Values[20]

ExxonMobil's core business values comprise "work flexibility, safety, and security, recognizing human rights, integrity and diversity, and inclusion." ExxonMobil exploits these guiding principles to bring affordable energy to new global markets.

2.9.2 Royal Dutch Shell[19]

Company Name	Company Logo	Status	Headquarters	Revenue Billion US $
Royal Dutch Shell		Public IOC	The Hague, Netherlands	265

Vision

We make the difference through our people, a team of dedicated professionals, who value our customers, deliver on our promises, and contribute to sustainable development.

Mission

Royal Dutch Shell is committed to safely market and distribute energy and petrochemical products while offering innovative value-added services.

Values

Not available.

2.9.3 British Petroleum[21]

Company Name	Company Logo	Status	Headquarters in	Revenue Billion US $
BP	bp	Public IOC	London, UK	223

Vision

One of the seven supermajors, BP, is determined to have the best competitive corporate, operating, and financial performance. It commits to improve and to be accessible, inclusive, and diverse.

Mission

The company seeks to display integrity, honesty, mutual respect, and advantage, and contributing to human progress in all their activities.

Values[22]

BP's corporate values include safety, respect, excellence, courage, and one team spirit.

2.10 Private Sector, Small-Size Public Limited Companies

Anadarko, Apache Petroleum, and JAPEX

2.10.1 Anadarko[23]

Company Name	Company Logo	Status	Headquarters in	Revenue Billion US $
Anadarko	Anadarko	Public	The Woodlands, Texas, USA	11

Vision

Not available.

Mission

Anadarko's mission is to provide a competitive and sustainable rate of return to shareholders by exploring for, acquiring, and developing oil and natural gas resources vital to the world's health and welfare.

Values

Anadarko promises:

- To maintain full integrity and trust and act with the highest ethical standards to honor their commitments and obligations to work, family, faith, and community.
- To demonstrate servant leadership by placing others' success above its own and pledge to exhibit personal humility and professional courage.
- To recognize and reward strong performance and respect diversity in thought, practice, and culture.
- To safeguard the long-term interest of their shareholders and maintain high HSE standards.
- To encourage open communication and allow constructive debate.

2.10.2 Apache Petroleum[24]

Company Name	Company Logo	Status	Headquarters in	Revenue Billion US $
Apache Petroleum	Apache Exploring What's Possible	Public	Houston, Texas, USA	6

Vision

Apache Petroleum has the vision to be a premier exploration and production company, contributing to global progress by meeting its energy needs.

Mission

The company's mission is to grow in an innovative, safe, environmentally responsible, and profitable manner for its stakeholders' long-term benefit.

Values

The following core values govern the company's business activities and operations:

- Safety is not negotiable and will not be compromised
- Expect top performance and innovation
- Seek relentless improvement in all facets
- Drive to succeed with a sense of urgency
- Invest in the company's greatest asset: its people
- Foster a contrarian spirit
- Treat company stakeholders with respect and dignity

- Take environmental responsibility seriously; and
- Conduct the business with honesty and integrity

2.10.3 Japan Exploration Petroleum Company[25]

Company Name	Company Logo	Status	Headquarters in	Revenue Billion US $
JAPEX	JAPEX	Public	Tokyo, Japan	3

Vision & Mission

JAPEX contributes to society through a stable supply of energy and solves social issues to realize the following sustainable development agendas:

- Explore, develop, produce, and deliver oil and natural gas in Japan and overseas.
- Enhance the natural gas supply chain consisting of the domestic infrastructures.
- Develop and commercialize new technologies to solve problems for a sustainable society on energy and climate change.
- Achieve sustainable growth and maximize corporate value.

Values

Japex corporate values are listed as follows:

- Comply with national and international laws and regulations.
- Accord top priority to HSE.
- Keep stakeholders' interests in mind and contribute to the welfare of society.
- Respect the human rights of all people.
- Engage in fair and free competition and maintain appropriate business practices.
- Have no relationship with any anti-social elements.
- Protect confidential information and personal/clients' data.

2.11 General Comments on the Vision, Mission, and Values of the Companies

The corporate vision, mission, and values of the NOCs and IOCs have a very subtle difference. It is probably so because all the selected companies are well-established public limited companies and have to live up to international standards.

As expected, NOCs place a greater emphasis on social/national value creation, whereas IOCs quite understandably endeavor to maximize the value for their shareholders. This difference will be steeper if the vision, mission, and values are compared with the private sector, private limited companies. This ideological difference between the two has an impact on the business guidelines and operating philosophy of the private sector and public sector companies engaged in the management of E&P operations.

2.12 Effect of Operating Company's Strengths on Reservoir Management

The IOCs have long ranked as the largest oil and gas producers worldwide. They have delivered a significantly higher return on capital than national oil companies of similar size and operations. The petroleum sector is unique due to its multi-pronged character with shades of social, economic, and geopolitical considerations and their impact on host nations. However, to take advantage of these resources, they need investment capital, technology, and know-how that they have in limited measures. Consequently, the NOCs have to look for external support available from IOCs on specific terms and conditions.

IOCs and NOCs have their strengths and weaknesses, and they are well aware of them (Table 2.3). As a result, their collaboration through strategic alliances for exploration, development, and management of reservoirs is an inevitable and welcome recipe for success.

Quite clearly both IOCs and NOCs have the opportunity to complement each other to overcome their limitations. Together they can form a formidable combination and leverage each other's strengths in the upstream and downstream sectors to create a win–win situation.

2.13 Commercial Agreements

Oil- and gas-bearing countries have three options to explore and develop their petroleum resources:

a. Create their national oil companies, for example, Saudi Arabia, China, Venezuela, Iran, Kuwait, India, and so forth

TABLE 2.3

Strengths and Weaknesses of IOCs and NOCs in the Petroleum Sector

No.	Criterion	Private Sector IOCs and Others	Public Sector (State-Owned NOCs)
1	Access to Reserves	Low	High
2	Technical Competence	High to Medium	High to Low
3	Technological Capabilities	High	Medium to Low
4	Regulatory & Commercial Influence	Low	High
5	Company Culture	Adhocratic	Bureaucratic
6	Decision making	Efficient	Relaxed
7	Project Management	Good	Average
8	Performance	Good	Average
9	Profit Margin	High	Low
10	Project Size Handling	Small to Large	Large
11	Redundancy	Low	High
13	Compensation	Performance-based	Seniority-based
14	Job Security	Medium to Low	High

b. Allow private sector companies to undertake E&P operations, for example, the US, the UK, Russia, Brazil, Canada, and so forth
c. Use a mix of the above two systems, for example, Nigeria, Azerbaijan, Kazakhstan, India, and so forth

Implementation of options "b" and "c" requires the host nation to identify an interested and capable company that can meet its investment and technology aspirations. The next step is to hold negotiations with this company to develop agreed terms and conditions. A detailed, self-sufficient document providing details of the commercial and legal framework is then prepared and signed by the two sides. This arrangement between the host nation and the operating company that is selected to support the E&P objectives is commonly referred to as a commercial agreement and can be classified into four categories:

2.13.1 Concessions

Concessions are probably the oldest kind of long-term contract. The host governments have been using this instrument to grant third parties the right to explore and extract petroleum reserves on their land or offshore. A self-contained document called "concession" contains the terms and conditions of this agreement. It may also be called a "license" or "permit" for exploration and "lease" for exploitation purposes.

The agreement conditions could vary and are subject to the negotiations between the host nation and the Private Oil Company (POC). Generally, the POC bears the risk in exploration plus the cost and taxes of all operations. Should exploratory efforts materialize in a commercial discovery, the POC reserves the right to produce and sell oil while the host government receives royalty and income tax.

It is a simple and straightforward type of agreement if the host nation adopts a transparent bidding system. Most of the oil and gas reserves in the Middle East, US, France, Russia, Brazil, and several other countries were explored and developed under the concessionary system. However, this system's inherent disadvantages left both parties-host governments and POCs with a bitter taste. In most cases, the pace of exploration and development by POCs did not match the expectation and urgency of host nations. As a result, national governments created NOCs to work with POCs in a Joint Venture (JV), which had more equal terms and conditions for both sides.

A summary of the concessionary system with its advantages and disadvantages to the host nation is presented below:

Contract Type: Concession
Participant: IOC/POC
Ownership: IOC/POC
Control: IOC/POC

Advantages to the host nation:

- Full discretion and control for awarding the field/area
- IOC/POC bears operational risks
- Receipt of royalties and income tax on commercial production without any investments
- Flat and fixed royalties without consideration of the level of production or oil price

Disadvantages to the host nation:

- Ownership and control of its resources compromised
- The potentiality of the oilfield not known
- Traditional arrangement granted rights to IOC/POC to decide market and price
- Inordinate delays in exploratory work due to risk perceptions and long-term engagements
- The blocked potential of natural resources, no provision for relinquishment of non-explored areas

- No or low signature bonus
- Terms were frozen for the life of the contract

Revenue Sharing under the Concessionary Model

Revenue sharing under the Concessionary system is explained with a simple example (Table 2.4). Gas revenue is ignored. Assumptions are made for the oil price, royalty, and tax rates.

In this example, the final revenue split between the host government and POC is in a ratio of approximately 51:49%.

2.13.2 Joint Ventures

JVs simply bring together the host nation and the POC with an explicit desire and commitment to pursue a joint undertaking of oil and gas exploration and exploitation. Jenik Radon likens the arrangement of this undertaking to a "premarital understanding of the two interested parties. There is no attempt to solve the material issues before entering the JV. Instead, it proceeds with the spirit of collaboration, forbearance, and sportsmanship to make it work."[26]

The guiding principle of a 50:50 deal of the JV is "unanimity." Based on this understanding, the two parties agree to work together to achieve their common goals. However, a JV in the petroleum sector can have a long list of complex issues that arise from the human nature of looking at "strengths before marriage" and the "weaknesses" after. The changes in business guidelines, hidden agendas, national or international relationships, gain or loss of market share, unpredictable fluctuations in oil and gas price, trust deficit, and geopolitical conflicts can trigger unforeseen problems for the JV between the host nation and participating POC. Jenik Radon's comparison of the success rate of modern-day marriages with JVs is quite a nice example.

The brief highlights of the JV system are presented as follows:

Contract Type: JV

Participant: NOC and IOC/POC

Ownership: Host government/(NOC) unless otherwise stated in the contract

Control: Host government/(NOC) and IOC

Advantages to the host nation:

- Retains ownership of its resources
- Joint responsibility for a project and decision-making by the host nation and IOC (experts in the business)
- The host nation can count on the expertise of the IOC
- Profit is shared on top of royalties and taxes

TABLE 2.4

Revenue Sharing between the Host Government and POC under a Concession

Head	Unit	Value	Distribution	
			Govt. Take	POC Take
A. Wellhead Price of Oil	$/bbl	40.00		
B. Royalty = (A*Royalty Rate)	%	10	4.00	
C. Net Revenue = (A-B)	$/bbl	36.00		
D. Deductions towards Operating Cost and DDA, IDCs, etc.*)	$/bbl	6.00		6.00
E. Taxable Revenue = (C-D)	$/bbl	30.00		
F. Provincial Taxes = (E*Tax Rate)	%	10	3.00	
G. Net Taxable After Provincial Tax = (E-F)	$/bbl	27.00		
H. Federal Income Tax = (G*Tax Rate)	%	50	13.50	
I. Net after Royalty, Operating Cost, and Taxes = (G-H)	$/bbl	13.50	0.00	13.50
Total ($/bbl)			**20.50**	**19.50**
			51.25%	48.75%

DDA-> Depletion, Depreciation, Amortization

IDCs-> Intangible Drilling and Development Costs refer to various expenses incurred during the location and preparation of wells used for production. IDCs include wages, drilling equipment repairs, hauling costs, and supplies, among other expenses associated with the production process. For federal tax, IDCs are treated as deductions.

Disadvantages to the host nation

- Discretion and control are significantly reduced
- Risks and costs to be shared
- Direct participation makes it responsible for potential liability on projects causing pollution or environmental damage
- Constant media glare and publicity

2.13.3 Production-Sharing Agreements or Production Sharing Contracts

This commercial arrangement is quite complicated and is built on the experience of the two other mechanisms discussed in the section "Concession" and JVs. Production-Sharing Agreements (PSAs) or their revenue-equivalent version of hydrocarbons called "Profit-Sharing Agreements" came into vogue after Indonesia's successful experimentation and experience in Indonesia.

Under PSAs, the host nation or its NOC can retain ownership of the oil and gas resource. It compensates the POC, which manages the exploration or development operations by sharing the profit in proportion to the agreed terms based on the following formula:

Profit Oil = Total Profit – Cost Oil

In this arrangement, financial risks of exploration and development are mostly borne by the POC. POC is entitled to recover the cost of operations from the total profit. Production from the oil or gas field specified in the PSA is divided between the host nation and POC based on their share. Some JVs may keep the NOC-POC split of profit oil share constant irrespective of the production level, while others may vary this split with a change in output level. The host nation or NOC may participate in the venture as an interest holder in the PSA. It may contribute some of its future profit as "share capital" to the consortium assigned with the development of the area granted under the PSA. All these provisos and other commercial, technical, financial, and legal issues that can make or break the PSA are subject to negotiations and agreement before it is signed. A detailed PSA document is then scripted and signed by the participating companies.

The brief highlights of the Production Sharing Contracts (PSCs) are presented as follows:

Contract Type: PSC

Participant: NOC and IOC/POC

Ownership: Host government/ (NOC) unless otherwise stated in the contract

Control: Host government/ (NOC) and IOC/POC

Advantages to the host nation:

- Retains ownership of its resources
- Receipt of signature bonus, discovery bonus, production bonus, and royalty. It may sometimes be traded for a more significant future profit share.
- All financial and operational risks rest with the IOC/POC
- IOC/POC is required to pay taxes on its share, often waived by the host government and accommodated in the agreed percentage split
- The host nation has the option to bring another company if participating IOC/POC does not comply with PSC or continues to fail in objectives

Disadvantages to the host nation

- Recovery of production costs
- Profit oil (strategic commodity) split with IOC/POC
- Loses an opportunity if an exploration or development project fails (time value of money or oil)

- Higher the risks to the investments, the higher the share of profit demanded by companies.
- The host government might introduce new policies and tax structures which can impact project economics

Example of Revenue Sharing under a PSC

Revenue sharing arrangement under a PSC agreed between a government and POC is explained with a simple example (Table 2.5). Gas revenue is ignored for simplification. Assumptions are made for the oil price, royalty, and tax rates.

For the PSC example, the final revenue sharing between the host government and POC works out in a ratio of approximately 68:32%.

2.13.4 Service Contracts

Service contracts are a convenient work-around for host nations to access POC/IOC expertise, technology, capital, and brand reputation. Such agreements provide a straightforward contractual framework to acquire specialized oilfield services or consultation without compromising or controversies. Sometimes service contracts may also be used to retain a reputed IOC as a shelter for organizational inefficiencies, mishandling of sensitive projects, or constant failure to achieve targets.

Master Service Agreements provide a long-term ongoing relationship between large oil companies and service providers. Drilling contracts, Seismic Licensing Agreements, and Production Well Testing Contracts best describe the nature of Service Contracts. Some drilling service

TABLE 2.5

Revenue Sharing between the Host Government and POC under a PSC

Head	Unit	Value	Distribution	
			Govt. Take	POC Take
A. Wellhead Price of Oil	$/bbl	40.00		
B. Royalty	%	10%	4.00	
C. Net Revenue After Royalty = (A-B)	$/bbl	36.00		
D. Operating Cost	$/bbl	5.00		
E. Depletion, Depreciation, & Amortizations; IDCs etc.	$/bbl	2.00		
F. Total Cost Recovery = (D+E)	$/bbl	7.00	0.00	7.00
G. Profit Oil = (C-F) shared in 60:40 ratio	$/bbl	29.00	17.40	11.60
H. Federal Income Tax @50%	%	50%	5.80	–5.80
Total ($/bbl)			**27.20**	**12.80**
			68.0%	**32.0%**

companies are better equipped than others to deal with the problems posed by offshore, horizontal, multi-lateral, or high-pressure high-temperature wells. Seismic Licensing Agreements may be required for a seismic shoot or non-exclusive use of acquired seismic data. Production well testing contracts may specify the nature and number of jobs to be performed every day or as required. Well-Service contracts usually include elements of both services and equipment.

Some important considerations for the Service Contracts are as follows:

- **Payment of fee:** Fees may be paid on Turnkey, Day Rate, or any other basis. Turnkey well drilling may imply drilling wells to a certain depth for a fixed fee. Day Rate structure of payment may represent payment based on "days on the job" with inclusive or exclusive provision for mobilization or demobilization of drilling rigs. Seismic Licensing Agreement rates may be fixed on the "Line Kilometer" basis.
- **Compensation for service-related risks:** Service contracts may be "Risk Exclusive" or "Risk Inclusive" contracts. Risk exclusive contracts are risk-free contracts with no liabilities of any kind to the oil company. In the "Risk Inclusive" deals, operational risks are borne or shared by the oil company. In the case of NOC being in charge of exploration and drilling operations carried out by a service company, NOC absorbs the risk of wells not achieving their objective. However, under the provisions of the Turnkey or Day Rate structure, the Drilling Service Company may have to incorporate drilling risks. In the case of IOC being in charge of exploration, it bears all the exploration costs in Risk Service Contracts.
- **Indemnification (claims for unforeseen loss of time, equipment, property, injury, or life)** The simplest way of dealing with indemnification is where both parties, that is, Oil & Gas Company and Service Company, assume all risks associated with their equipment and personnel.

The commercial arrangements discussed earlier might have varying characteristics; the core of the matter is how smartly the host nation and the participating POC or its consortium share the cost and risks and divide the profits.

And since contracts must be signed well before work can start, neither the host nation/NOC nor POC/Service Provider is fully geared to forecast this accurately. NOC has the advantage of knowing its reservoirs but cannot predict the turn of events during operations. POC/Service Provider has the experience but not the knowledge of reservoirs. Interestingly, both sides keep their cards to the chest until they come to the negotiation table to discuss the most loaded items: cost, profit, and risks.

2.14 Reservoir Development Objectives & Early Planning Considerations

The discovery of oil and gas is excellent and welcome news for everyone in a country. It portends hope for a better future in underdeveloped countries and reassurance of energy supplies in developed countries. Discovery gives rise to a flurry of pre-development activities since operating companies are usually interested in fast-tracking oil and gas production for the following reasons:

- Generate a quick cash flow for the company and support the economy
- Complement the energy supply to the national or international market
- Develop local infrastructure and strengthen the local market
- Generate employment, training, and development opportunities for people

It is a good idea for the development team to be proactive and think through the challenges it can face in developing an oil or gas field. Critical management and operational decisions need to be made at each stage relating to schedule, cost, drilling, production, technology, training, and so forth. Undoing an action is impossible; it can only be corrected with associated consequences in time delays, cost overruns, and lost capital and opportunities.

Short- to long-term impacts of the development activities must be considered thoroughly. Large-scale corrections to the development plans can result in wastage of money and inventory, missed targets, and reduced credibility. Since development and production activities for primary and secondary recovery may continue for 30 to 40 years and involve huge capital expenditure, it is good to split the plan into phases. Each phase would require a detailed development plan that can be prepared with a reasonable degree of confidence about activities and expenditure and lower the risk to investments. These Development Plans would usually be operative for 10 to 12 years and are discussed below.

2.14.1 Management Questions and Objectives

Even before the implementation of the very first phase of development starts, the company must be in a position to address the following fundamental questions:

- Is the organization capable of managing the project in terms of capital, technology, and human resource?
- How will the major oilfield operations (seismic surveys, well drilling, oil and gas production, processing, and transportation) be carried out?
- What commercial agreements/contracts will be required for oilfield services and the sale of oil or gas, and by what time?
- What pace of development is necessary? The following will determine it:
    - The urgency of oil or gas requirement
    - Reservoir depth, type, and number of wells to be drilled
    - Number of drilling rigs and equipment/material availability
- What technologies are necessary, and how the gap, if any, will be bridged?
- Is the company fully aware of petroleum legislation, environmental regulations, local labor laws, and taxation systems?
- Does everyone involved in the project understand HSE policies and issues clearly?
- Does the company have financial resilience to absorb the CAPEX and OPEX until the payout period?
- How deeply is the company going to be involved in fulfilling social objectives?

Once a positive answer to these questions is ascertained, the stage is set to move forward with the Development Planning.

2.14.2 Conceptual Development Plan

A Conceptual Development Plan (CDP) is based on data acquired from wells drilled in the exploration and delineation phase and an outlook of development activities for the next 5 to 10 years.

Key Technical Considerations:

- Type of hydrocarbon (oil, free gas, or condensate)
- Type of reservoirs: clastic or carbonate, single or multiple, sheet, stacked or staggered
- Size of the resource and uncertainty associated with the estimates (P90, P50, or P10)
- Size and strength of the aquifer and gas cap if present
- Depletion policy

  - At what rate the fluids will be withdrawn from the reservoir
  - Production capacity of the wells and reservoir(s)
  - Natural operating drive mechanism and its efficiency
  - Assessment of pressure depletion
  - The need and timing of pressure maintenance
  - Oil, gas, and water production profiles with estimates of recovery
  - Well-completion strategy (technical, economic considerations)
  - Number of wells (production, injection, and disposal), well-type (vertical, horizontal, deviated, multi-lateral), completion type (openhole, cased hole, single or dual)
  - Estimated Ultimate Recovery Factor (EURF) under primary recovery and well spacing
  - Facility capacity required to handle oil, gas, and water production
  - Options and strategy for disposal of produced water
- Well-hook up arrangements, contracts, and so forth
- Possible use and discussion of Artificial Lift (AL) methods
- Crude storage and dispatch, transportation to export/sale point
- Possible risks associated with the CDP, and the mitigation plan

An essential piece of information for the CDP is the pressure and production variation with time, also known as pressure-production history. All this knowledge is compiled and used to prepare the CDP.

2.14.3 Initial Development Plan

An Initial Development Plan is based on the data acquired in the early stages of the development phase. It presents an outlook of development activities for the next 5 to 10 years. It is an update/revision of the CDP after a thorough analysis of what went right or wrong in the early development phase.

Key Technical Considerations:

- Construction of a geological model with structure, faults, properties, and fluid distribution
- The revised deterministic and probabilistic estimates (P90, P50, P10) of hydrocarbon-in-place volumes of oil and gas
- Revised well-completion strategy based on the experience gained in the earlier phase
- Review of the production capacity of the field/reservoir(s)

- Confirmation of aquifer/gas cap (if present) size and strength from reservoir performance data
- Evidence for the need and timing of pressure maintenance
- Construction of a reservoir simulation model and its calibration with dynamic data
    - Revised EUR with and without pressure maintenance
    - Assessment of possible development options and selection of the optimum variant
    - Required number and type of production wells, completion strategy, well spacing
    - Drilling schedule, plateau rate, timing, and plateau period
    - Injection strategy (peripheral or pattern) and injector locations
    - Water/gas injection schedule and rate forecast for maintaining the target reservoir pressure
    - EUR with and without pressure maintenance
    - Long-term and Life of Field (LOF) oil, gas, and water production forecasts
- Application of standard/classical reservoir engineering methods to crosscheck in-place volumes, recovery factor, decline rates, and so forth
- Availability and compatibility of injection fluids (water or gas), injectivity testing
- Pilot test proposal for pressure maintenance and outlook for expansion of the project
- Type of drilling and workover rigs required, an estimated load of the drilling and workover rigs, rigless workover activities
- Type of AL, timing, and details of the AL plan
- Effluent Water Disposal (EWD) plan: It is a critical piece of the field development plan. This plan must identify a reservoir for disposal, provide an estimate of the size and capacity of the reservoir to accept EW, have a blueprint of the transmission system from the gathering center to disposal wells, and provide the number of EWD wells, their locations, and their average disposal capacity. This plan must present a future outlook for disposal or reinjection into the reservoir for pressure maintenance, decide about the quality and quantity, makeup volumes, need for treatment before disposal/reinjection, and many other pertinent questions.
- Possible risks associated with the initial development plan and the mitigation plan

2.14.4 Mid-Term Development Plan

A Mid-Term Development Plan is based on a review of activities carried out during the Initial Development Plan phase. The idea is to check what went according to the plan and what didn't. It also presents an outlook of development activities for the next 10 to 15 years.

Key Technical Considerations:

- Revise deterministic and probabilistic estimates (P90, P50, P10) of hydrocarbon-in-place volumes
- Review of well-completion strategy based on the experience gained during the Initial Development Plan phase
- Design and test one or more water/gas injection pilots, evaluate results
- Assessment of initial water/gas injection performance and implementation schedule of the full-fledged waterflood development plan according to the injection scheme, revised EUR, and production profiles
- Review of production and injection capability of the reservoir(s) with water/gas injection.
- Review of EW disposal/utilization strategy. Continue with EW water disposal into the subsurface or utilize for reinjection?
- Revisit the makeup volumes and the need for treatment before disposal.
- Determine suitability of water for injection from other sources, need for filtration and treatment. In the case of pressure maintenance by gas injection, ensure gas availability and need for processing, if any.
- Justify water/gas injection costs based on incremental oil recovery obtained by coreflood studies and well logs (S_{orw} or S_{org}) before starting the water/gas injection.
- Need for correction/refinement of anything not doing well during the Initial Development Phase
- Refine static and dynamic models with newly acquired data and finetune the simulation results -
    - To confirm the techno-economic suitability of the development plan
    - To finalize the development plan locations
    - To estimate EUR with and without pressure maintenance
    - To forecast the long-term and Life of Field (LOF) oil, gas, and water production and water/gas injection forecasts

- Implement the AL plan if necessary
- Estimated load of the rig and rigless workover activities
- Assessment of after-effects of EWD, risk mitigation if necessary
- Techno-economics of the project
- Screening for EOR methods

2.14.5 Final Development Plan

A Final Development Plan is based on the activities carried out until the end of the Mid-Term Development Plan. This plan should generally visualize and freeze all further capital expenditure on development activities required to be completed under the secondary recovery phase. All preparatory work to find a suitable EOR method that started at the end of the previous phase must be pursued further.

Key Technical Considerations:

- Need for correction/refinement of anything not doing well during the Mid-Term Development Phase
- Final estimates of hydrocarbon-in-place volume (P90, P50, P10) and reserves
- Well calibrated reservoir model with consistent volumes of oil-in-place and EUR determined by other methods
- Final development plan with well locations and production profiles
- Review of expenditure made on wells, flowlines, AL, oilfield services, and other items
- Estimates of spending on future wells, flowlines, facility upgrades/modifications/construction to handle oil, gas, water, and other projects
- Techno-economic parameters and estimated cost of the primary and secondary oil
- Assessment of EOR potential based on the results of the screening study
- Recommendations for laboratory studies to decide the most suitable EOR method
- Collect information for a feasibility study on EOR-
    - The laboratory efforts must establish the incremental oil recovery in the lab based on the corefloods
    - Identify and test wells to determine Sor with the proposed EOR agent
    - Best time to implement the EOR scheme

- Logistics of EOR project, handling of EOR production and injection streams upstream and downstream of wells, need for the refurbishment of facilities
- Estimated cost of the EOR oil
- Likely size and configuration of injection production patterns

Final Development Plan is not the end of the development planning, as the name may imply. This plan is also subject to further revision(s) based on new information or surprises such as a further extension or even shrinking of fields.

This plan should nearly sum up the pressure maintenance/secondary recovery plan and expenditure to be incurred on the field development. It is also a good time to initiate a feasibility study for EOR, depending on the positive results of the pre-work.

EOR methods may require slight modification in the quality of injection water or a completely new injectant in place of water as established by the laboratory experiments. This consideration is, however, dependent on the economic viability of the process.

Notes

1 EY Poland Report; Short-termism in business: causes, mechanisms and consequences
2 IFA Financial; https://ifafinancial.com/oil-gas-law/
3 Lanre Aladeitan; Ownership and Control of Oil, Gas and Mineral Resources in Nigeria: Between Legality and Legitimacy, Thurgood Marshall Law Review, Vol. 38, Spring 2013 No. 2, A Publication of Texas Southern University; Thurgood Marshall Law review. 38.
4 Arfaa, Noora; Tordo, Silvana; Tracy, Brandon S. 2011; National oil companies and value creation (English). World Bank working paper;no. 218 Washington, D.C.: World Bank Group. http://documents.worldbank.org/curated/en/650771468331276655/National-oil-companies-and-value-creation
5 Wikipedia The Free Encyclopedia; https://en.wikipedia.org/wiki/Uzbekneftegaz
6 Wikipedia The Free Encyclopedia; https://en.wikipedia.org/wiki/SOCAR
7 Wikipedia The Free Encyclopedia; https://en.wikipedia.org/wiki/List_of_oil_exploration_and_production_companies
8 Website; http://www.kmg.kz/eng/
9 Wikipedia The Free Encyclopedia; https://en.wikipedia.org/wiki/Naftna_Industrija_Srbije
10 Trend news agency website; https://en.trend.az/casia/turkmenistan/2478389.html
11 Paul Stevens, "International Oil Companies The Death of the Old Business Model"; Research Paper, Energy, Environment and Resources, May 2016; https://www.chathamhouse.org/publication/international-oil-companies-death-old-business-model

12 Ownership and control, AmosWEB Encyclonomic WEB*pedia, http://www.amosweb.com/cgi-bin/awb_nav.pl?s=wpd&c=dsp&k=ownership+and+control
13 Wikipedia The Free Encyclopedia; https://en.wikipedia.org/wiki/United_Nations_Convention_on_the_Law_of_the_Sea
14 Wikipedia-the free encyclopedia; https://en.wikipedia.org/wiki/United_Nations_Convention_on_the_Law_of_the_Sea
15 Robert Churchill, Encyclopedia Britannica, Law of the Sea (1982); https://www.britannica.com/topic/Law-of-the-Sea
16 Woo, L.M., 2005, 'Summer circulation and water masses along the West Australian coast', PhD thesis, University of Western Australia, Perth
17 COMPARABLY website; https://www.comparably.com/companies/saudi-aramco/mission
18 CNPC website; https://www.cnpc.com.pe/Quienes%20Somos/Pages/Misión%20Visión%20y%20Valores-en.aspx?ord=2&lang=en&mord=2
19 Alar Kolk, President at European Innovation Academy; Visions missions of fortune global 100 https://www.slideshare.net/openinnovation/visions-missions-of-fortune-global-100/
20 EXXONMOBIL website; https://mission-statement.com/exxonmobil/
21 Comparably website; https://www.comparably.com/companies/bp/mission
22 BP Website; https://www.bp.com/en/global/corporate/who-we-are/our-values-and-code-of-conduct.html
23 Website; https://www.anadarko.com/content/documents/apc/Responsibility/Anadarko-CRDeck.pdf
24 Website; http://www.apachecorp.com/About_Apache/Mission.aspx
25 Website; https://www.japex.co.jp/english/company/vision.html
26 J Radon; "The ABCs of Petroleum Contracts: License-Concession Agreements, Joint Ventures, and Production Sharing Agreements," in S Tsalik & A Schiffrin (eds), *Covering Oil: A Reporter's Guide To Energy and Development*, Open Society Institute, New York, 2005

3

Reservoir Management Requirements

3.1 Introduction

The modern world is divided into three categories: developed, developing, and underdeveloped based on development in a country or region. Despite disagreement on a specific set of social and economic criteria to decide the development level, energy consumption is unarguably one of the common indicators of progress. Practically all development activities in the social (domestic, food, water, nutrition, education, and healthcare) and economic segments (agriculture, transportation, industrialization, manufacturing, and infrastructure) are energy driven, and that is why the consumption of energy is one of the leading indices of development.

Energy is the ability to do work. The man himself has a limited physical capacity, but he continues to expand it by harnessing mechanical, electrical, chemical, and digital power. While the future annual energy consumption in OECD (Organization for Economic Co-operation and Development) or the so-called "developed" countries is more or less stable, the energy requirement for non-OECD or "developing" countries shows a significant upward trend (Figure 3.1).

Primary sources of energy include petroleum (crude oil and natural gas), coal, nuclear, and renewable energy. Presently, crude oil meets about 33% and natural gas 22% of the world's energy demand. The inset picture in Figure 3.1 presents the relative share of the consumption of petroleum liquids and natural gas until 2040 for all the countries. Projections of the International Energy Outlook (IEO) indicate that petroleum will continue to meet more than half of the world's energy demand until 2040.

3.2 Significance of Crude Oil

The heat value represents the importance of energy fuels like crude oil and natural gas. It is determined by the amount of heat released during their

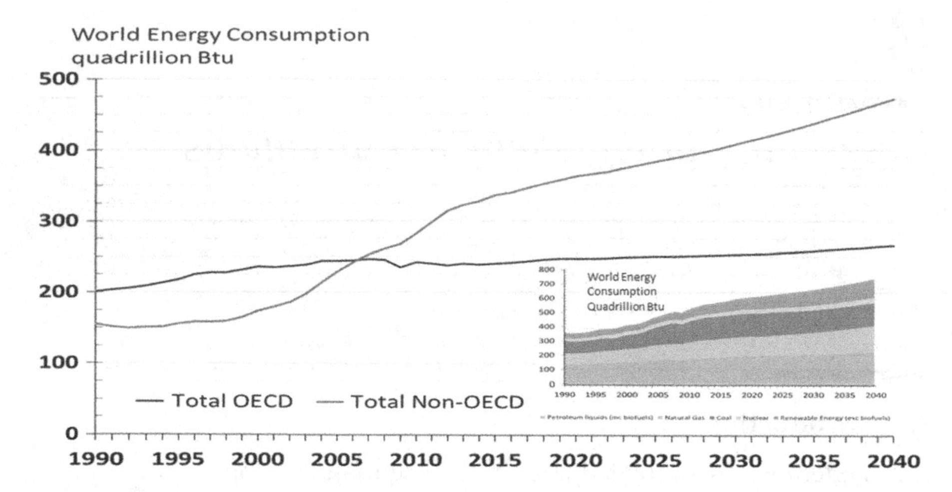

FIGURE 3.1
OECD and Non-OECD Countries Energy Consumption -Past, Present, and Future[1] (*Source: US Energy Information Administration, International Energy Outlook 2018*).

combustion or calorific value. The heat value of standard fuels, including petroleum, is presented in Table 3.1 for comparison.

Crude oil and other petroleum derivatives have higher heat values, except nuclear, and are, therefore, superior energy fuels than firewood and coal.

The word petroleum includes crude oil and natural gas. The significance of crude oil is determined by its products for human use and comfort (Figure 3.2). Natural gas is a cleaner, recognized energy source used for heating, cooking, and electricity generation. It is also used as a fuel for vehicles and as a chemical feedstock to produce fertilizer and generate

TABLE 3.1
Heat Value of Common Fuels

Energy Fuel	Heat Value MJ/Kg	Heat Value KgOE	Heat Value KCal
Firewood (Dry)	16	0.036	362
Hard Black Coal	24	0.054	543
Crude Oil	45	0.102	1,018
Natural Gas	49	0.111	1,109
Petrol/Gasoline	45	0.102	1,018
Diesel Fuel	44	0.010	996
Liquefied Petroleum Gas (LPG)	49	0.111	1,109
Natural Uranium in Fast Neutron Reactor	28,000,000	63.367	633,673,405

Key: MJ = 10^6 Joule, Kg → Kilogram, OE→Oil Equivalent, K→1000, Cal→Calories

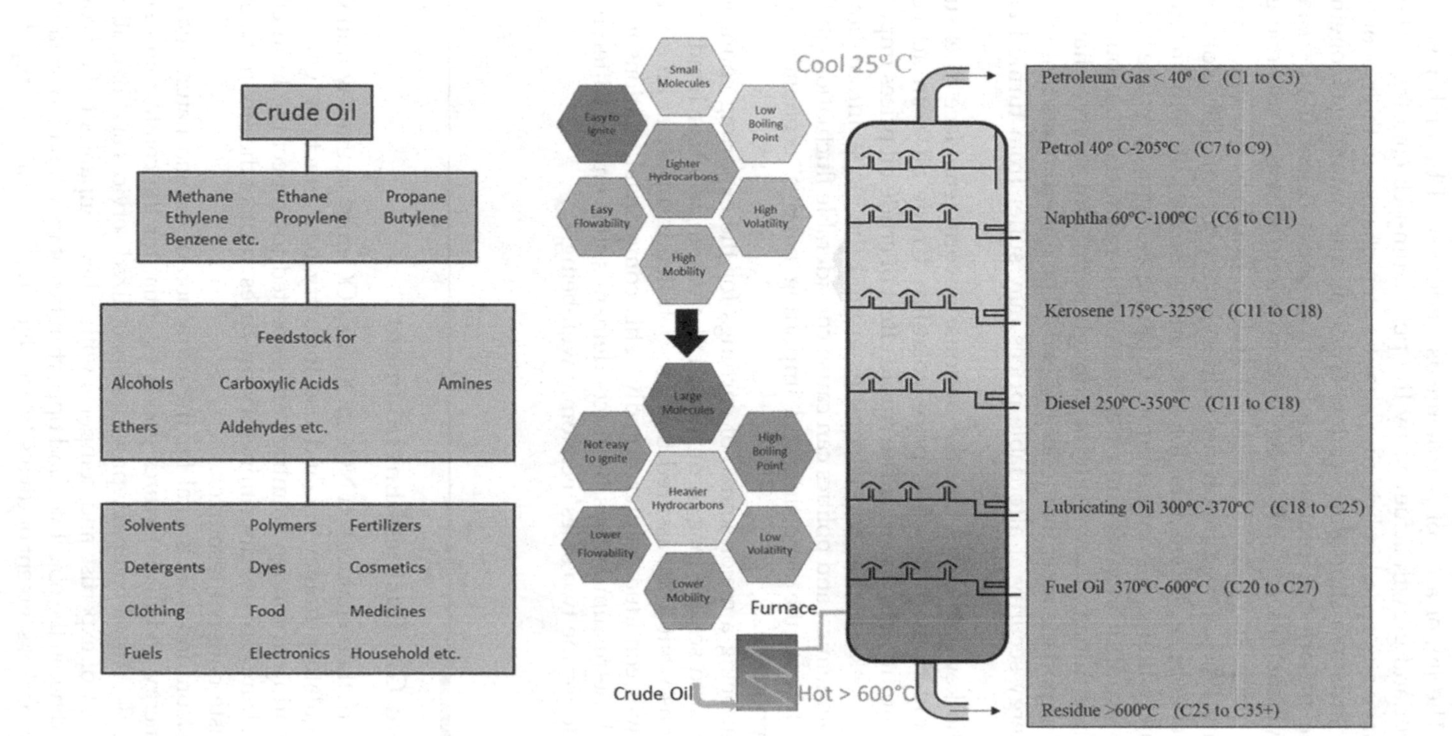

FIGURE 3.2
Significance of Crude Oil for the Modern World.

electricity. The importance of petroleum is underscored by high levels of development and growth achieved by the petroleum-rich countries. These countries have succeeded in bringing about a step-change in their people's living standards in a short time. Today, crude oil and natural gas together provide the majority of products and conveniences essential to sustain modern civilization's life. All peace and wartime activities use energy, whether cooking, travel, office work, laboratory research, manufacturing, farming, any kind of war involving military, navy, or air. Therefore, assurance of energy security is vital for all countries so that life and business can continue without any interruption. Countries surplus in crude oil or natural gas export it in raw form or as petroleum products. The countries that are not fortunate to own any or enough petroleum reserves have to depend on imports to meet their energy needs. The governments that fail to provide energy security at affordable prices may suffer from turmoil and backlash.

One of the key challenges that the national governments face to ensure energy supply from crude oil is price volatility. Crude oil is an actively traded commodity in the world. The ripple effect of crude oil prices impacts the price of stocks, bonds, and currencies around the globe. International incidents, pandemics, and politics can cause considerable fluctuations in the oil price, affecting the exporting and importing nations' economies and lifestyles oppositely.

Petroleum being a major source of earnings for the exporting countries, they are keen to seize a long-term market share. On the other hand, consumer nations want to be assured that the petroleum supply at an affordable price will continue uninterruptedly. This commercial relationship to buy and sell petroleum and its products demands that mutual differences be cast aside for the two sides' economic well-being.

3.3 Oil and Gas Business Environment

Irrespective of their character, a NOC, IOC, or POC must operate within the national regulatory framework and use industry best practices. Oil companies operating in foreign countries are expected to respect the local customs and culture, not use discriminatory practices, protect the environment, and not waste national resources.

The petroleum industry is vital for the extraction of oil and natural gas. It produces energy and manufactures goods essential for the comfort and survival of people. An oil company provides specialized services and operations in its core area of expertise and competes with other companies to be profitable. Depending on the objectives and opportunities, it may limit its operations to a specific market segment or diversify activities over the entire supply chain.

The oil industry is divided into three main sectors: upstream, midstream, and downstream. The upstream sector involves activities relating to Exploration and Production (E&P). The midstream sector includes transportation by road, rail, and sea. It also includes the storage, marketing, and distribution of crude oil and its products. The downstream industry comprises crude oil refining, where crude oil is refined into more useful products such as naphtha, gasoline, diesel fuel, kerosene, liquefied petroleum gas, jet fuel, and so forth. All three sectors define the entire supply chain of petroleum, starting from its exploration to the sale/distribution. A company that operates over all three segments is called a vertically integrated oil company.

Upstream, midstream, and downstream industries are subject to operational and market risks. Operational risks are typical of a particular industry. However, market risks are common to all sectors. They may arise from social unrest, political turmoil, fluctuations in interest rates, natural disasters, terrorist attacks, and geopolitical events or conflicts at a local, regional, or international scale. Any national or international incidents that can affect the global balance of demand and supply can trigger market risk.

Domestic conflicts can be of social, political, environmental, or regulatory nature. Work strikes and social unrest due to poor working conditions, demand for higher wages, and benefits can result in loss of work hours and profitability. These unrests could take on the dimensions of criminality and give rise to oil thefts and artisanal refining.

In the age of globalization, where business operations and infrastructure of oil and gas companies are spread over several countries, national events have regional and international fallouts. Profits and losses of the petroleum exporting and importing countries can soar and slump with the rise and fall in oil prices. History of the political standoff between the US and Iran on nuclear testing, escalation of the Iran-Saudi proxy war in Yemen, political conditions in Iraq, Syria, Libya, and Venezuela go on to prove the fragility of the world geopolitical order in respect of oil prices and threats to oil supply. The fact that oil and gas exporting countries are far-flung from the importing countries, huge sea tankers or pipelines must supply the oil and gas. In 2015, nearly 60% of the world's total supplies were routed through the three famous chokepoints, namely, the Suez Canal, the Strait of Bab el-Mandeb, and the Strait Hormuz and the Suez Mediterranean (SUMED) Pipeline in the Arabian Peninsula.[2] Chokepoints are narrow maritime seaways that are critical for global oil cargo (Figure 3.3). Geopolitical risks to energy supplies are real and a significant source of abrupt changes in oil prices. Any kind of conflict or tension in the neighboring countries around these chokepoints can have a devastating impact on oil prices and the energy security of the consumer nations.

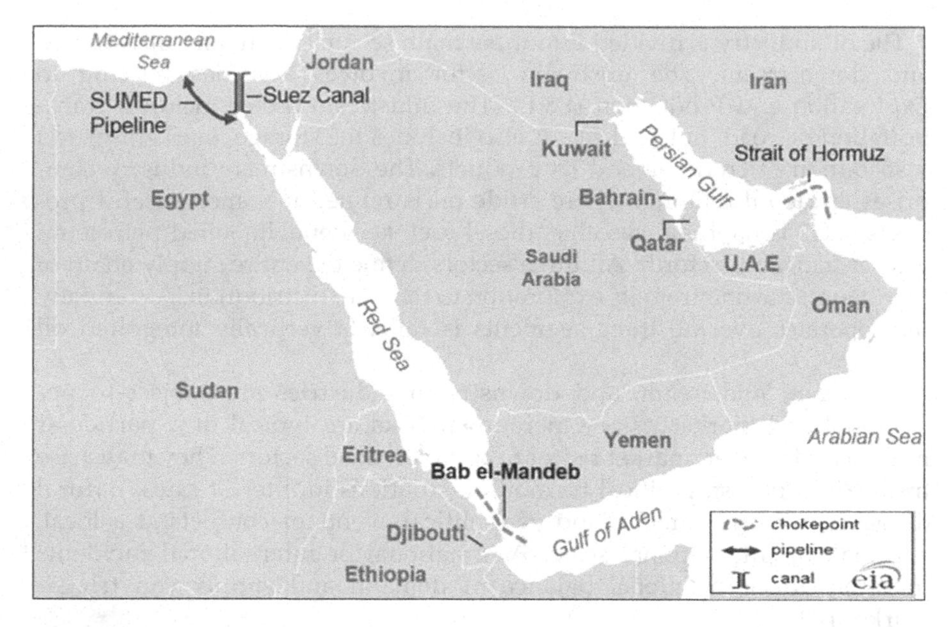

FIGURE 3.3
Maritime Chokepoints around the Arabian Peninsula. (*Source: US Energy Information Administration; Aug 2017*).

Pipelines transport most of the natural gas across the globe. The construction of gas pipelines is not only based on the economic and topographical factors alone. Geopolitics and the gas market along the route of the pipeline are equally important factors in the decision-making.[3] Since a gas pipeline may transit through several countries, the agreements with the concerned states for the pipeline's safety and security can take time besides the legal and commercial negotiations.

Petroleum, as a source of energy fuel, is an enabler of globalization and economic growth. The economy of one country affects other economies. Assuming that petroleum may be one reason for interdependence between producer and consumer nations, even a perceived threat of disruption or delay in its supply can cause a panic among energy importing countries. The sourcing of oil and gas by international cargo or end-to-end pipelines is vulnerable to piracy, sabotage, embargos, and attacks in local or regional conflicts and war times. If such a situation develops and persists for long, the scarcity of petroleum resources can manifest itself in more than one way. As oil price goes up, the oil demand is significantly reduced. Oil importing countries record a weak industrial output cascading into a drop in the manufacturing of durable goods. Efforts to control expenditure and reduce costs result in workers' layoffs, a high unemployment rate, and a steeper inflation rate. In an inflationary economic environment, people curtail spending and increase savings. High-interest rates limit liquidity or the amount of money available to invest. Global shares take a hit, and stock markets tumble, affecting investments. It triggers an economic slowdown and lowers the GDP. This economic slowdown can snowball into a worldwide recession. A tension-free world and stable demand can hold the oil price in check.

Because oil is such a vital commodity today, an oil-producing or consuming nation must be aware of the environment to maximize benefits. Oil-producing nations focus not only on controlling the cost of indigenous oil production; they also look for opportunities to seal the long-term cheap oil deals in the international market. An oil company's performance is affected by internal and external factors discussed below:

3.3.1 Internal Factors

Internal factors mostly deal with organizational capability and resources and are addressed by the company management. Some of these factors are:

3.3.1.1 Expertise

Expertise relates to skills and knowledge in a particular field. It is a game-changing attribute that has a telling effect on problem-solving and winning the competition. Expertise exists in a company in the form of the employees' knowledge and experience about their reservoirs, company strategy, business guidelines, short- and long-term plans, sources, digital and analog databases, workflows, processes, and so forth. A manager's challenge is to harness this knowledge coherently and productively. By working together, it is possible to identify the Subject Matter Experts (SME) who can expedite decision-making. An SME is a competent professional who addresses a question with a deep understanding and wealth of experience in his area of discipline to meet the project requirement.

3.3.1.2 Performance

In the context of an organization, performance means converting a plan into action and deliver results. Clear and precise communication is the key to achieve performance targets. The management must align the organizational vision, mission, values, and aspirations to create a sound performance culture. The corporate structure should not be a barrier; it should facilitate interaction and collaboration. Organizational performance is a participative process that involves teamwork, quality, and cost consciousness, load sharing, and owning of responsibilities. A fair and consistent performance measurement system reflects these elements to reward and recognize high achievers. The organization must prepare a personnel development plan for all staff members based on identifying their training needs. It will benefit both the organization and the employees with improvements in the quality of service and results.

3.3.1.3 Cost

Oil and gas development is expensive, and a net positive cash flow may take several years. The company must have sound financial credentials to sustain these operations and resilience to absorb setbacks. Cost is an amount of money that has to be paid or spent to produce oil and gas. However, the word "cost price" also includes other outlays necessary to produce oil and gas, such as property costs, materials, power, worker wages, etc. Most companies treat the cost price as a piece of confidential information for several reasons. It influences the cost-consciousness of customers who prefer to buy goods or services within a specific price window. Competitors can strategically manipulate the price of their goods/services to secure a higher market share to improve their profitability. Cost price can also determine investors' profits and decision-making into either buying or selling the tradable securities. Costs affect profit, and they're used to make decisions for both small and large businesses.

3.3.1.4 Supply

The word supply of oil and gas is an overarching function that includes oil and gas production at the in-country point of export/sale and then further transportation to the consumer country by rail, road, ships, or pipeline. Under the internal factors, the reference is made only to the in-country supply. Oil and natural gas are extracted from individual reservoirs and wells. It is then brought to gathering centers, processed to separate oil from gas and water, and other impurities. Both oil and gas are saleable. Crude oil should conform to agreed specifications of bottom sediment and water (BS&W), crude API, and sulfur content at the export/sale terminal. Natural gas should be free from heavier hydrocarbons and meet the heating value specifications. Heavier hydrocarbons removed from the gas, also known as natural gas liquids (NGL), can be sold separately or mixed with crude oil.

The natural gas supply must comply with emissions regulations and pipeline gas specifications in terms of water content, H_2S, CO_2, O_2, N_2, gross heating value, and delivery temperature and pressure. Oil processing at the gathering centers can achieve the target specifications for delivery. Free or associated natural gas may require additional processing to bring it to the desired specifications.

3.3.2 External Factors

External factors are beyond the control of the organization and its management. Therefore, the organization must continually monitor national and international events to decide its business strategy and action (Figure 3.4).

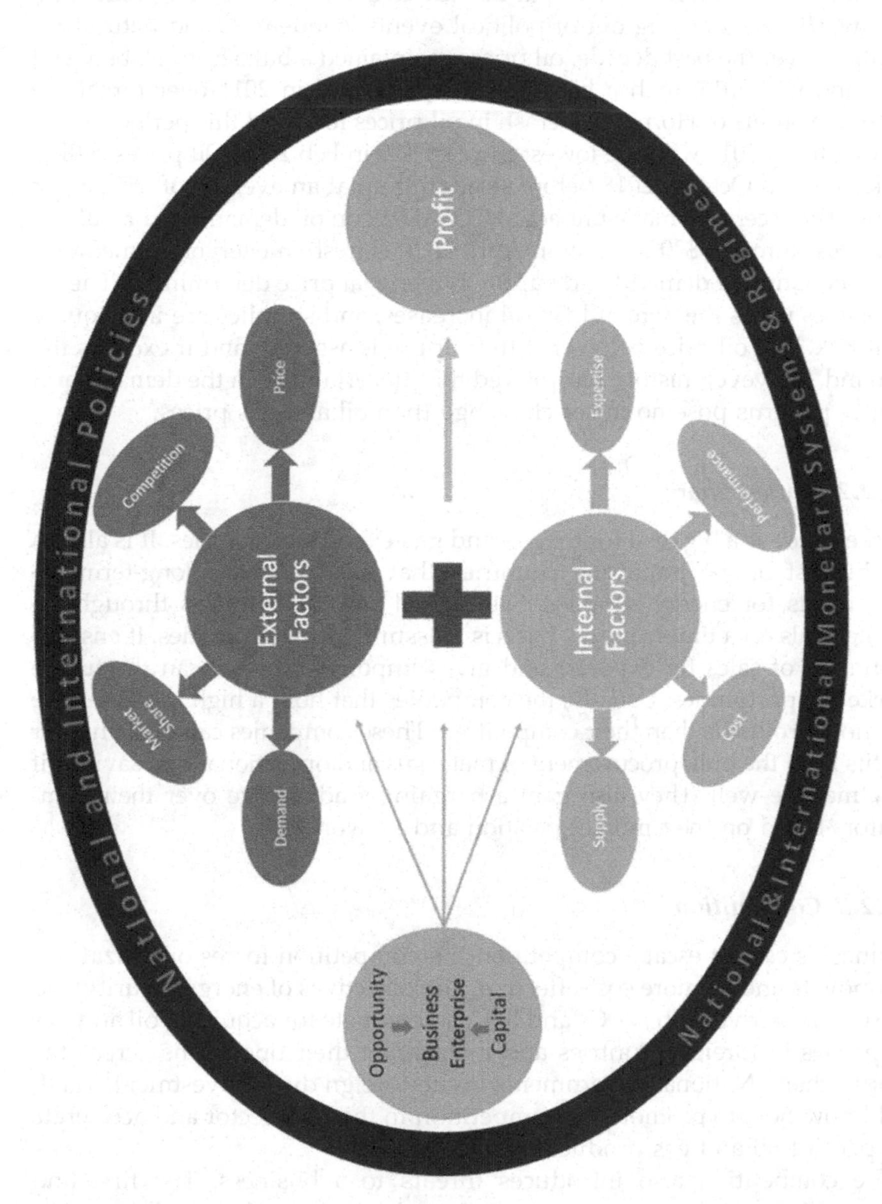

FIGURE 3.4
Business Environment of a Petroleum Company.

3.3.2.1 Demand

Price plays a significant role in the profitability of an oil and gas company. It is controlled by several factors that include demand and supply, production cost, world crises arising out of political events, pandemics, and natural calamities. Over the past decade, oil prices maintained a bullish streak between 2011 and mid-2014, with a high of $112 per barrel in 2011 over threats to choke the Straits of Hormuz. A crash in oil prices followed this period due to the oil glut in 2014 with the lowest price of $29 in Feb 2016. Oil prices rallied back to $76 in October 2018 before steadying up at an average of $55-56 per barrel. The recent dramatic impact of COVID-19 on oil demand and resulting oil prices slump to $20 a barrel in April 2020 is fresh in everyone's memory.

The equation of demand and supply is a crucial price determinant. The oil price goes up as the demand for oil increases, and supplies are inadequate or at risk. The oil price is lower if the supply is assured, and it exceeds the demand. However, history has proved that uncertainties in the demand and supply patterns pose no lesser challenge than oil and gas prices.

3.3.2.2 Market Share

Market share is a big deal for the oil- and gas-exporting countries. It is also in the interest of the importing countries that wish to make long-term arrangements for energy supplies. Settling oil and gas supplies through bipartite deals on a time-window basis is reassuring for both parties. It ensures continuity of sales for exporter and gives importer time to scan the future market: opportunities. Usually, the companies that hold a high market share are more profitable than their competitors. These companies can make higher profits from the bulk procurement of materials and operational cost savings if they manage well. They also gain a bargaining advantage over their competitors based on their market position and network.[4]

3.3.2.3 Competition

Businesses cannot escape competition-the competition forces organizations to innovate and be more cost-effective. The objectives of energy security and profitability drive both NOCs and IOCs to compete for acquiring oil and gas properties in foreign countries and integrating their operations across the supply chain. National governments invite foreign direct investments (FDI) and know-how to promote the competition in the E&P sector and accelerate the pace of oil and gas production.

The competition also introduces threats to a business. The first and foremost is the threat of new entrants into the market with possibly higher talent, desire, and capability to compete. However, the new entrants have to overcome resource-ownership barriers, large capital requirements, and scale of operations that come in their way of success.

Another risk to oil and gas comes from disruptive technologies. Disruptive technology is a concept, invention, product, or technology that disrupts conventional paradigms. It is an "out of the box" idea that can develop into a new commercial product or industry. Disruptive technologies can change the lifestyles and behaviors of people. For oil and gas, transition via unconventional resources is gaining momentum. Recovery of hydrocarbons from shale, tight carbonates, coalbed, and EOR methods is being pursued seriously. Oil price shocks and climate change initiatives have forced the world to expedite alternative energy methods. The fuel cell, solar, wind, geothermal, and other technologies are being tested and expanded. The oil and gas companies of today are aware of these challenges. They evaluate the potential of the new technologies to diversify their portfolio with an eye on the future.

The competition also drives exporting countries to form oil cartels. Oil cartels are the alliances of powerful suppliers. They control oil production, supplies, and prices to ensure the profitability of their organizations. Similarly, influential buyers use their position to secure a reduction in oil price or demand better quality for the same price.

Investors and financial markets are pushing oil companies to make more efficient and prudent use of energy capital. Companies will have to think of plans and strategies to reduce shale oil and EOR production costs to compete with secondary oil production.

3.3.2.4 Oil Price

It is essential to understand the microeconomic and macroeconomic implications of oil prices. Microeconomics is a study of the behaviors and interactions of individuals and companies as scarce resources get reallocated under the new price regime.

The common man feels the first impact of higher oil prices in the form of increased gasoline prices. Gasoline prices affect the supply cost because transportation and distribution of goods become costly. That leaves a significantly smaller share of the budget for other household activities. On the other hand, higher oil prices tend to increase production costs because operations and services become expensive. It has been observed that oil and gasoline prices are closely linked, and therefore, a surge in oil prices affects the common man and businesses alike.

Macroeconomic factors deal with the health of national or regional economies defined by gross domestic product (GDP), inflation rate, unemployment rate, and so forth. An increase in oil price is generally associated with a higher inflation rate due to an increased cost of goods and services. It reduces economic activity and growth. Based on the experience of the oil shocks of the 1970s, it can be concluded that a sustained increase in oil prices suppresses the economy.

3.4 International Monetary System

Knowledge of the international monetary system is essential to find the capital for financing oil and gas projects and oil and gas trading. The objective of trade is to transfer commodities between the buyers and sellers. Money transfers take place between investors who have surplus funds and borrowers who have the need. Financial markets facilitate the transfer of funds between investors and borrowers and the proceeds between buyers and sellers. Globalization of the economy has facilitated the massive cross-border flows of excess funds available with different countries to be diverted to the needy states to develop energy, infrastructure, construction, and other sectors.

The international monetary system is the framework that allows various countries to trade and make payments across political borders. It is required because the trading countries' currencies are not at par, and their exchange rates vary. International trade helps the participating nations to improve their "balance of payment" position and reduce the trade deficit.

In the very early days, the business was carried with the help of gold and silver. Between 1876 and 1913, countries adopted the Gold Standard during the free-trade period, setting a gold value of their currency. Way back in history, the two trading countries, the United States and Britain, pegged their money to an ounce of gold that became the exchange rate for conversion of US$ into British £:

$$\frac{\$20.67/\textit{ounce of gold}}{£4.2474/\textit{ounce of gold}} = \$4.8665/£$$

Governments of the other countries also used the Gold Standard to calibrate their currencies with their trading partners. This practice continued until World War I (1914–1918). However, in the post-World War I era, the member countries failed to revive this system due to the severe economic recession because of small gold reserves' holdings. Also, the need for recalibrating the gold to the US dollar became apparent. It was done in 1934 by revisiting the gold price from $20.67 to $35 per ounce.[5]

Austria's banking system collapsed in 1931 as it did not receive any support from the banking institutions in the US, Britain, and France. The crisis deepened as Germany too withdrew its money from Austria, and ultimately the Gold Standard had to be disbanded.

In July 1944, as World War II (1939–1945) approached an end, representatives of 44 countries aspiring to introduce the new financial order met at Bretton Woods, New Hampshire, in the US. Their efforts were fructified with the International Monetary Fund's creation to facilitate trade based on the member countries' reserves in proportion to their financial status and business need. Each member country contributed 25% of its gold share and the remaining 75% in its national currency.[5] Subsequent modifications to the system required the US to maintain a reserve of gold, and other countries to maintain a reserve of US dollars.

The Bretton Woods system lasted until 1971. A subsequent decline in the value of the US dollar, followed by the oil crisis of 1973, forced a revision in the gold price from $38 to $42.22 per ounce. In the post-1973 period, the role of gold is somewhat moderated. However, it continues to be a vital asset to the IMF. Today, the IMF holds around 90.5 million ounces (2,552, 908.5 kilograms, or 2,814.1 metric tons) of gold. The value of this gold at the current market price is about $140 billion (using end–December 2019 exchange rates).[5] The international currency-based system has since been revised many times to address the concerns and inconsistencies mostly arising from exchange rate issues. Despite the IMF's various actions to equilibrate international currencies, most oil and gas sector contracts continue to be in US dollars.

In its present avatar, the IMF is an international organization of 189 member countries that

> *promotes international financial stability and monetary cooperation. It also facilitates international trade, promotes employment and sustainable economic growth, and reduces global poverty. The primary mission of the IMF is to ensure the stability of the international monetary system-the system of exchange rates and international payments that enables countries and their citizens to transact with each other.*[6]

The IMF has three leading roles: economic surveillance, capacity building, and money lending. Its economic surveillance activities include collecting data on global trade and economics and updated economic forecasts periodically. These national and international forecasts are published in the World Economic Outlook. Under Capacity Building, the IMF trains and develops people to support its surveillance activities. IMF's lending programs aim at extending loans to member countries in financial distress. IMF funds are often conditional on recipients making reforms to increase their growth potential and financial stability.[7]

Other prestigious financial institutions help the developing and third-world countries in their energy and poverty alleviation programs. The World Bank is one such international financial institution and a sister organization of the IMF born in 1944 out of the Bretton Woods agreement. It comprises the International Bank for Reconstruction and Development (IBRD), which provides loans and grants to poorer countries to pursue capital projects.[8] Some other banks/institutions, which provide financial support in the form of loans, grants, guarantees, and equity investments

TABLE 3.2

Active Oil and Gas Project Financing by World Bank Group 2014–2018 (million US$)

Energy Type	Equity	Loans	Guarantees	Grants	Total
Gas	529	7,558	3,336	0	$11,422
Oil & Gas	499	1,071	1,527	0	$3,097
Oil	218	1,187	24	0	$1,429
Total Oil & Gas	1,246	9816	4,887	0	**$15,948**

besides technical assistance for the development of energy and poverty alleviation programs, are as follows:

- Inter-American Development Bank (IADB)
- Asian Development Bank (ADB)
- African Development Bank (AfDB)
- European Bank for Reconstruction and Development (EBRD)

All these institutions have similar mandates to support their member countries, and they use various financial instruments on a short-, medium-, and long-term basis. An example of financing of 675 fossil fuel projects by the World Bank Group between 2014 and 2018 is presented[9] in Table 3.2. However, these commercial and concessional loans and grants can often be conditional, based on social reformist themes, environmental, or economic importance. For the loan seeking state and its people, these conditions may be hard. This practice of conditionality is often criticized for its financial wisdom, and sometimes also for its lawfulness.

3.5 Reservoir Management Framework

The primary goal of a business is to make a profit and enhance asset value for its stakeholders and society. They attain this goal by exploiting petroleum reservoirs at an optimum rate and target recovery. Focus on profit entails the organization to be innovative and conscious of costs. Objectives of achieving optimum rate and recovery require a deep insight into the reservoir characterization and its dynamic behavior. Continuous optimization of production rate, oil recovery, and profit is the ongoing endeavor of a reservoir management plan.

The scope of Reservoir Management is all-inclusive. The key aspects of the reservoir management framework are presented in Figure 3.5.

i. Strategic

Modern companies have to think about the growth and value creation from oil and gas than a mere revenue collection from the sale of these products in the market. The objective is to maximize the strategic value of petroleum by devising a plan under a set of political, environmental, social, and technological (PEST) factors. This plan may cover a specific business segment or the entire chain of operations, as applicable.

Usually, most oil and gas companies have a Strategic Planning team that reports to the company management to manage its strategic elements. This team is supported by SMEs and team leaders who control and manage the activities in their respective areas. There can be various questions that the strategic planning team must be prepared to address. Some of these questions relate to the organization's response to market share, competition, resource allocation, capacity build-up, and so forth. Of equal importance to the management are issues concerning organizational structure and efficiency. The strategic team must have the capability to advise the company management on critical questions such as:

- How should the company split its resources between exploration and development assets?
- What reserves to production ratio (R/P) should the company maintain?
- What are the most appropriate time and sizes of investments in the oil price cycle?
- How should oil and gas production react to changing oil prices?
- How can the company manage the price elasticity of demand to its advantage?

Figure 3.6 shows a simple reservoir management strategy to maximize the profit from unstable oil prices. The oil company produces at lower rates when the oil prices are low and higher production rates when the oil prices are high.

ii. Technical

For an asset in the development/production phase, this element deals with core issues of development planning, operations, and reservoir management. The most critical consideration in development planning is time. Time sets the pace of development, which decides the allocation of money,

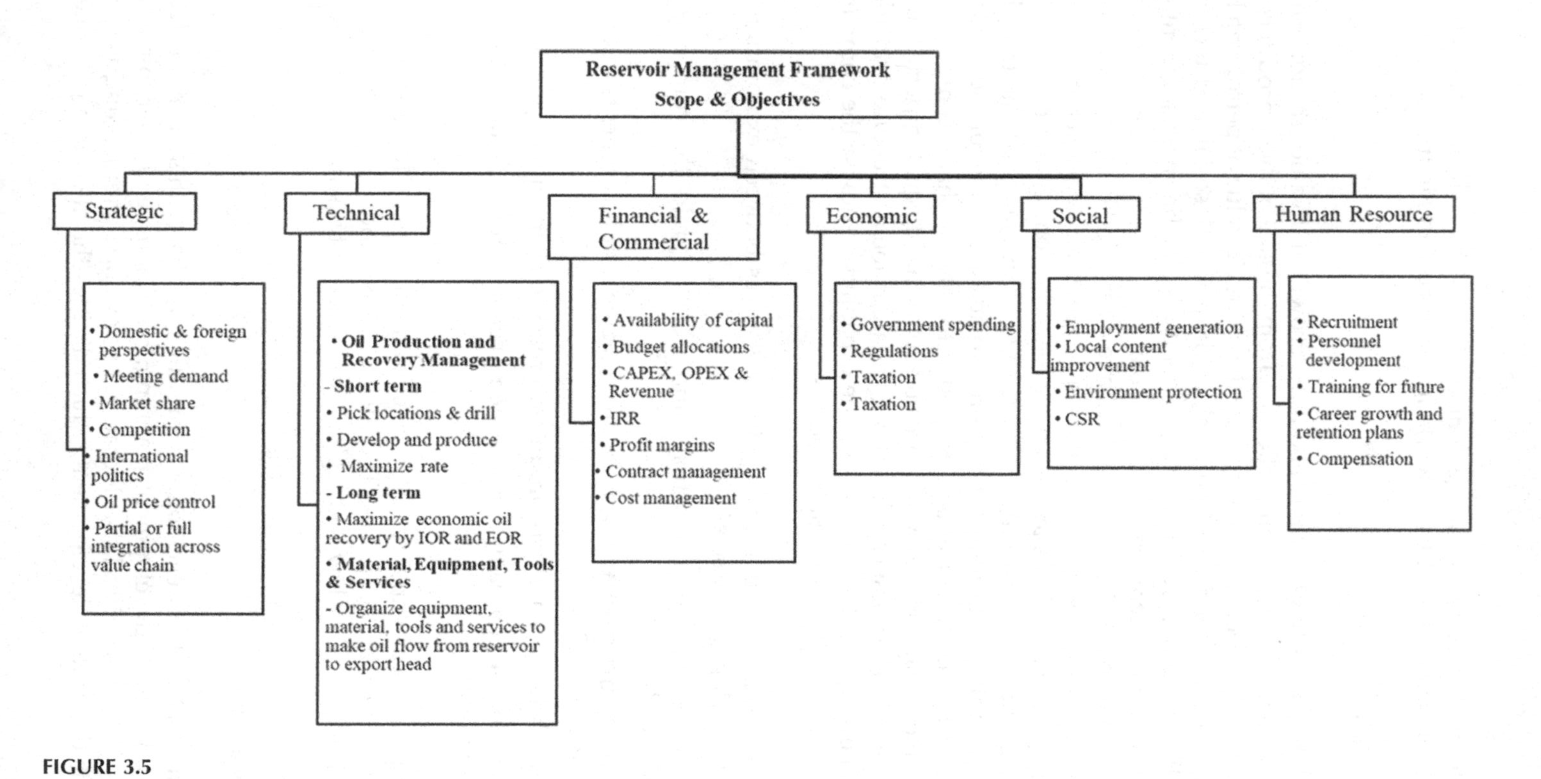

FIGURE 3.5
Key Aspects of Reservoir Management Framework.

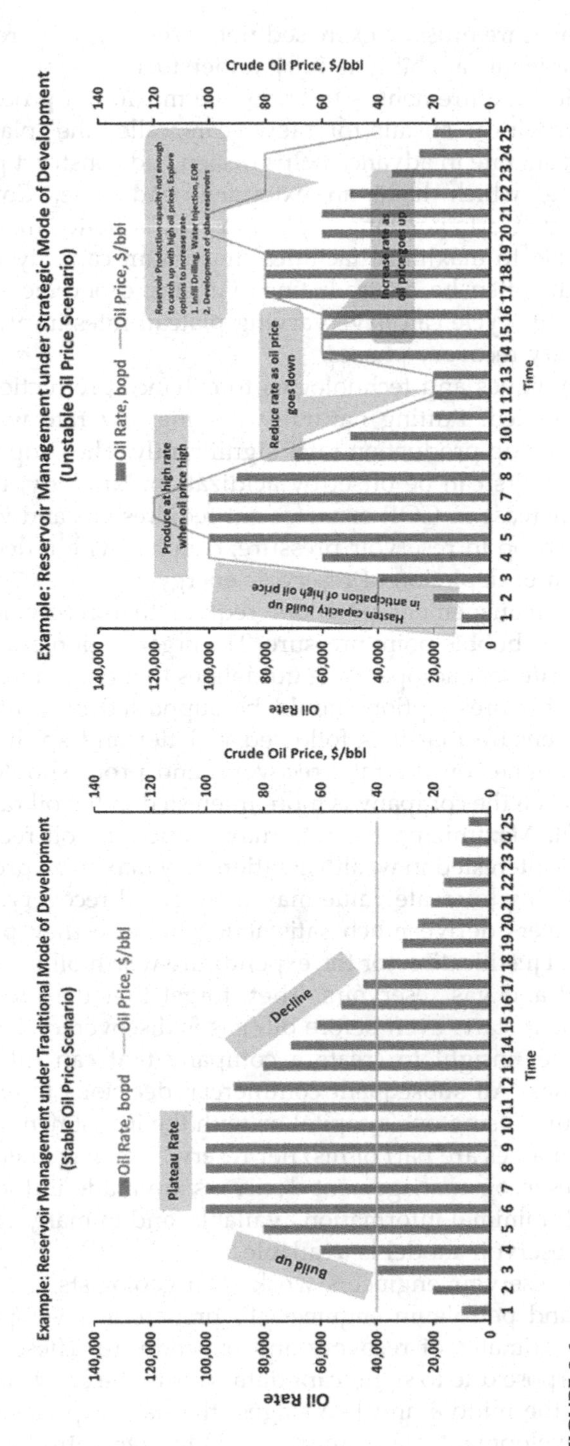

FIGURE 3.6
Reservoir Management Strategies.

men, and material. Care must be exercised not to set a very aggressive pace until the organizational capability is fully understood.

Another technical requirement is to know the maximum production potential that a reservoir can sustain for a few years, called the "plateau" rate. Knowing the plateau rate in advance helps design and construct production handling facilities, which have an extended lead time. Conventional wisdom dictates that a lower longer plateau rate is better than a higher shorter plateau rate to maximize facilities' maximum capacity utilization. However, it might often be contradicting with the economic analysis. A reservoir over its life-cycle can have varying plateau rates in primary, secondary, and tertiary recovery phases.

The role of techniques and technologies to enhance production must be continuously evaluated. Putting naturally flowing low rate wells on the artificial lift can step up production rates significantly. The drop in oil rates in carbonate reservoirs can be offset by acidization. Similarly, the oil rate decline with an increase in GOR or water cut requires gas and water shut-off methods. The drop in reservoir pressure, coupled with a decline in oil rate, is a symptom of depletion of reservoir energy.

Good reservoir management practices require that reservoir pressure should not go below bubble point pressure. The organization must establish specific policies, criteria, and operating guidelines to maintain its reservoirs in good health. This prescription should be supported by a healthy surveillance plan to ensure that it is followed in letter and spirit. Reservoir management guidelines on the rate, recovery, and profit should be clear. There are times when the company is more interested in the oil rate than the recovery or profit. Maximizing the rate may reduce the oil recovery and profit. Companies interested in wealth creation may maximize profits, while the companies aiming to create value may increase oil recovery.

Reservoir engineers derive much satisfaction because they provide the basis and technical justification for the expenditure worth billions of dollars in developing oil and gas reservoirs. They forget that the process of reservoir management starts even before oil/gas is discovered. The resource owner uses his/her insight to create a company that can fulfill his/her business aspirations. All subsequent commercial decisions to explore and develop the resource using one's capital or with the investment and know-how of other companies are part of his/her reservoir management strategy. Many strategic reservoir management decisions are made in the company board rooms with minimal information available, and in many cases, much before a proper reservoir model is available.

Only later, the reservoir engineers work with geologists, geophysicists, petrophysicists, and petroleum engineers to prepare a development plan used to support estimates of recovery and expenditure. These plans may only be fit-for-purpose due to significant data gaps in the early stage of field development. In the middle and late stages, the data gap closes, and the quality of the development plans improves. However, introducing a new recovery process such as the secondary or EOR process in the middle or late stages tends to increase complexity in terms of reservoir behavior and open up data gaps, affecting the quality of development plans.

The typical reservoir development process is presented in Figure 3.7 as it progresses through a phased development in time. The reservoirs are managed based on a conceptual model tested for its accuracy against the observed reservoir. The level of understanding improves as more data is collected, analyzed, and validated by field observations. Uncertainties and risks associated with the development are substantially reduced with improved knowledge of the reservoir and its behavior. When reservoir behavior is different from expected in real life, frequent updates to the development plans become necessary.

a. **Conceptual Development Plan (CDP)**

Development projects for large reservoirs can be broken into smaller phases to manage risks, uncertainties, investments, and delivery better. CDP is mostly prepared to highlight the magnitude of initial investments under primary recovery operations based on early estimates of oil-in-place, wells' productivity in different parts of the reservoir, and the number of wells required to be drilled in this phase. The CDP must provide practical solutions to manage effluent water production and excess gas production that is not contracted out for sale. A process needs to be put in place to ascertain well integrity and manage the full cycle of water/gas production, injection, and disposal or reinjection of produced water.

CDP is revisited periodically based on new geologic information, petrophysical, and well test data. Newly drilled wells are constantly monitored to map any changes in faults, folds, formation tops, pay thickness, and petrophysical properties to reduce uncertainty in oil-in-place estimates. Long-term production tests and time-lapse reservoir pressure measurements determine variations in Productivity Indices (PIs) with time. All this information is used to update and match production history.

This history-matched model forms the basis for generating a long-term production forecast with reservoir pressure and recovery factor. The rate of decline in reservoir pressure is an indication of the size of the reservoir. Reservoir simulation models or analytical methods properly calibrated with actual field tests are used to justify the investments in development activities. Due to the limited understanding of reservoir behavior, uncertainty in production forecast and development risks are high.

b. **Initial Development Plan (IDP)**

IDP is more detailed than the CDP. Its objective is to maximize the production from primary recovery and do the groundwork to apply pressure maintenance or secondary recovery methods. IDP's framework is similar to

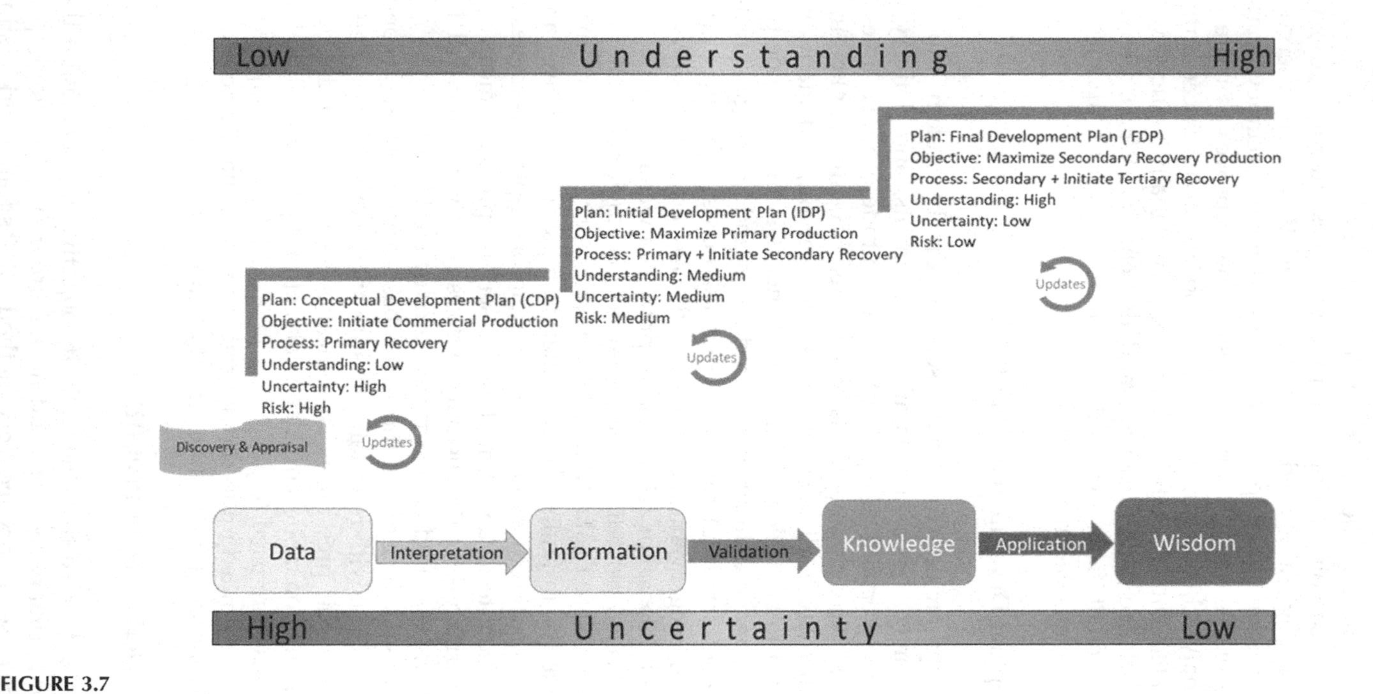

FIGURE 3.7
Reservoir Development in a Constrained Environment.

that of the CDP, except that IDP has a more refined and more extensive data set. IDP's scope is certainly broader, and its insights deeper than the CDP because it is based on the firmer estimates and categories of resources and reserves. The IDP has a good handle on investments as development risks are significantly reduced due to an improved understanding of the reservoir and its dynamic behavior.

In case water injection is chosen as the IOR method, then this is the new focus area of the study for the IDP. Much pre-work has to be done before it can be considered for field implementation. Asset subsurface teams usually do the preparatory groundwork based on their observations of reservoir pressure decline. It includes doing coreflood studies with water, Step Rate Tests to determine fracture pressure gradient, water compatibility studies, and injectivity tests to estimate sustained injection rates. A waterflood pilot can be designed and implemented in collaboration with surface teams at a suitable location in the reservoir to assess the response time and impact.

Water injection has the potential to increase oil rate and recovery significantly. Many waterflood options such as line drive, peripheral, pattern waterflood, or even their combination are evaluated in detail. A simulation model can perform this kind of task in a relatively short time. An economic evaluation of technically preferred scenarios is carried out. The option with the most attractive economic indices is recommended for field implementation. This approach provides approximate estimates of the anticipated oil production rate, recovery, CAPEX, OPEX, earnings, payout period IRR, and the project's profitability.

It is essential to regularly monitor wells and reservoir performance to look for unexpected deviations, which may be the tell-tale sign of something not working as expected. The IDP must consider and address the issue as soon as possible through an updated document to reflect the change.

Mid-term development may be necessary in case of significant deviation between actual and predicted performance by the IDP. It may also be required by a field that must go through the pressure maintenance/secondary recovery program. A Mid-term Development Plan is built on a stronger foundation of the improved assessment and more certain oil-in-place and reserves estimates supported by an enhanced understanding of reservoir performance.

c. FDP

FDP is prepared after the reservoir has been mostly drilled, and its performance was observed for a considerable period. By this time:

i. The Asset Subsurface Team has a good understanding of the reservoir characterization and performance of various sectors
ii. Artificial Lift methods are in extensive use, and their performance as well as problems are appreciated

iii. The reservoir has matured, and water cut and GOR trends are stabilized and
iv. The field-wide implementation of the water/gas injection method is in use

The objective of the FDP is to maximize the recovery from water injection/waterflood methods. FDP sums up the final well drilling requirement and if any significant inputs are necessary to maintain the production rate and recovery. Well-integrity and water management might pose a substantial challenge in this phase of development. FDP needs to be backed up by a master surveillance plan for maximizing the benefits of water injection. A review of surface facilities in terms of their modification or upgrade may also be necessary.

The FDP is designed after the development team has all the essential information. The static and dynamic reservoir models have been calibrated adequately against the actual reservoir performance, and the confidence in oil-in-place, production profiles, and reserves are high. Under these conditions, the FDP presents a good view of future investments in the development and the remaining risks. The name "FDP" does not imply the end of development planning. It will undoubtedly require further updates and fine-tuning if there is a significant gap between actual and predicted reservoir performance.

iii. Financial and Commercial

The financial aspect deals with the optimum allocation of capital or funds for business activities that bring the highest possible returns. Corporate finance is a very specialized subject today. It aims to manage funds to maximize its wealth, thereby increasing company stock value in the market.

Capital is the most crucial ingredient that is required to start and sustain a business. It is like oxygen for a human being; its lack can choke the company or even kill it prematurely. The exploration phase might use a vast amount of capital without any return. Business owners have to ensure that enough funds are available to carry out essential business activities in the payout period.

Capital is always finite or limited. It is further allocated to the projects based on the priority of the plans. Corporate finance's role is to ensure that they make the most efficient use of available financial resources without impeding its progress and performance.

The selection of a reservoir management proposal is very much dependent upon its economic viability decided by the following criteria:

- Capital Expenditure (CAPEX)
- Operating Expenditure (OPEX)

- Price of oil and gas
- Discounted cash flow
- Payback period
- Rate of return on investment
- Net present value
- Internal rate of return

Companies with limited financial capacity may use CAPEX or OPEX to screen the projects. Large-size companies with no budgetary constraints can choose the plans with the least payback period or a high rate of return on their capital investments.

The commercial aspect is vital because oil- and gas-producing companies have short-term and long-term agreements to deliver crude oil and gas to the national and international refineries and customers. They can also have joint venture agreements with other oil companies to explore and develop oil and gas resources. E&P companies also have commercial contracts with a variety of service providers for the supply of goods and services. The majority of oil companies employ drilling, logging, perforation, well-hook up, testing, and many other functions on a contract basis. These service providers can provide only a specific "service" or "service with material" as agreed and defined in the contract. Commercial contracts are in writing, signed, and dated by both parties. A general format for such agreements is given below:

- **The goods/service being sold:** This clause mentions the specific name of the goods or services with quality specifications, if applicable.
- **Delivery of goods/service:** Include place, schedule, and options for the delivery. How the goods will be transported, who will bear the cost of transportation, and when, before or after delivering the goods/service. Liability for damages or loss of products during transit rests with whom.
- **Contract price:** This is the negotiated total payment agreed between buyer and seller, the basis of payment, for example, in time delivery without damage, etc., payment in installments or lump sum, details of currency/currencies.
- **Terms of payment:** This clause includes if payment terms are linked to performance, states mechanisms for payment, and if there are contract conditions that can suspend the obligation to pay.
- **Implications of non-performance:** This clause details the ramifications of delivering a sub-standard product/sub-optimal performance through warranties, indemnification, termination, and liquidated damages.

- **Force Majeure:** It comprises a list of acceptable events to the buyer and seller for the delay in delivering goods/services and the payment.
- **Dispute Resolution:** the contracts generally give details of the agreed procedure to settle any contract disputes.
- **Suspension, Termination, and Settlement of Accounts:** This clause clearly states which party can suspend or terminate the agreement and under what conditions. The next step clarifies the payment and performance obligations for both parties if the contract is terminated. In the case of services contracts, the service provider can terminate/suspend the agreement if the buyer refuses to make payments.

Commercial agreements can also cover other business aspects, including wages, leases, loans, hiring, and employee safety.

iv. Economic

Economics is quite different from finance, and these two should not be confused with each other. Finance deals with investments, budgeting, expenditure, income, money flow, time value of money, rate of return and profit, and so forth. The goal of economics is to create, optimize, and preserve value. It aims to analyze the interplay of attendant economies globally and make economic decisions within the international trade market. It evaluates the trends of production, distribution, consumption of goods or services, transfer of wealth, and related factors to indicate how a country's economy performs. It explains how limited resources are appropriately allocated between many sectors and efficiently utilized at the national level.

At the corporate level, many vital aspects are controlled by the national governments, directly impacting its balance sheet. Some of these are terms and conditions of allocation of concessions, government spending in the oil sector, petroleum regulations, setting up Joint Ventures, the role of NOC vis-à-vis IOC or any other players in the field, tax structure, exemptions, and so forth. Other factors are outside the national governments' domain, such as a sudden change in oil price and an upward or downward revision in the US dollar value that impacts international trade and the global economy. High oil prices may result from high demand, low supply, OPEC quotas, and a decline in the US dollar value. Seasonal fluctuations in demand for oil and gas are part and parcel of "business as usual" and mostly absorbed by traditional development practices within the reservoir management framework of "Stable Oil Price." The unpredictable price fluctuations prompted by geopolitical factors necessitate quick changes in reservoir management programs.

Oil-exporting countries peg their national currency to the US dollar since most oil contracts worldwide are managed in US dollars. A drop in the dollar value causes oil revenues to go down and costs to go up. A jump in inflation rates results in reduced consumer spending and a lack of investments in the projects. Such market conditions can sometimes be exacerbated by the onslaught of pandemics, which can plunge the economy into a recession.

NOCs, under the instruction of their governments, tend to control the production of oil to balance internal and external factors. Oil-exporting countries with excess production capacity control their production to regulate oil prices, maintain or increase their market share of petroleum supply, and secure attractive economic or strategic returns. On the other hand, oil-importing countries try to maximize their production and introduce conservation measures when the oil price is high to protect their economy. In times of low or average oil prices, some countries may resort to maintain an oil reserve by way of oil storage above or underground. These conditions also usually apply to IOCs or other private companies operating in an oil-producing country subject to their commercial arrangements.

v. Social

The social aspect of the Reservoir Management Framework is commonly referred to as Corporate Social Responsibility (CSR). For IOCs and POCs, this might be limited to a couple of items on the corporate BSC, but for NOCs, the CSR has expanded objectives.

The CSR for an ideal oil and gas or any other business organization is characterized by Archie Carroll's Pyramid (Figure 3.8), which has four crucial layers of responsibilities: economic, legal, ethical, and philanthropic, in that order.

a. **Economic Responsibility:** The essence of any business outfit is to earn money to make a profit. By staying profitable, it can create long-term value and maintain a growth profile. Risk management and business practices are crucial parts of an organization's economic responsibility. A business, particularly oil and gas by its very nature and significance to the national economy, is an integral player in nation-building. The financial component forms Archie Carroll's CSR pyramid's foundation because only if the business can stay financially sound and wholesome can shoulder the following three responsibilities.
b. **Legal Responsibility:** Carroll described the legal obligation as "codified ethics." Operating companies are obliged to conduct their business according to the laws of the land. Good understanding and compliance with local regulations are essential to avoid litigation and conflicts. Companies can avoid legal issues by setting company policies and guidelines that are fair to all the stakeholders or creating a standard contract where necessary.

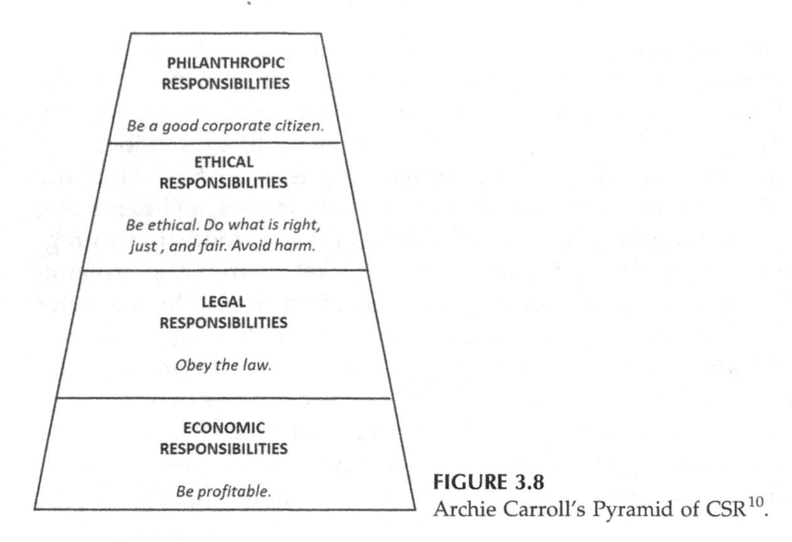

FIGURE 3.8
Archie Carroll's Pyramid of CSR[10].

c. **Ethical Responsibility:** Ethics are broadly the set of rules, written and unwritten code of conduct, and practices that govern our own and others' behavior. Most ethical codes ensure human behavior such as honesty and integrity, objectivity, carefulness, openness. Some of the ethical issues, such as discrimination based on race and religion, protection of the environment, intellectual property, and confidentiality, seemingly overlap the company's legal responsibility. Ethical responsibilities also include not harming the business, colleagues, society, property, climate, forests, rivers, oceans, living beings, and so forth.

d. **Philanthropic Responsibility:** Philanthropy may be described as love for humankind. It is an integral part of a civilized society and is motivated to drive social change. Wealthy sections of the community, including corporations, have a moral responsibility to the less fortunate people. It may include financial contributions, time, and resources. Some common examples of philanthropic programs are providing education or healthcare to the poor, building libraries, museums, scientific research facilities, etc.

vi. Human Resource

A business organization needs employees to carry out a large number of tasks in a fixed timeframe. Most employees do not have the necessary knowledge base or skill set to accomplish the job they are hired for at the time of hiring. They have to be educated by their employer, trained, and groomed suitably for delivering a quality output. Generally, the task assigned to an employee is repetitive, and by doing it repeatedly, the employee can excel in delivery.

For both employers and employees, the business may initially be a means of survival and sustenance and growth later. Both the parties, that is, the employer and employees, are particularly hungry for success in the early phase because it opens the gateway of opportunities to prosperity for all the stakeholders.

Modern business organizations recognize that employees are valuable resources. Highly motivated teams are capable of delivering phenomenal results in terms of productivity. In technical and specialized industries such as petroleum, employees have a more critical role to play. Proper staff training and development programs ensure that employees can carry out risky and costly oilfield operations. These programs also help avoid failures that can result in loss or suboptimal use of resources. Dynamic organizations also look at their employees as agents of change and manage transformation essential to keep pace with new developments. Change management refers to re-engineering the old processes, while transformation entails a fundamental cultural change in achieving positive growth.

Human Resource Management (HRM) is a well-established system comprising well-structured policies and procedures in most modern organizations. It is inclusive and integrated, covering the full cycle of hiring, training, development of skills, career growth, motivation, rewards and punishment, retention, and exits. By offering opportunities to employees for advancement through training and education, which is now more or less a standard practice, this system has brought the following benefits to all stakeholders:

- Increased employee job satisfaction, recognition, compensation, and job stability
- Improved efficiencies in processes, resulting in financial gain
- Enhanced potential to adopt new technologies and methods
- Increased innovation in strategies and products
- Enhanced company image

3.6 Reservoir Management Requirements

Reservoir management is an overarching process that controls the systematic conduct of crucial oil and gas activities relating to rate and recovery management, project and cost management, technology management & HSE management (Figure 3.9).

3.6.1 Rate and Recovery Management

This objective requires developing petroleum resources by drilling necessary wells and creating facilities to handle the production and injection of fluids. Rate and recovery management relates to the withdrawal of reservoir fluids at a rate prescribed by the development plan. For reservoirs under pressure maintenance, a corresponding injection rate of water or gas must also be maintained. This objective requires drilling of a certain number of production and injection wells at pre-identified locations.

Production from wells must be separated into constituent oil, gas, and water streams by the processing facilities before dispatch to the sale or export terminal. Over time, the production and injection wells may suffer from loss of productivity, which can be restored with the rig or rigless workover/operations.

An oil reservoir must contribute at a commercially attractive rate to pay out for the costs. Daily rate allows the company to generate revenue for sustaining its day-to-day activities and is also an indicator of reservoir health, long-term recovery, and profitability. The new reservoir management code underscores the need to maintain petroleum reservoirs in good health, adhere to aggressive project management strategies, and employ best practices across all operations and activities to maximize benefits.

3.6.2 Project and Cost Management

Drilling wells, creating oil and gas separation/treatment facilities, oil storage, transport infrastructure, and so forth, takes a lot of capital expenditure. It starts with the project definition when the project scope must be precisely defined at the conceptual stage. At the Front End Engineering Design (FEED) stage, the project-specific objectives, requirements, and data must be openly shared with all concerned so that all project stakeholders are on the same page. In the contracting phase, an appropriate contracting model and contractors must be selected. There should be no ambiguity of any kind when price structure and contract terms are finalized.

All those involved in project definition and execution must be conscious that any revision in the project scope implies a review of its cost and schedule. Therefore, appropriate project management measures must be in place to control project time, price, and quality.

Cost management is a necessary process of planning and managing the budget of a project. It involves comparing and analyzing the project's actual time and cost against the estimates and controlling them by eliminating waste and performance loss. The purpose of this cost management process is to keep the overall project budget under control. This process also allows preparing the revised estimates of budgets and predicting future prices more accurately.

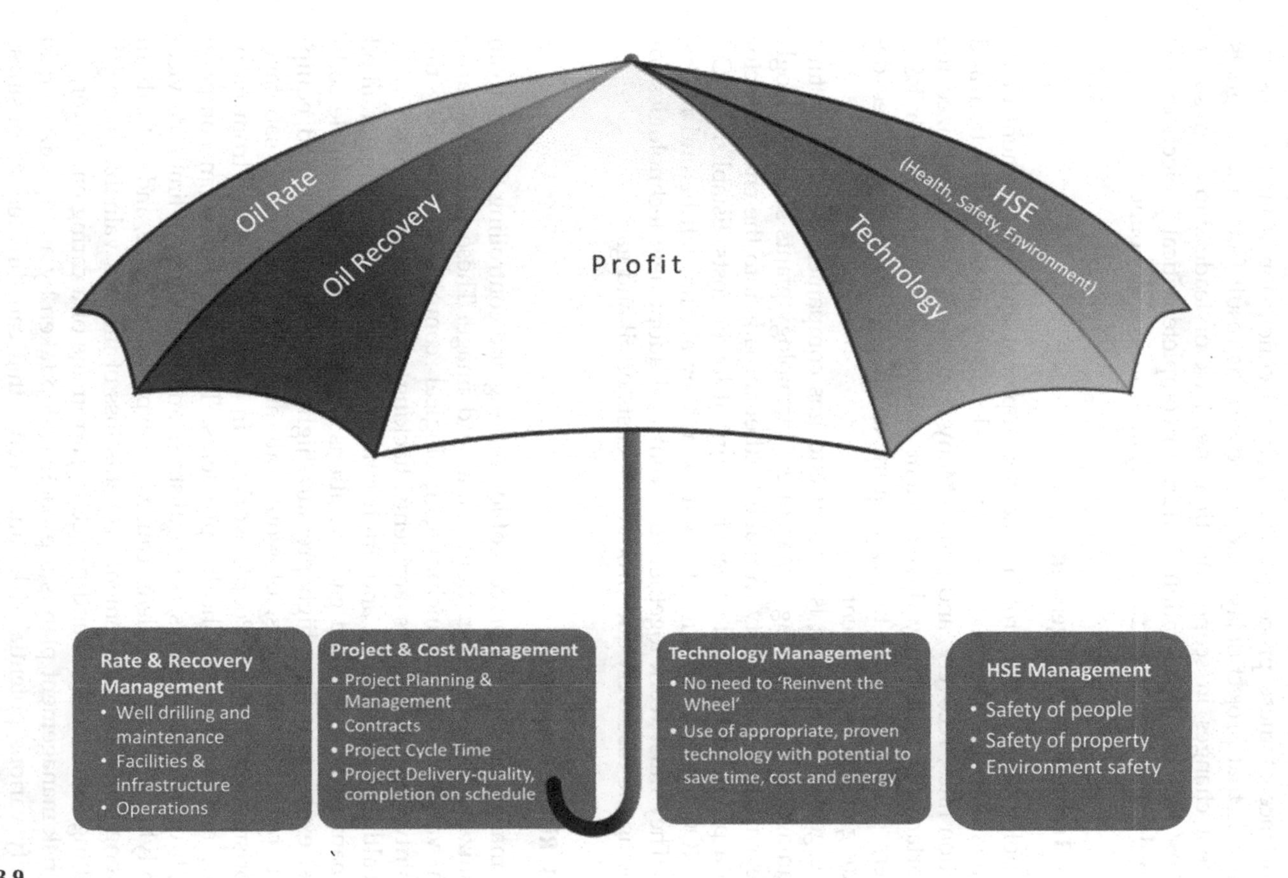

FIGURE 3.9
Reservoir Management Requirements.

Project management and cost management are crucial to ensure the on-time delivery of projects within the planned cost. Companies recognize the importance of these processes. The general conclusion from the lessons learned is that project delays and cost overruns result from unclear goals, frequent changes in scope and timeline, lack of leadership, and stakeholders' lack of consideration. Failure to complete critical projects on the schedule does impact oil rate, recovery, and project economics.

3.6.3 Technology Management

Technology plays a key role in the development and management of reservoirs. It helps to reduce cost, lessen drudgery, boost production and injection rates, recovery, and profit. Many companies have recognized the importance of technology development and invest heavily in their R&D centers. Those who don't may acquire technology by paying a fee depending on the requirement.

The general perception is that oil and gas companies are slow to take advantage of the digital age. Many new technology giants such as Google are eyeing the opportunity to make further inroads into the energy sector with a promise to provide cleaner energy at lower costs. Established IOC and NOC heavyweights must ensure that they are not left behind in this race. They must work together to invent and adopt new technologies for reducing the discovery and production costs of oil and gas.

3.6.4 Risk Management

Risk management is the process of identifying and controlling threats to an organization's assets, operations, data, and image. These risks may arise from various sources, including geological complexity, financial uncertainty, natural disasters, accidents, hacking, and strategic errors.

Health, Safety, Security, and Environment (HSSE) is a focus area of all oil companies today. Oil and gas operations include the risk of vehicle collisions, explosions, fires, high pressure-high temperature lines and equipment, radiation, oil spills, blowouts, air and water pollution, sabotage, terrorism, and so forth. People working in offices may suffer from ergonomic and occupational health problems. These problems can comprise cuts, broken bones, sprains, physical stress and strain, loss of limbs, cervical spondylosis, hearing problems caused by exposure to noise, and so forth. In the long run, risk management prevents loss of life and valuable property resulting from accidents and projecting an image of a caring employer.

A risk management plan is a proactive step taken by an organization to identify various potential risks and events that can threaten its business before they occur. Such a plan gives the organization time to prepare for any eventualities and reduce their impact should they occur.

3.7 Reservoir Management Framework

Reservoir management requires many important technical and investment decisions to be made in the company at various levels of hierarchy, depending on the significance and size of the investment. Some of these decisions have a long-term impact and some immediate to short-term. For example, critical decisions about the strategy of development, construction of facilities, mode of crude transportation, sourcing, and water treatment before injection or disposal must come with the company management's approval. Less significant decisions on reservoir management issues may be made by line managers, team leaders, and the SMEs in line with their roles and responsibilities.

Decision-making becomes difficult due to alternative solutions to a problem and the attendant risks. Each option has its share of cost and consequences. Clear objectives and cost–benefit analysis provides useful support for making crucial decisions, as explained with the following two examples:

3.7.1 Development of Multiple Stacked Reservoirs

In prolific hydrocarbon basins, multiple reservoirs may be stacked in layers, one above the other. This kind of subsurface setting of reservoirs with different reservoir pressures and fluid properties poses a fundamental development question. The operator needs to decide if such reservoirs should be developed individually with a separate (segregated development) or with a shared (integrated development) set of wells Figure 3.10.

The integrated development results in commingled production (Figure 3.10). Ease in implementing well integrity and flow assurance solutions. Commingling in oil and gas wells refers to the simultaneous production from multiple reservoirs through a single tubing. Such completions restrict the productivity of the individual zones.

A good reservoir development practice prescribes that oil and gas production from distinct reservoirs or pools must remain segregated for the following reasons:

- Avoiding crossflow of reservoir fluids that may adversely affect ultimate oil recovery
- Proper allocation of production to different reservoirs can affect reservoir management decisions
- Pre-empting any concerns about fluid compatibility (e.g., sweet and sour crude)
- Maintaining the ability to gather data on an individual reservoir for reservoir management

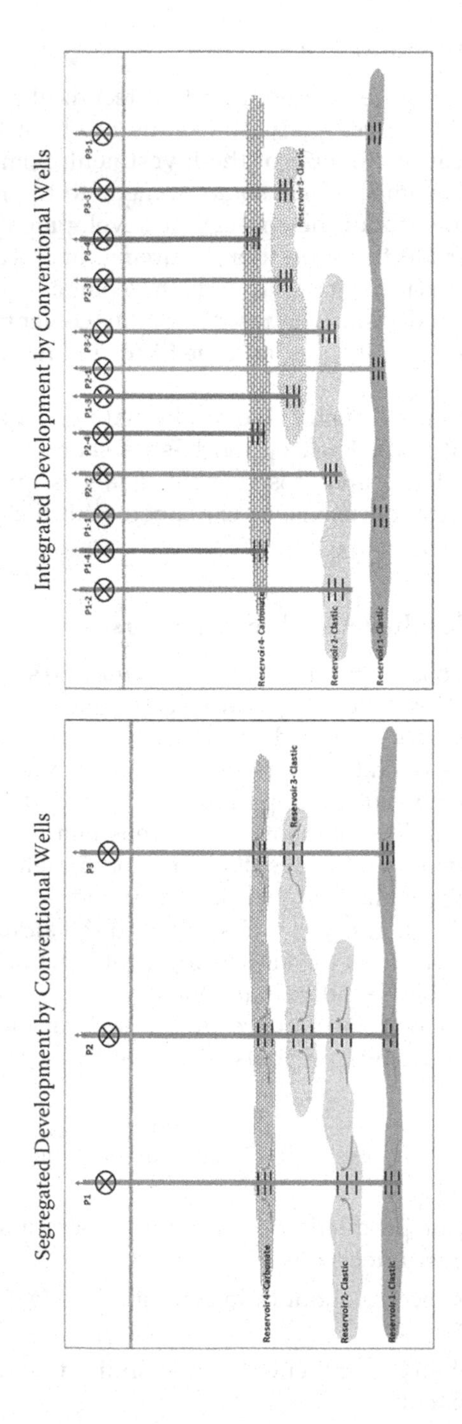

FIGURE 3.10
Well Requirement for Development of Multiple Stacked Reservoirs (Integrated versus Segregated Development by Conventional Wells).

- Facilitating secondary recovery operations such as water flooding targeted towards specific intervals
- Ease in implementing well integrity and flow assurance solutions

So technically, the commingling of production is not a preferred option. However, it has a tremendous potential to increase Net Present Value (NPV) for the following reasons:

- A higher production plateau and faster cash flow
- Accelerated production, many layers can produce from the same wellbore
- Comingling delivers economically more attractive oil rates than single reservoirs
- Fewer wells, less infrastructure, lower capital costs, and lower operating expenses
- Smaller surface footprint due to fewer wells and less infrastructure

A vertical well can target only one sweet spot in a reservoir. It will be nothing more than a coincidence if the same well encounters sweet spots at the same location in other reservoirs. Therefore, each vertical well location is likely to be optimal for only one reservoir in terms of productivity. Experience has shown that commingled production causes severe reservoir management constraints as an asset moves from primary to secondary recovery.

3.7.2 Well Allowables

In petroleum engineering, the term "Well Allowable" refers to the allowable daily oil production rate from an oil well that it can safely make. This term is equally applicable to water injection, gas production, and gas injection wells and can express the permissible water or gas injection/production.

Setting production and injection allowable is essential to control excessive reservoir withdrawals or injection, which can be harmful to reservoir health (Figure 3.11). Maintaining production or injection "Well Allowables" within prescribed limits ensures a long, trouble-free, and healthy productive life with minimal intervention.

Setting well-allowables can be compared with the speed limits that the traffic department imposes on vehicular traffic. Despite the opportunity to achieve higher on-road speed, drivers are advised to observe the speed limits to avoid accidents and improve road safety. It not only aids in traffic control and management; it also assures a smooth driving experience to riders and safe passage to pedestrians.

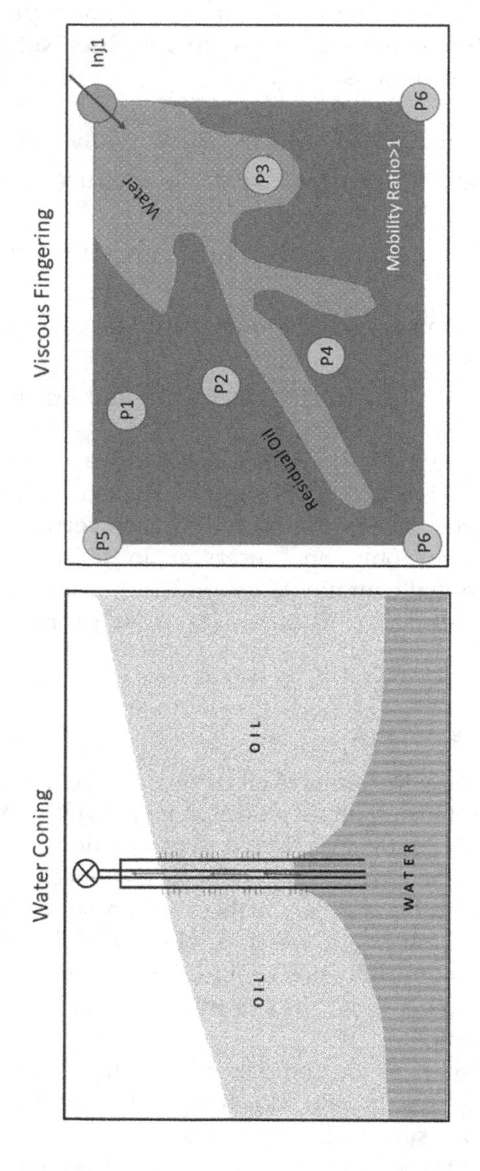

FIGURE 3.11
Consequences of Exceeding "Well-Allowables".

"WellAllowables" permit a proactive approach to reservoir management. If practiced sincerely, well-allowables can save a lot of unnecessary effort and expenditure on costly well-interventions to bring reservoir performance back on track.

Good production practices require that displacing fluids displace oil in a slow piston-like movement with a uniform front. Excessive and haphazard withdrawals cause a non-uniform flow of fluids in the reservoir due to coning/cusping or viscous fingering, which often causes a premature breakthrough of fluid that is the primary source of energy in the system; solution gas in the depletion drive, aquifer water, or gas cap gas. Exceeding allowable rates can aggravate undesired gas or water production at the cost of oil rate and recovery. It hits the operator resulting in poor economics. Consequently, the operator suffers from loss in oil revenue and has to incur extra cost to handle and treat gas and water.

Reservoir simulation is an excellent tool to determine production and injection rates provided the model has a representative geological description and calibrated well test data. In the absence of a simulation model, well allowables are determined based on the following factors:

- Lithology and stratigraphy of producing horizon
- Type of well (vertical, deviated, horizontal, multi-lateral)
- Reservoir layering and formation permeabilities
- Reservoir drive mechanism
- Results of the multiple-bean test
- Drainage unit size and target area for a proposed location
- The distance of Oil Water Contact (OWC)/Gas Oil Contact (GOC) from the perforations
- Bubble point and Solution GOR
- Water cut status and history
- Oil rate requirement

Exceeding well allowable for a long time has severe economic consequences due to increasing water/gas production rates and continuously declining oil rate and recovery.

3.8 Impact of Reservoir Management

The scope of reservoir management is overarching. It includes all phases of the petroleum lifecycle and encompasses the strategic, technical,

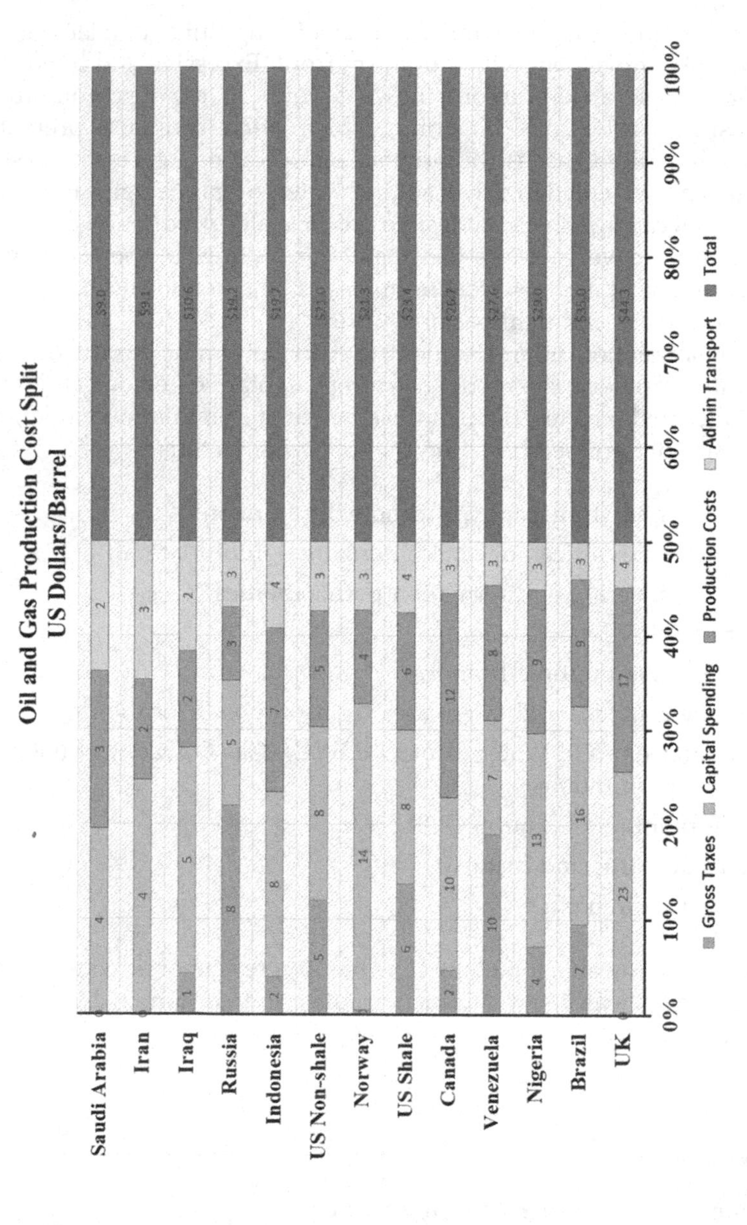

FIGURE 3.12
Production Cost Break Up of a Barrel of Oil and Gas (2016)[11].

commercial, financial, social, legal, and human resource aspects of the business. A well-designed reservoir management program can outreach the company objectives and integrate with the national mission of the governments of employment generation, human resource development, and overall growth of the national economy.

Reservoir management is not the responsibility of an individual or a team. It is everyone's shared responsibility to contribute to its objectives. The government, owner, and company management undoubtedly play a significant role in designing the strategy to develop the oil and gas resources, arrange the capital, and keep the company business afloat. Employees contribute through their knowledge and skills. They are responsible for doing all technical, operational, managerial, and other kinds of spadework. Owners, management, and employees have exclusive roles and functions as they cannot replace each other in work or position. Overall, within the organization, they rely on mutual output and strengths, yet they are functionally autonomous.

Modern reservoir management guidelines recognize that oil and gas extraction should not undermine the community's interest, climate, and environment. Archie Carrol's CSR pyramid is no longer a fine-print statement; it is now an explicit mandate in bold letters. This new code underscores the need to maintain petroleum reservoirs in good health, adhere to aggressive project management strategies, and employ best practices across all oil and gas field operations and activities to maximize benefits. These requirements assume enormous significance today when most energy-starved nations compete to secure their oil and gas share despite their side effects on the climate.

An effective organizational structure is a prerequisite to the seamless integration of its four elements: organization, capability, policy, and governance. It ensures the blending of authority with the delegation, know-how with skills, tools with technology, and interdependence with independence. The cost of production can be significantly reduced by enhancing efficiency coming from excellent teamwork and capitalizing on individuals' talent and potential in the organization.

Reservoir management has made rapid strides due to accumulated knowledge acquired in diverse conditions and continuous improvements towards developing new tools, technologies, and hardware. IT software has further advanced data acquisition, processing, interpretation, and management of all available data. Reservoir simulation is a powerful technology that supports reservoir management. When coupled with pre- and post-processing tools, it can enhance modeling efficiencies and provide deep insights into reservoirs' static and dynamic behavior. Development strategies can now be designed, evaluated, and compared faster to select the best development option. These advancements help the economic optimization of oil and gas recovery and lower oil and gas production costs.

Figure 3.12 highlights the cost of producing a barrel of oil by some of the world's top producers. The split of the total cost of production indicates that capital and operating expenditure is the lowest in middle-east countries. In general, these countries have distinct advantages of oil-rich sediments, conventional crude oil quality, economies of scale, and coastal environments. When combined with a zero or concessional tax regime, these factors help keep the cost of oil and gas production lower.

Reservoir management has its role cut out for the future as the oil and gas operators intensify their efforts to explore and extract hydrocarbons from unconventional resources, remote locations, problematic oil, and harsh environments. It can make a massive impact on the world energy scene if it can improve the reserve replacement ratio, increase the recovery factor, and reduce the cost of exploration and production. It assumes special significance at a time when energy-hungry nations are in a rat race to secure the largest share of hydrocarbons and manage their prices. Let us hope that crises like the COVID-19 will bring the governments, operators, investors, producers, and consumers together to design the new fiscal regime that meets everybody's basic need for energy at affordable prices.

Notes

1 US Energy Information Administration; International Energy Outlook 2018; https://www.eia.gov/outlooks/ieo/
2 Lejla Villar, Mason Hamilton; US Energy Information Administration (EIA); "Three important oil trade chokepoints are located around the Arabian Peninsula;" Africa Oil and Gas Report https://www.eia.gov/todayinenergy/detail.php?id=32352
3 Dr. Idris Demir; Ahi Evran University, Faculty of Economic and Administrative Sciences Department of International Relations, "Strategic Importance Of Crude Oil And Natural Gas Pipelines"; Australian Journal of Basic and Applied Sciences, 6(3): 87-96, 2012, ISSN 1991-8178
4 Robert D. Buzzel, Bradley T. Gale, and Ralph G.M. Sultan; "Market Share—a Key to Profitability"; Harvard Business Review; January 1975
5 MBA-H4030, International Business Finance; UNIT – I, International Monetary and Financial System; http://www.pondiuni.edu.in/storage/dde/downloads/finiv_ibf.pdf
6 International Monetary Fund; https://www.imf.org/en/About/Factsheets/IMF-at-a-Glance
7 Website Investopedia; https://www.investopedia.com/terms/i/imf.asp
8 Wikipedia The Free Encyclopedia; https://en.wikipedia.org/wiki/World_Bank

9 Heike Mainhardt; "World Bank Group Financial Flows Undermine the Paris Climate Agreement: The WBG contributes to higher profit margins for oil, gas, and coal" Urgewald; March 2019
10 Carroll, Archie. (1991). The Pyramid of Corporate Social Responsibility: Toward the Moral Management of Organizational Stakeholders. Business Horizons. 34. 39-48. 10.1016/0007-6813(91)90005-G.
11 Wikipedia The Free Encyclopedia; https://en.wikipedia.org/wiki/Price_of_oil

4

Significance of Reservoir Health and Its Impact on Reservoir Performance

4.1 Introduction

The formation of petroleum is attributed to the burial and compression of the remains of dead plants, animals, or other organisms at the end of their life cycle under millions of tons of sediment. Sediments and rocks are made up of atoms. Atoms form elements; elements form minerals, and minerals lead to the formation of rocks. Geological events leading to folding or faulting create conditions for accumulation and entrapment of oil and gas that migrate from high to low-pressure areas until it encounters an impermeable cap rock.

Formation pore pressure is the primary source of energy that drives fluids through the formation. Over geologic time, forces in the reservoir caused by the overburden (mass of rock and soil above a reservoir), temperature, pore pressure, capillary, molecular diffusion, and thermal convection equilibrate[1]. During this time, gravity plays its role in the segregation of fluids according to their density. Gas being the lightest goes to the top, water being the heaviest, settles at the bottom. The oil stays in the middle, overlain by gas and underlain by water. Drilling and completion operations connect the reservoir with the surface via wellbores allowing passage of fluids from high pressure to low pressure. The wellbore receives the fluids from its neighborhood, causing a pressure sink at the perforations. This disturbance in the pressure equilibrium creates a pressure drawdown, which forces reservoir fluids to move from areas farther away towards the wellbore. The effect of continued withdrawal of reservoir fluids in this manner is ultimately felt at the boundary, and reservoir pressure experiences a decline (Figure 4.1).

4.2 Reservoir Drive Mechanisms

In the primary recovery phase, a reservoir uses its inherent energy manifested in the form of pressure. The entire setup of reservoir rock, fluids,

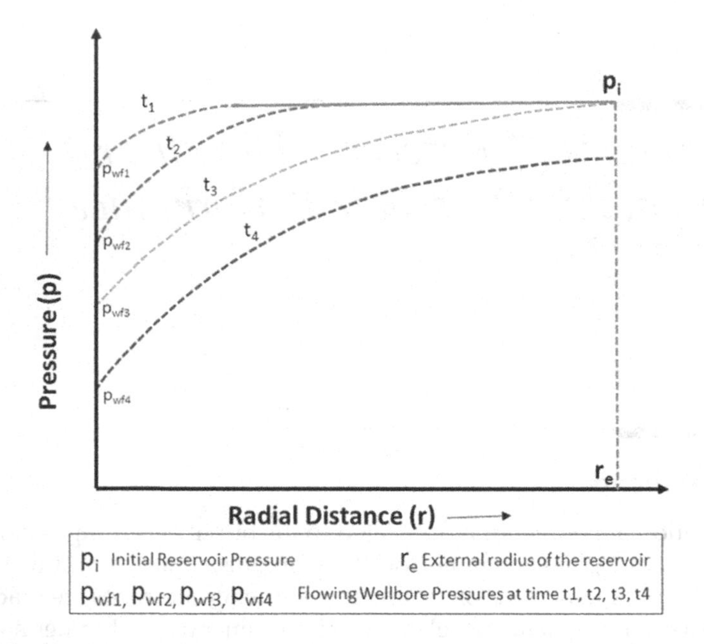

FIGURE 4.1
Pressure Profile in the Well Drainage Radius.

and geological conditions forms a drive mechanism, which pushes the fluids out of pore spaces in the rock to the surface via a wellbore. The drive mechanism is identified with the primary reservoir energy source as the dominant player in oil and gas recovery.

4.2.1 Solution Gas Drive Reservoirs

Solution gas drive reservoirs (Figure 4.2) are also known as depletion drive, dissolved gas drive, or internal gas drive reservoirs. Such reservoirs are entirely bounded by sealing faults and impermeable formations. Crude oil, without any gas dissolved in it, is called "dead oil" and cannot be recovered without external pressure support or intervention. However, reservoir oils contain natural gas, which is dissolved in oil. The expansion of reservoir rock and fluids aided by the pressure gradient drives hydrocarbons out of the pore spaces. The dissolved gas in oil also reduces the oil viscosity, thereby improving its mobility.

The reservoir pressure at which the gas just comes out of solution and stays as a bubble in the gas phase is known as the bubble point pressure. By implication, if the reservoir pressure is higher than the bubble point pressure, a mixture of oil and gas would be in a single-phase, that is, liquid phase, and this mixture of oil and gas is called "under-saturated" oil. At the bubble point pressure and below, the mix of oil and gas is "saturated" with liberated gas, which can now coexist as a separate

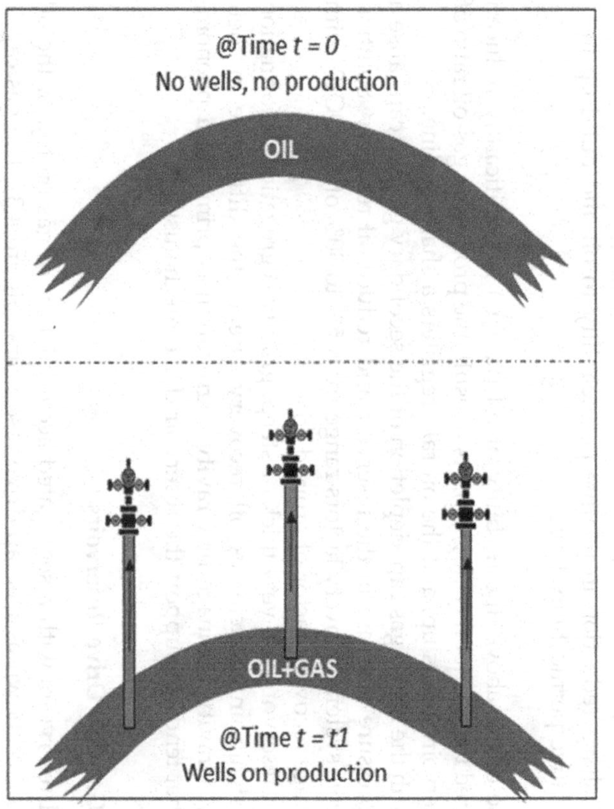

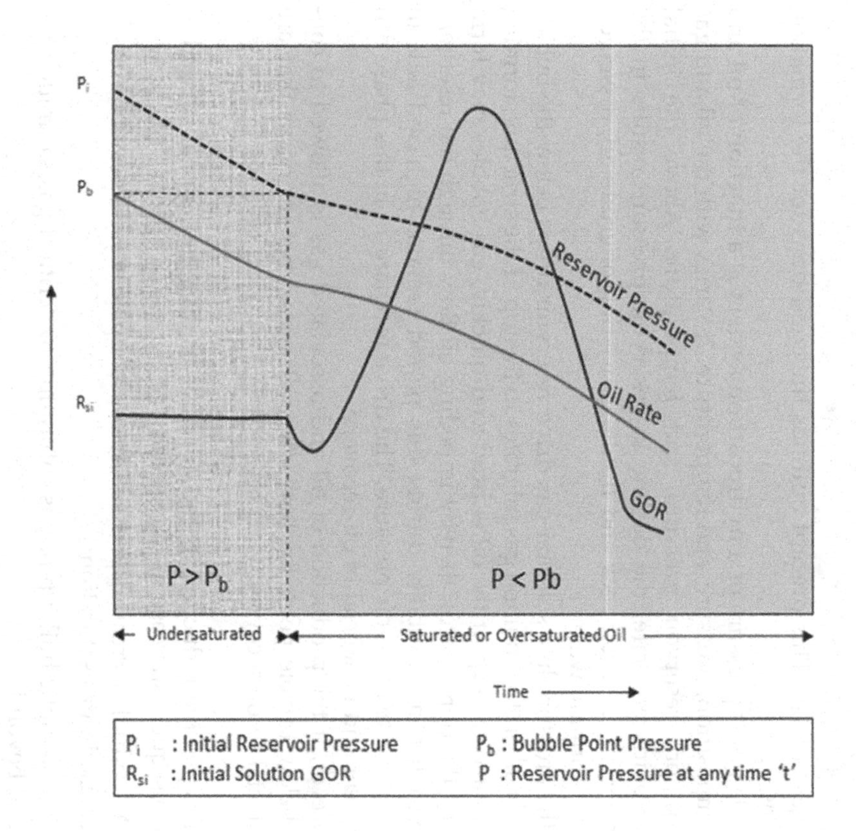

FIGURE 4.2
Typical Performance of a Solution Gas Drive Reservoir.

"free gas" phase. This precept outlines the conditions for the formation of a Gas Cap.

Bubble point pressure of a hydrocarbon system is a function of oil and gas composition, pressure, and temperature. It varies with the oil and gas composition, reservoir pressure, and temperature. Above the initial bubble point pressure, the decline in reservoir pressure results in the expansion of reservoir rock and fluids, consisting of crude oil, dissolved gas, and connate water.

As the pressure in a solution-gas drive reservoir drops below the bubble point, the liberated solution gas migrates to the top of the structure, forming a secondary gas cap. If the oil is produced quickly, gas bubbles may form and clog the pore spaces, thereby impeding oil flow through the reservoir. The system compressibility during this period remains high and aids in maintaining the reservoir pressure. The drive mechanism in this phase can be efficient and beneficial in oil recovery.

However, if the production of oil and associated gas is allowed to continue below bubble point for long, the following changes occur that are detrimental to oil recovery:

- Progressive reduction in the quantity of dissolved gas in oil and consequent increase in oil viscosity
- A gradual decrease in oil saturation coupled with a resulting increase in gas saturation
- A systematic buildup of gas saturation to "critical gas saturation" and beyond
- A sharp reduction in relative permeability to oil and build-up in relative permeability to gas

Because of the above, the mobility of oil (k_{ro}/μ_o) is significantly reduced compared to gas mobility (k_{rg}/μ_g). As a result, the producing gas-oil ratio of the reservoir shoots up, and the oil rate registers a sharp decline.

As with the initial gas cap, depletion of the secondary gas cap can cause a rapid pressure decrease in the reservoir and reduce oil recovery. Solution Gas Drive system recovery factors range from 5% to 30% of the OOIP, with an average recovery factor of around 15%.

When reservoirs are very thick or steep, gravity segregation of formation fluids significantly increases oil recovery due to the difference in their density. Gravity Drainage or Gravity Segregation is primarily a phenomenon that tends to support the reservoir drive mechanisms.

4.2.2 Gas Cap Drive Reservoirs

The oil reservoirs with a segregated accumulation of gas on top of the oil column are known as gas cap drive reservoirs (Figure 4.3). The gas cap is

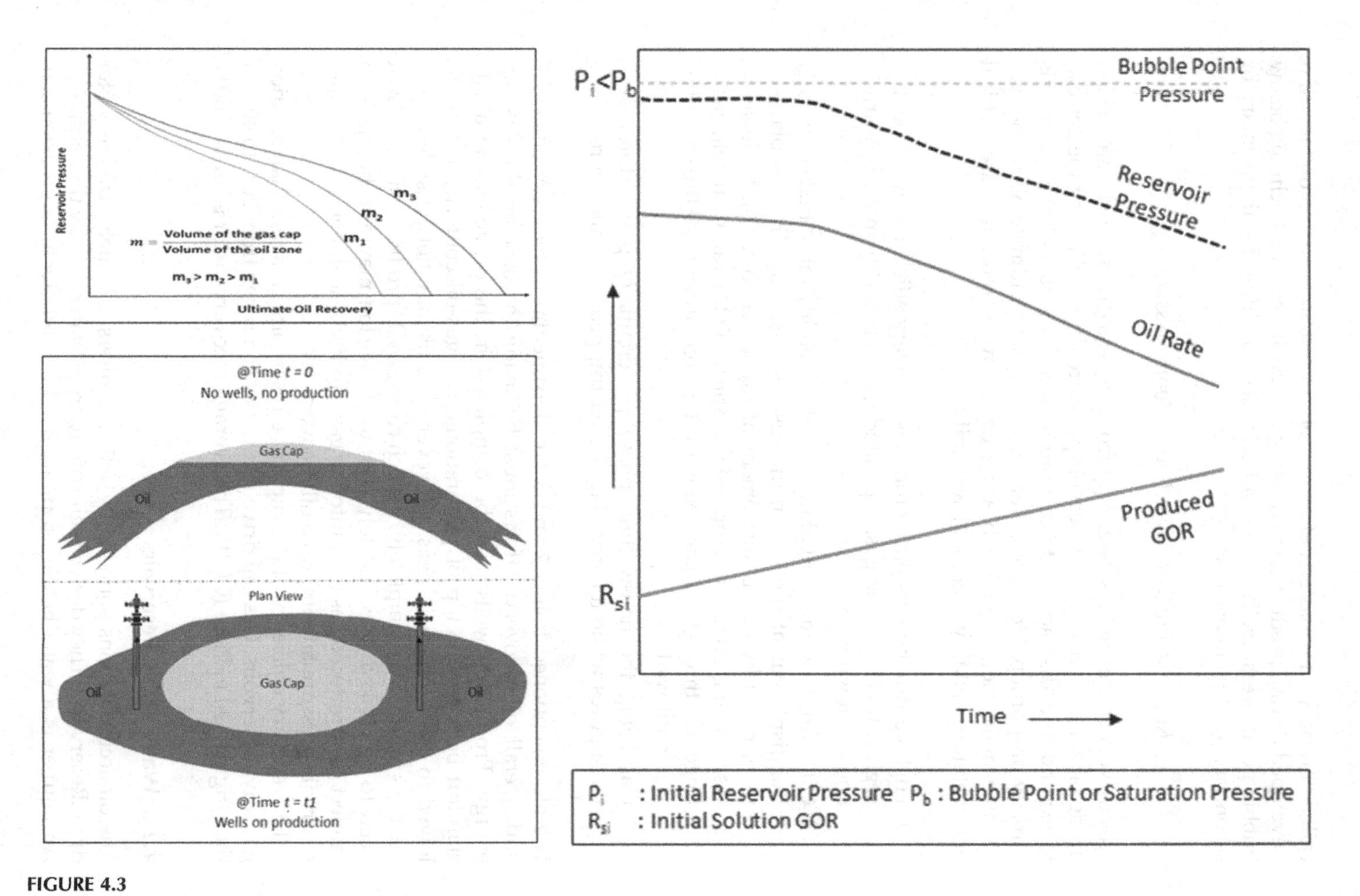

FIGURE 4.3
Typical Performance of a Gas Cap Drive Reservoir.

called "primary" if it existed before any oil production and "secondary" if it developed during production because of reservoir pressure declining below bubble point pressure. Therefore, a necessary condition for the primary or secondary gas cap to exist is:

$$\textit{Reservoir Pressure } (P) < \textit{Bubble Point Pressure } (P_b)$$

This drive mechanism gets its energy from the gas cap. As wells are drilled in the oil zone to recover oil, reservoir pressure drops, allowing the gas cap to expand and displace oil from pore spaces. The size of the gas cap is directly proportional to the recovery efficiency and ultimate oil recovery.

Three main factors with a bearing on recovery efficiency are vertical permeability, mobility ratio, and wettability.

1. High vertical permeability promotes recovery efficiency by efficient segregation of oil and gas. Separated gas can quickly move upward and oil downward.
2. Oil viscosity is an essential input to the mobility ratio[2] that controls the displacement process, in the case of a gas-cap drive displacement of oil by gas. Displacement of higher viscosity oil by lower viscosity gas gives rise to an unfavorable mobility ratio, causing gas fingering through oil and leads to the premature breakthrough of gas in oil wells.
3. Gas being the non-wetting phase preferentially passes through large pores without contacting the oil trapped in smaller pores.

The aforementioned Points 2 and 3 are detrimental to recovery efficiency and overall oil recovery. Efforts must be made to conserve the gas cap energy. Production wells must be completed in the oil zone, and a safe standoff between the top of perforations and gas-oil contact must be allowed to avoid producing gas. However, it is usually not possible to prevent gas production completely. In such cases, wells producing at high GOR need to be choked or completely shut-in. Provision may also be made to reinject the produced gas back into the gas cap if viable. It helps to maintain reservoir pressure and improve oil recovery.

The recovery efficiency of gas cap drive reservoirs is higher than solution gas drive reservoirs. Gas cap drive recoveries can range from 20% to 40% of the Original Oil in Place (OOIP). The average recovery factor is around 30%.

4.2.3 Water Drive Reservoirs

The oil accumulations bounded by active aquifers are categorized as water drive Reservoirs. They derive their energy from water held by the "aquifer." An aquifer is a water-bearing formation with a relatively large extension.

It can transmit water into the oil reservoir to displace oil from pore spaces as wells are drilled and produced. Aquifer location, size, strength, and permeability govern its behavior and reservoir performance. The aquifer's size and strength determine if it is a finite (limited and weak) or infinite (unlimited and strong) aquifer. A finite aquifer may be confined by impermeable rock such that the two, aquifer and oil reservoir, form a single hydraulically sealed unit. It may also be unconfined and connected with another water source such as rain, river, or sea. On the other hand, an infinite aquifer is vast and can be more than ten times larger than the reservoir.

An oil reservoir with an aquifer attached to the bottom is classified as a bottom water drive, and the one attached to the flanks is termed as the edge water drive system (Figure 4.4). In a bottom water drive reservoir, the aquifer directly communicates with the overlying oil reservoir. The water from the aquifer moves vertically upward into the oil zone as oil is produced. In an edge water drive system, the water moves upward along the reservoir dip.

Another important consideration in the classification of aquifers is their activity or strength. The size of the aquifer is only a qualitative indicator of the strength of the aquifer. Its actual capacity is judged by the dynamic behavior and performance of the reservoir when on production. The aquifer's strength is directly proportional to its ability to replenish the voidage created by fluid withdrawals from the reservoir and maintain the reservoir pressure. The aquifer size, permeability, expansion of the rock-water system, and the pressure drop imposed by reservoir withdrawals are the major factors that control this behavior.

At relatively low withdrawal rates, a large and efficient aquifer can quickly maintain reservoir pressure by filling the production voidage with water. If the reservoir withdrawals are high, then a small aquifer with low permeability will not efficiently replenish the voidage, and the reservoir pressure will fall.

In general, the operators' focus is on the production of oil and gas. They usually work to get as much information as possible about the oil reservoir; investigation of aquifer size and strength is not their priority due to the time and cost involved. They instead use indirect methods such as the water cut or time-lapse pressure measurements in the oil reservoir to determine the aquifer's strength. The only direct information that a company has about the aquifer is through drilled exploratory/delineation wells, which may have unintentionally penetrated the aquifer. Even in these wells, detailed testing of aquifers is avoided to save costs. Sometimes, an aquifer might be misinterpreted as dormant or completely absent due to the aquifer's delayed response to the reservoir withdrawal. Late aquifer kick-off is not uncommon and can happen when the aquifer does not immediately experience a pressure drop caused in the reservoir.

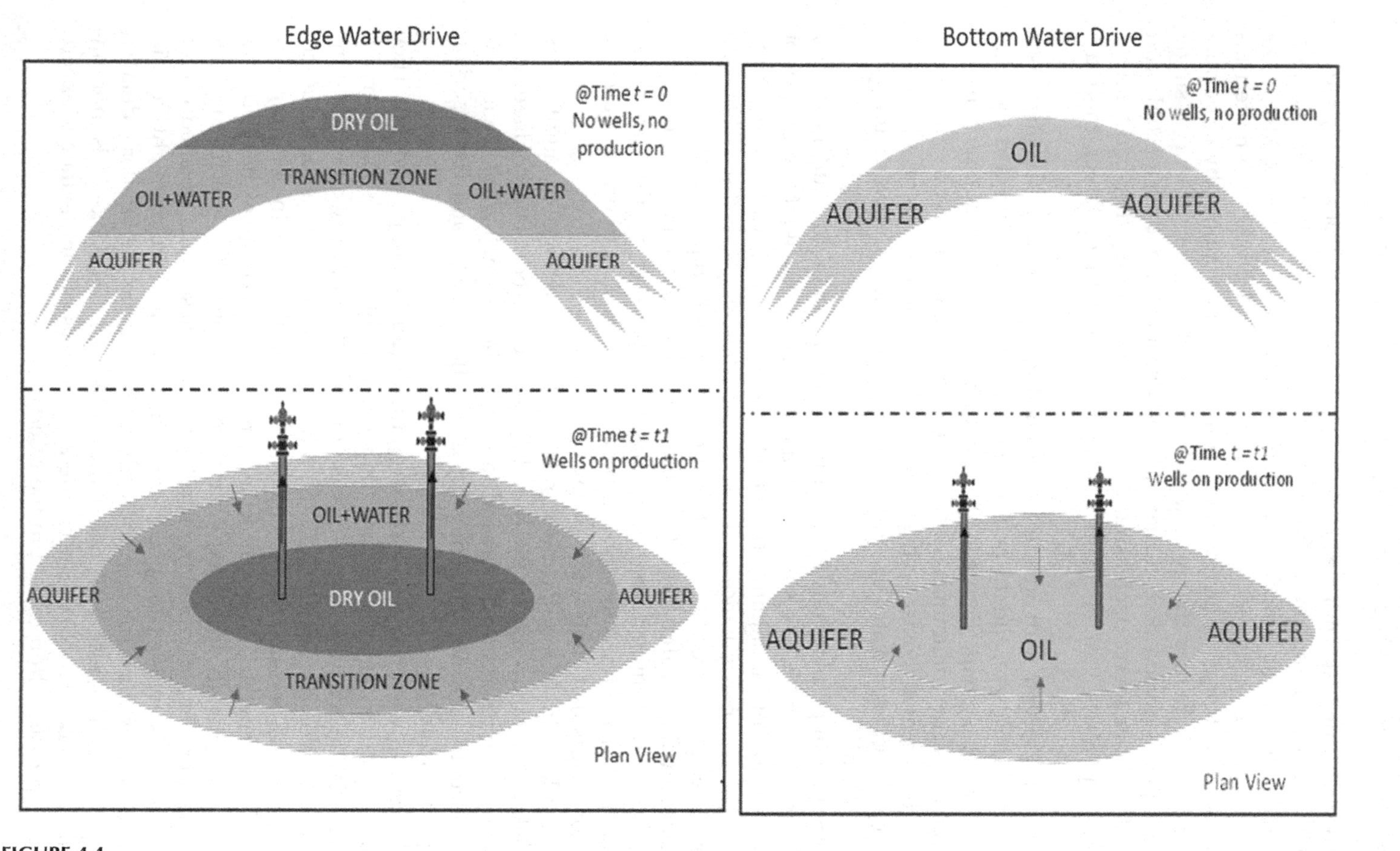

FIGURE 4.4
Water Drive Reservoir.

Water production is an expected outcome in the case of water drive reservoirs. However, its timing depends on the aquifer's size and strength, production offtake, and reservoir management strategy. Ideally, water production can be contained by perforating the wells in the clean oil column, keeping a safe distance away from the Oil Water Contact (OWC), and maintaining relatively low production rates. However, it may often not be a practical option for the company, usually interested in higher production rates to quickly recover their investments. It may instead choose to manage water if the facilities can handle the produced water volumes. Controlling the production of formation water in the later phase of exploitation is not easy when the water cones can form an extended transition zone where both water and oil are mobile. In this phase of production, each barrel of oil is accompanied by several barrels of water. In such cases, operating costs and oil prices decide the economic cutoff for production.

A water drive reservoir (Figure 4.5) is characterized by stable reservoir pressure and a nearly constant production (oil + water) rate. The production GOR for water drive reservoirs tends to remain stable as long as the aquifer can maintain the reservoir pressure above the bubble point pressure.

Water drive reservoirs constitute some of the world's most prolific oil fields based on their production rates and recovery. Their EUR depends on the aquifer's characteristics, mobility ratio, and sweep efficiency. However, reservoir heterogeneity can significantly lower oil recovery due to erratic water movement, causing its premature breakthrough. Edge water drive reservoirs generally perform better because water coning in bottom water drive

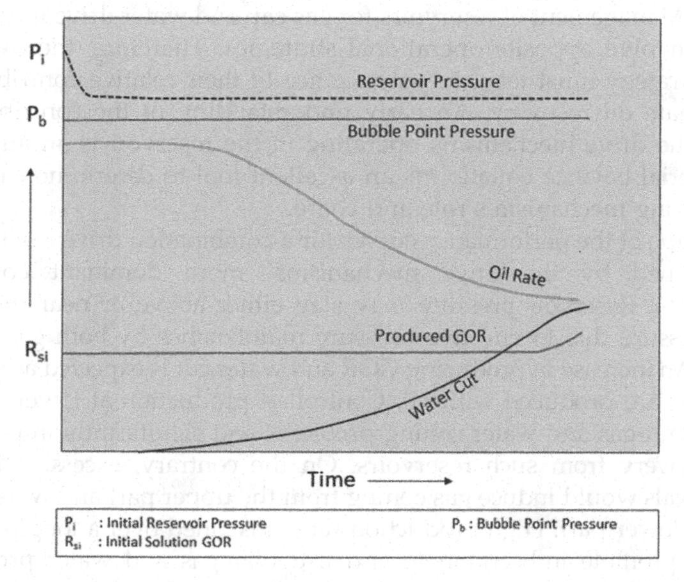

FIGURE 4.5
Typical Performance of a Water Drive Reservoir.

reservoirs can substantially reduce oil recovery. In general, the recovery factors from water drive reservoirs are 30% to 50% of the initial oil-in-place.

Under the most favorable conditions, a large aquifer in communication with a medium or light oil reservoir (mobility ratio $M < 1$) can effectively maintain the reservoir pressure. It helps to retain the dissolved gas in oil, thereby maintaining excellent oil mobility and displacement efficiency. In such cases, oil recovery may well go beyond 50% and sometimes even exceed 60% of the OOIP. High oil recoveries can be achieved by controlling the reservoir withdrawals to match the aquifer's ability to maintain reservoir pressure. If not, the aquifer must be appropriately strengthened by additional water injection.

4.2.4 Combination Drive Reservoirs

A reservoir may incorporate more than one drive mechanism. Combination drive reservoirs, also referred to as mixed drive reservoirs, are those reservoirs where the gas cap and water drive mechanisms operate simultaneously.

It is a very efficient recovery mechanism, particularly in the early development phase. The reservoir benefits from the dual recharge of energy from the gas cap and aquifer for pressure maintenance. As oil production is initiated, the aquifer underneath expands and pushes the oil column upward. At the same time, the gas cap also expands and pushes the oil downward. The middle and late phases of developing the combination drive reservoirs can be quite complicated in tracking fluid movement and workover operations. Management of reservoirs for gas cap and water drive mechanisms usually involve opposite operational strategies. Therefore, the overall reservoir strategy must take due cognizance of their relative contribution to the ultimate oil recovery. An early understanding of the contribution of each of the drive mechanisms operating in the reservoir is an advantage. The material balance equation is an excellent tool to determine each operating driving mechanism's role and share.

The shape of the performance curves for a combination drive reservoir will be governed by the drive mechanisms' more dominant constituent (Figure 4.6). Reservoir pressure may stay either above or near the bubble point pressure due to effective pressure maintenance by both gas cap and aquifer. An increase in producing GOR and water cut is expected as some gas and water are produced with oil. Controlled production at lower rates can help contain gas and water coning problems and significantly improve ultimate recovery from such reservoirs. On the contrary, excessive reservoir withdrawals would induce gas coning from the upper part and water coning from the lower part. High production rates sustained over a long period are likely to promote indiscriminate and irregular gas and water production, resulting in the waste of reservoir energy. Many infill wells would be subsequently required to recover the bypassed oil by gas and water.

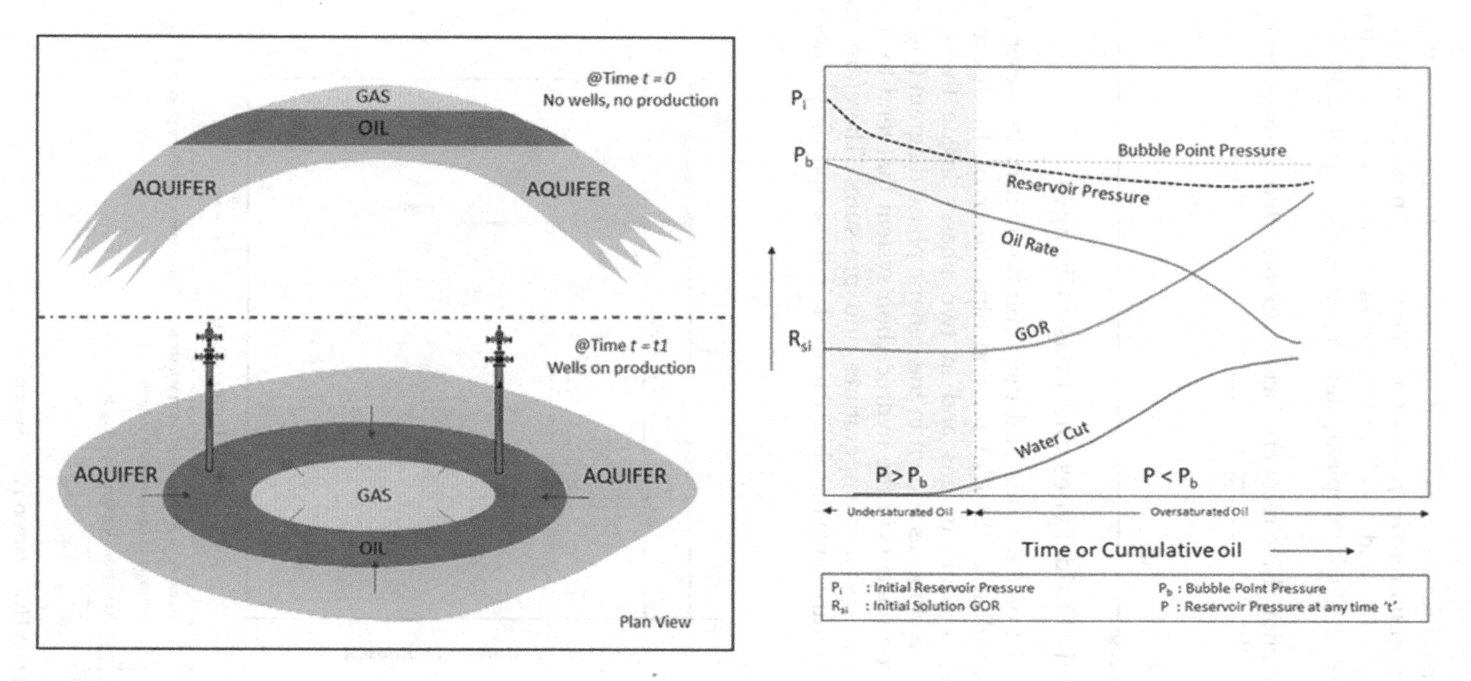

FIGURE 4.6
Typical Performance of a Combination Drive Reservoirs.

Reservoir models are an excellent tool to estimate the impact of gas and water coning on oil recovery. It is possible to design production strategies that will help GOC and OWC to remain uniform as oil production continues. However, field implementation of such a plan can be quite challenging.

The combination drive recovery factors are reported to range from 5% to 70% of the OOIP. The average recovery factor is around 50%.

4.3 Reservoir Fluid Types and Phase Changes

The state of petroleum fluids and their composition in the reservoir and surface is controlled by the PVT relationships. Reservoir fluids can be in a single phase as liquid or gas and in two phases inside the pressure-temperature envelope, as shown in the generic phase diagram (Figure 4.7).

There are components in the hydrocarbon system that tend to condense and vaporize at specific temperatures and pressures. Therefore, the full phase diagram must have the ability to explain these specific phenomena as

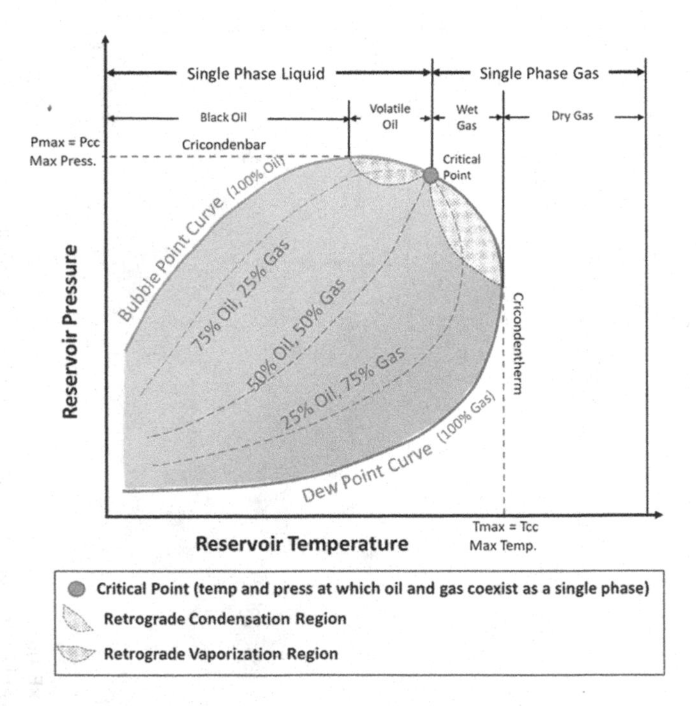

FIGURE 4.7
Phase Diagram of a Hydrocarbon Fluid System.

the process of extraction of oil and natural gas involves significant variations in pressure and temperature. The P-T phase envelope delineates the monophasic and diphasic regions of the hydrocarbon fluids separated by the bubble point curve and the dew point curve.

The bubble point curve separates the single-phase liquid region from the two-phase region. It differentiates between the crude oils that are saturated or undersaturated with natural gas. Any oil and gas mixture above this curve will stay in a liquid phase. This oil is referred to as "undersaturated" because it can dissolve more gas if it were present.

However, on the bubble point curve, the hydrocarbon mixture is 100% liquid with an infinitesimal amount of gas in equilibrium. It represents the evolution of the first gas bubble as the mixture can not dissolve any further gas. Therefore, the oil and gas mixture at the bubble point is said to be "saturated."

The dew point curve represents a region of 100% vapor saturated with an infinitesimal amount of condensate liquid. It demarcates the single-phase gaseous region from the two-phase region. Any fluids below or outside this curve will be in the dry gas phase.

Critical Point is defined by the common meeting point of the bubble point and dew point curves of a multi-component hydrocarbon system. At this point, the liquid and gas phases coexist such that they are indistinguishable from each other. However, this is not valid for multi-component systems because the critical point neither represents the maximum pressure nor the maximum temperature for vapor-liquid to coexist.

Within the P-T phase envelope bound by the bubble point and dew point curves, the fluids are present in two phases- liquid and gas. Between the bubble point and the dew point curves, the liquid fraction in the mixture decreases while the gas fraction increases. Finally, the mixture contains a tiny amount of liquid saturated with mostly gas at the dew point curve. Any fluids beyond this curve are in the gaseous phase.

At this point, it is essential to introduce two new terms cricondenbar and cricondentherm to help define the retrograde phenomena that some oil and gas mixtures can experience during extraction. The cricondenbar and cricondentherm of the phase envelope of single-component pure fluids systems coincide with their critical pressure and critical temperature.[3] However, for multi-component hydrocarbon systems, cricondenbar (Pcc) is the maximum pressure (Pmax), and cricondentherm (Tcc) is the maximum temperature of the phase envelope. The word retrograde means "reverse" or "contrary to expectation." Retrograde condensation phenomenon is associated with the area inside the phase envelope where the gas in contact with the liquid may be condensed by an isothermal decrease in pressure; or an isobaric increase in temperature. This apparently reverse phenomenon of some gas condensing into a liquid phase instead of expanding or vaporizing when pressure is decreased is called retrograde condensation.

Similarly, some volatile oils in contact with the gas, having their pressure between critical pressure and cricondenbar, undergo vaporization by an isothermal increase in pressure or isobaric decrease in temperature. This phenomenon is known as retrograde vaporization or evaporation.

The phase diagram is an important means to classify petroleum reservoirs based on their fluid types, namely, black oil, volatile oil, condensate (retrograde gas), dry gas, and wet gas. An understanding of key differences in their phase behavior is essential to maximize the oil and gas recovery.

During extraction, oil and gas go through significant pressure and temperature variations, often causing a fluid phase change. Many fluid samples need to be collected and analyzed thoroughly in the laboratory to establish a representative phase envelope and determine PVT properties such as the initial bubble point pressure, oil and gas formation volume factors, solution GOR, oil and gas viscosity, oil and gas composition at reservoir and surface conditions, the gravity of the stock-tank liquid, and so forth.

Figure 4.8 displays four separate pictures of various fluid types at reservoir conditions, and their phase behavior changes due to pressure reduction in the subsurface and surface separators. Point A represents the initial pressure and temperature conditions of the reservoir (P1, T1). Points A, B, and C mimic the production path emulating the isothermal decline in reservoir pressure. The specific P and T coordinates define Points B and C's location on the P-T diagram; for example, the coordinates of points B and C are (P2, T1) and (P3, T1), respectively, and their position on the diagram decides the fluid quality and number of phases at the reservoir conditions.

Production path between the reservoir and the separator is represented by the line AS. Point A defines the initial reservoir conditions (P1, T1) and point S the separator conditions (P4, T2). The location of these points on the P-T diagram reveals the fluid quality and number of phases.

Black oils are dark in color due to the presence of heavy hydrocarbons. Their producing GOR increases as the dissolved gas in oil liberates inside the reservoir when the reservoir pressure falls below bubble point pressure (point B through C). Their reservoir temperature is lower than the critical temperature of the fluid.

Volatile oils are light brown or green in color, have initial API gravity over 40°, and high gas-oil ratios. They are rich in lighter hydrocarbons and have less heavy hydrocarbons. Their reservoir temperature varies over a narrow range, and its hydrocarbons are in the liquid phase because the reservoir temperature is lower than the fluid's critical temperature. Lower the gap between the reservoir and critical temperature higher the production of gas with lower oil recovery.

Gas condensate or condensate gas reservoirs are similar to volatile oils. Condensate may be colorless or have a light color like that of volatile oils with a gravity of 40° to 60°API. Reservoir temperature of a gas condensate reservoir is higher than the fluid's critical temperature; as such, its contents are in the gaseous phase.

During extraction, the pressure of a gas condensate reservoir must traverse through the dew point, resulting in the condensation of large volumes of

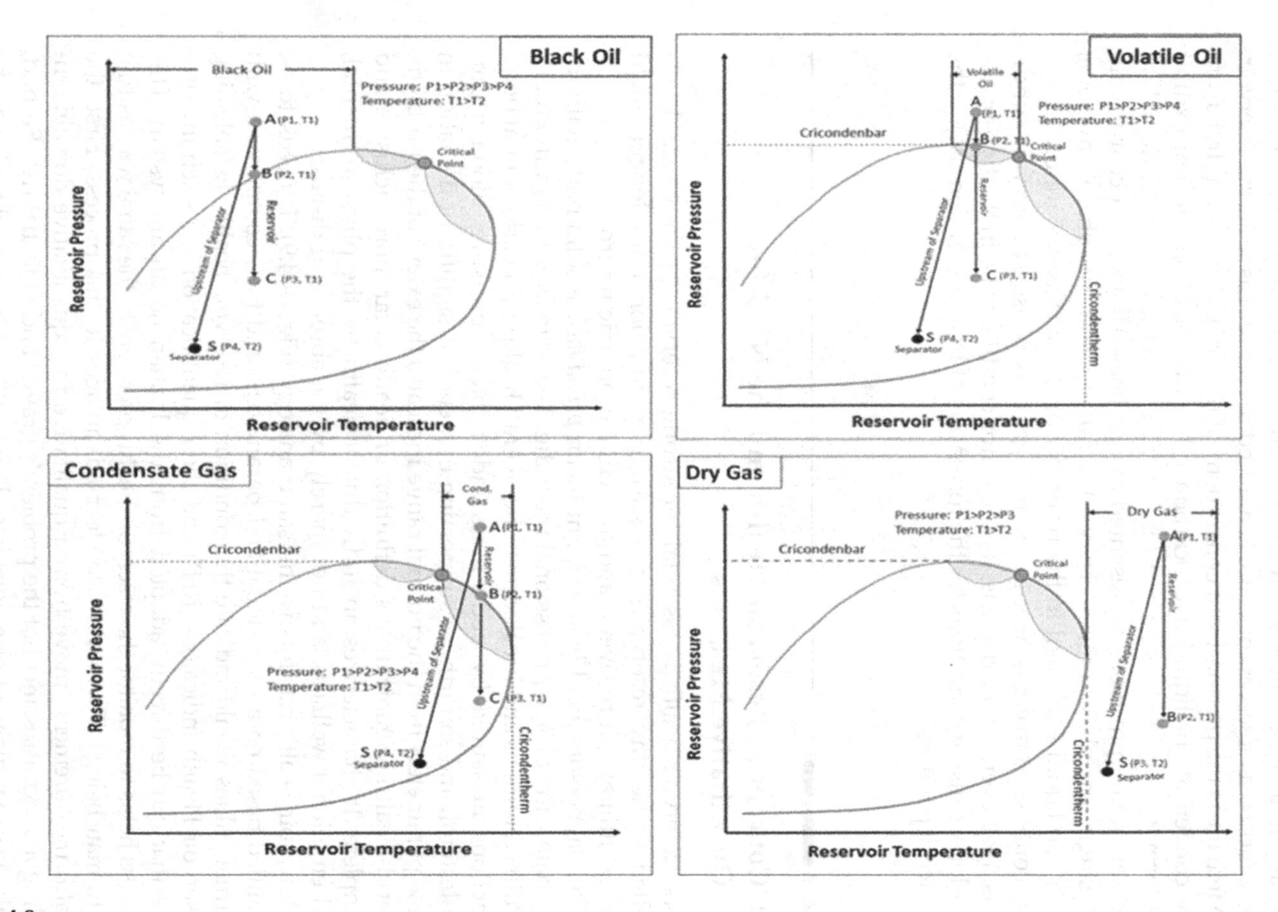

FIGURE 4.8
Phase Diagram and Reservoir Fluid Types.

hydrocarbons that can only be recovered by reinjecting dry gas to maintain reservoir pressures above the dew point for improving the condensate recovery. Volatile oils and gas condensate reservoirs can exhibit a retrograde condensation phenomenon as reservoir pressure is reduced to recover gas from the reservoir.

Methane is the predominant constituent of the dry gas reservoir, free from any condensate or liquid hydrocarbons. The hydrocarbons do not exhibit the formation of condensate in the subsurface or at the surface. Its depletion and separation processes represented by the lines AB and AS occur within the single gas-phase region. They also fall outside the phase envelope; hence, no liquid is formed in the reservoir or the surface separator.

In contrast to dry gas, wet gas is in a gaseous phase at reservoir conditions. However, unlike dry gas, its separator conditions lie in the two-phase envelope, causing conditions for the formation of liquids (condensate) in the surface separators.

4.4 Concept of Reservoir Health and the Need for Quantitative Measures

Good reservoir health is essential for strong reservoir performance. Wells drilled in healthy reservoirs can sustain production rates longer, yield higher ultimate oil recovery, and attractive economic returns.

To a layperson, good health might mean *prima facie* a solid body without any pain. However, a professional must dig deeper to assess an individual's health conditions and break down the overall health into various vital body functions measured by body mass index, blood pressure, blood sugar, cholesterol, and so forth, as supporting evidence. A significant deviation in these parameters from the normal range indicates the extent of the problem. World Health Organization's definition of health is far more inclusive and complete. It characterizes an individual's health by the physical, mental, and emotional wellness, and not merely by the absence of disease.[4]

Reservoir health can follow the same analogy (Figure 4.9). It needs to be qualified based on a set of identified parameters, and their comparison with normal values would indicate the condition of reservoir health. Establishing Reservoir Health Indicators (RHI) and their quantification is a scientific process that can be directly adopted from the human healthcare system. This process helps to monitor the reservoir health precisely in the same way as that of human beings. There are two clear advantages to this process. First, the health measurements move from a qualitative to a quantitative domain, enabling a direct assessment of the problem's gravity and potential risk. Second, it allows the ranking of reservoirs based on health ratings useful to decide the priority of intervention. It brings us to an essential question of the indicative parameters of good reservoir health and their typical values.

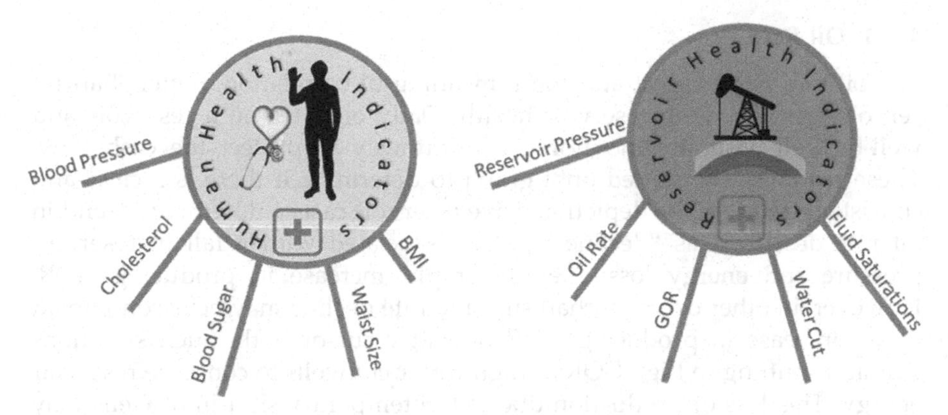

FIGURE 4.9
Analogy between Human and Petroleum Reservoir Health.

4.5 Reservoir Health Indicators (RHIs)

As for human beings, performance is a good albeit indirect measure of reservoir health. If a well or reservoir can perform efficiently, it is expected to have good health. Any fluctuation in reservoir performance reflects on reservoir health conditions.

The following five fundamental parameters that are widely used to assess the reservoir performance in the primary or secondary recovery phase may be adopted as RHIs.

1. Oil Rate
2. Producing GOR
3. Water Cut
4. Reservoir Pressure
5. Fluid Saturations

Apart from the above, the following two additional parameters can be used to determine the reservoirs' health conditions under the secondary recovery process.

1. Sweep Efficiency
2. Reservoir Souring

As established by time-lapse measurements, the change in these parameters with respect to the initial and previous values indicates the relative status of reservoir health.

4.5.1 Oil Rate

The oil rate has a direct and more meaningful correlation, better than reservoir pressure, with reservoir health. Daily oil rates on a reservoir and well-by-well basis are monitored continuously to protect the cash flow. These readings are plotted on a graph to determine if there is a clear and consistent trend. In the depletion drive reservoir case, a downward trend in oil rate described as "decline rate" is associated with a fall in reservoir pressure and energy loss due to a rapid increase in producing GOR. However, in other drive mechanisms, the rate decline may be accompanied by an increase in producing GOR or water cut or both. Such situations demand shutting-in high GOR or high water cut wells to conserve reservoir energy. This loss of production due to the temporary shut-in of wells may sometimes be compensated by other producers in the same layer or fault block.

Decline curves are a plot of production rate versus time on a Cartesian scale. The flatter curve is referred to as "exponential or constant percent decline" while hyperbolic and harmonic decline datasets plot as concave upward curves on an ordinary graph paper (Figure 4.10). Most oil wells tend to produce at an exponential or constant percent decline, which often plots as a straight line on a semi-logarithmic plot.

Semi-log, log-log, or any other plotting techniques often help replace a curvilinear shape into a straight line eliminating uncertainties in extrapolation and improving forecasting accuracy.

The decline rate is generally calculated annually, yielding the change in produced volume over the successive years. It is not necessarily constant and may vary with the wells' operating conditions or one reservoir drive mechanism dominating the other during exploitation. The decline rate is a positive number implying a decrease in production with time and is customarily expressed as a percentage. A negative decline rate represents an increase in production.

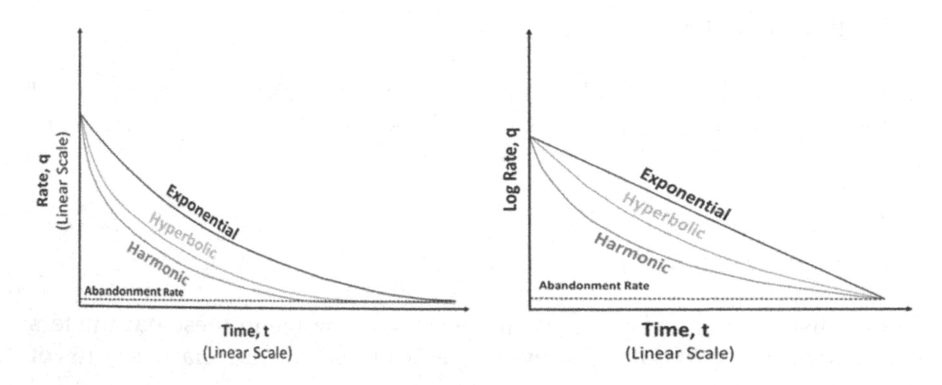

FIGURE 4.10
Types of Decline Rate Curves (Linear and Semilog Plots)[5].

$$Decline\ Rate_t = \frac{Production_{t-1} - Production_t}{Production_t} \tag{4.1}$$

In this equation, "*t*-1" and "*t*" represent two successive instants when the production rate is measured. If "t" is measured in months and the production rate is measured in bbls/day, then the decline rate is expressed in bbls/day/month.

The oil rate starts dropping when the reservoir boundary feels the effect of production. The decline in oil rate is the result of depletion due to production. High decline rates may be a symptom of reservoir maturity or inefficient use of drive mechanisms. In the first case, it is essential to reassess the reservoir potential and its economic development based on the incremental oil saturation (S_o-S_{or}). The latter situation can be corrected by modifying the production strategy.

DCA is a classical reservoir engineering method that has been widely discussed in petroleum literature. Oil companies extensively use it to forecast reservoir performance and estimate reserves. The technique relates the production performance of oil wells to reserves, assuming that all control factors remain unchanged. This assumption can severely limit the method's application due to ongoing essential workovers, infill drilling, well stimulations and artificial lift applications, and so forth. Sometimes the factors, such as politics, sabotage, changes in operating conditions of the wells, lack of investments, or inefficiencies, besides other factors, can also result in an artificial decline in rate. Modifications are made to account for such changes in operating conditions of wells or changes in reservoir behavior.

4.5.2 Producing GOR

Subsurface temperature and pressure are related to the depth and are higher than the standard temperature and pressure. When oil at reservoir temperature and pressure are produced and brought to the surface, it releases the dissolved natural gas causing the volume of oil to shrink and gas to expand. The producing GOR, at a particular instant, is the ratio of the total amount of gas production (solution gas + free gas) at standard conditions divided by the amount of oil production. It may be reported either in metric units as m^3/m^3 or in field units as ft^3/bbl. For a clear understanding of the behavior of the Producing GOR at any instant, it is essential to consider its variation with pressure. Reduction in pressure is representative of production history caused by depletion.

Figure 4.11 shows that the Producing GOR at any time "t" is the sum of two components made up of reservoir fluid and rock properties[6]:

i. Solution GOR (Rs) which is the amount of gas dissolved in oil at pressure P, and

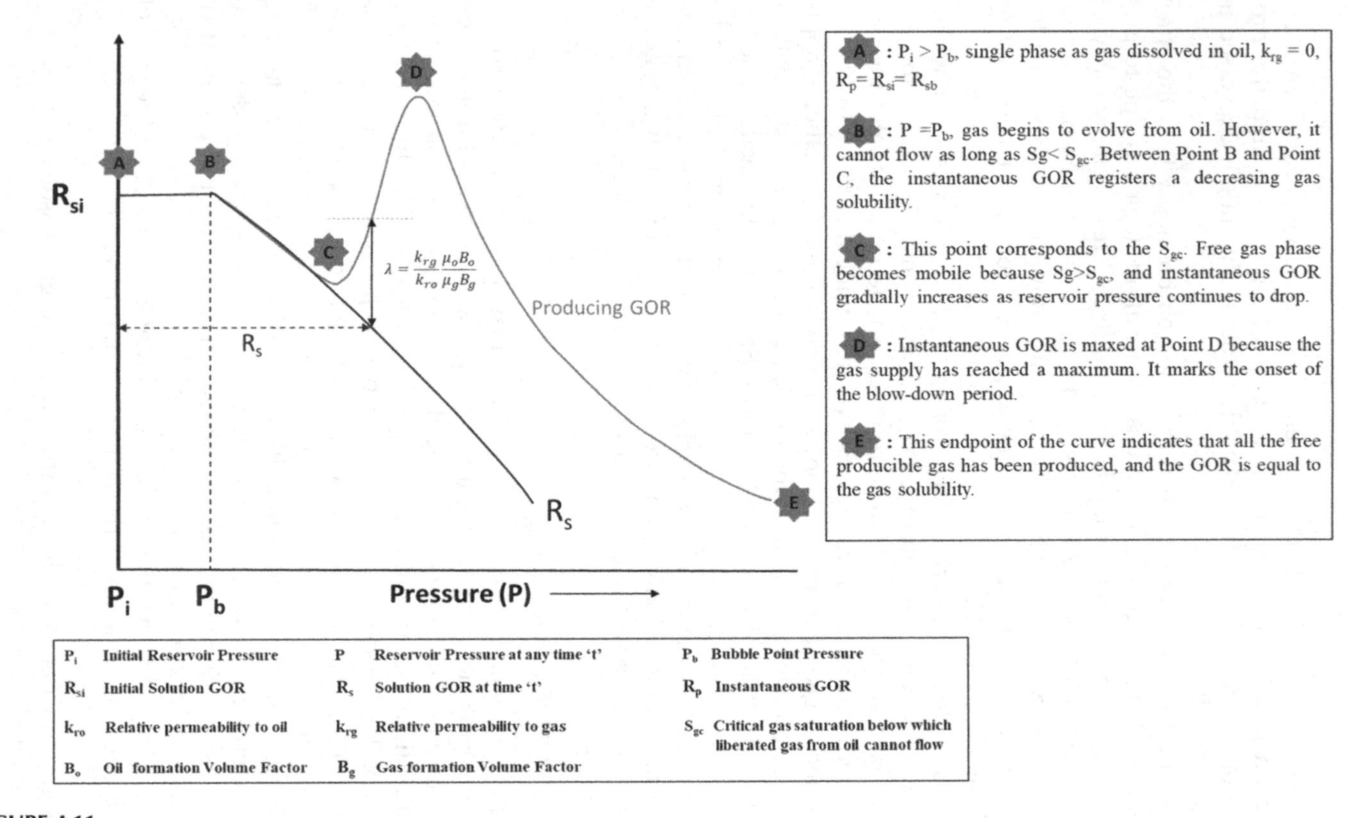

FIGURE 4.11
Producing GOR Behavior.

ii. $\lambda = \frac{k_{rg}}{k_{ro}} x \frac{\mu_o B_o}{\mu_g B_g}$

where k_{rg} is relative permeability to gas, k_{ro} is relative permeability to oil, μ_g is viscosity to gas, μ_o is viscosity to oil, B_g is Gas Formation Volume Factor, and B_o is Oil Formation Volume Factor.

Here, the solution R_s is a PVT property of the oil and gas system. It refers to the ability of gas to dissolve in oil. The reservoir oil can contain the maximum amount of dissolved natural gas at the bubble point pressure (P_b) and stay in equilibrium with it as a single phase. Any excess amount of gas would appear as a free gas phase. It implies that R_s for reservoirs with initial reservoir pressure (P_i) above bubble point pressure would be constant between P_i and P_b. However, as oil production from the reservoir proceeds below P_b, the gas dissolved in oil would evolve and accumulate as a free gas phase.

Producing GOR curve shown in Figure 4.9 can be broken into four segments marked by five points A through E on the curve:

Point A: At this point, which is the initial reservoir pressure $P > P_b$. There is no free gas in the formation, that is, $k_{rg} = 0$, the solution gas-oil ratio remains constant at R_{si} until the pressure reaches the bubble-point pressure at Point B. Therefore, between points A and B on this curve:

$$\text{Producing GOR } (R_p) = R_{si} = R_{sb}$$

Point B: Reservoir pressure $P < P_b$, the gas begins to evolve from the solution, and its saturation increases. However, this free gas cannot flow as long as the gas saturation $Sg < S_{gc}$, where S_{gc} is the critical gas saturation. Between Point B and Point C, producing GOR registers a decreasing gas solubility.

Point C: This point corresponds to S_{gc}. As the reservoir's gas saturation exceeds this value, the reservoir's free gas phase becomes mobile. The value of GOR gradually increases with the decline in reservoir pressure. During this phase of pressure decline, the producing GOR is described by the following equation:

$$GOR = R_s + \frac{k_{rg}}{k_{ro}} \frac{\mu_o B_o}{\mu_g B_g} \tag{4.2}$$

Point D: The maximum GOR is reached because the gas supply has reached a maximum. It marks the onset of the blow-down period.

Point E: This endpoint of the curve indicates that all the free producible gas has been produced, and the GOR is equal to the gas solubility, which may be nil or very small.

Producing GOR is a behavioral characteristic of different drive mechanisms. It helps to establish the type of drive mechanism operating in a reservoir (Figure 4.12). If the reservoir pressure stays above bubble point pressure in a water drive reservoir, the GOR remains constant at R_{si}.

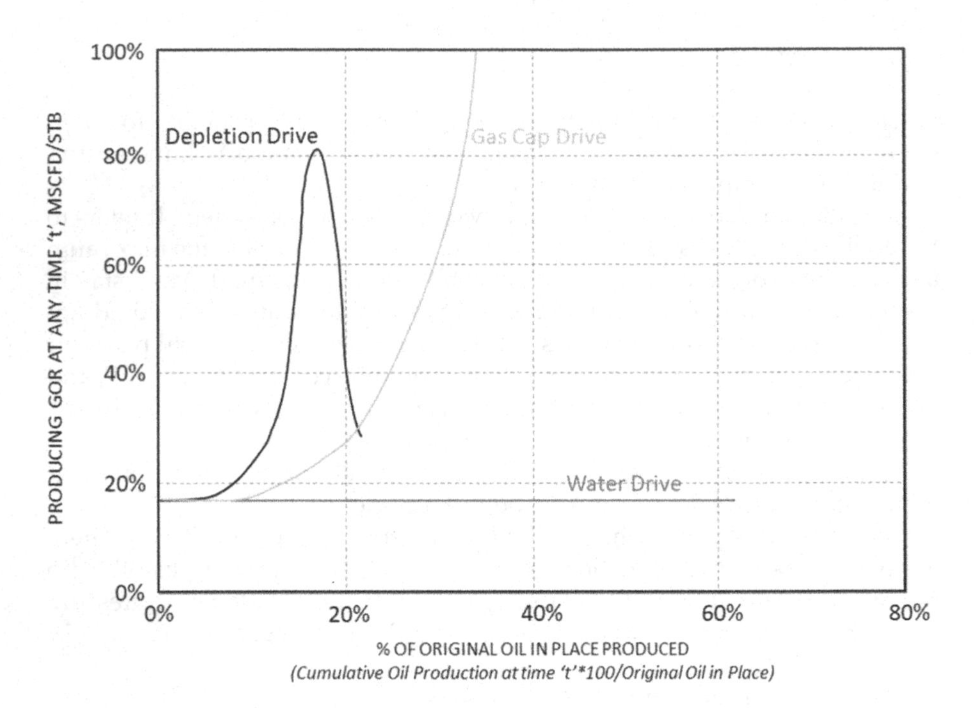

FIGURE 4.12
Variation of Producing GOR with Recovery Factor.

A depletion drive reservoir would register an increase in GOR for reservoir pressure $P < P_b$ and $S_g > S_{gc}$, as discussed earlier. For a dominating gas cap drive reservoir where reservoir pressure P < Pb, GOR would show a rising trend after $S_g > S_{gc}$.

4.5.3 Water Cut

Water cut is defined as the ratio or percentage of produced water divided by the volume of total produced fluids (crude oil + water). It may be calculated for a well, group of wells, zone, fault block or compartment, sector, reservoir, or asset. Reservoirs or wells can gradually turn from dry (0% water cut) to wet (more than 0% water cut).

Water cut is a behavioral characteristic of a water drive or waterflood reservoir. Sometimes in the early stages of production, a water drive or waterflood reservoir may not see any water cut. However, when the aquifer becomes active, water production is an expected outcome. The start of the water cut and its magnitude is dependent on the aquifer's size and strength, production offtake, and well locations.

Reservoirs producing by dominant natural water drive or waterflooding mechanisms, if not appropriately managed, can pose severe operational problems due to excessive water production, sometimes in the early phase of exploitation. The problem can be easily identified by an abrupt increase in water production on a plot of time versus water cut or water-oil ratio. Comparing problem wells/patterns with other wells/patterns in the neighborhood also helps differentiate between normal and abnormal water production. The problem of early water production is quite common in carbonate reservoirs owing to the presence of conductive faults, vugs, fractures, and fissures.

In water drive or waterflood reservoirs, two distinct types of water production, commonly referred to as "good" and "bad" water, can be recognized. The first type, that is, good water, usually occurs later in the life of a primary or secondary waterflood. This water is produced concurrently with oil because of the fractional flow characteristics of the reservoir rock. Any effort to reduce this water would reduce oil production. Each barrel of this water helps support oil production and recovery. In other words, this type of water production is inevitable, and the company must handle and manage this water.

The second type of water that does not support reservoir pressure or improve oil production is "bad water." It may appear in oil wells at any stage because of reservoir heterogeneity, cross-flow, adverse mobility ratio, water coning, high permeability streaks, fractures, fissures, and so forth. In the course of production, it may bypass oil and appear in oil wells prematurely. Water, like gas, is the source of energy. Excessive water production from a water drive or waterflood without adequate oil production amounts to a waste of reservoir energy.

Irregular, uneven waterfront movement creates pools of bypassed oil, which may be recoverable only through additional infill wells or new completions at extra cost. The undesired high water production not only increases both lifting and water disposal costs, but it also reduces oil production, shortens the productive life of oil and gas wells due to hydrostatic loading, migration of fines, and corrosion of tubular. Not only this, but it also adds to additional maintenance costs for surface facilities and requires downhole treatment for scale, bacteria, corrosion, and naturally occurring radioactive material (NORM). This type of water must be identified and controlled as soon as possible.

Water can appear in wells in the early stages of production or the normal course as reservoirs mature; the diagnosis of water production is central to solving the problems relating to it. Based on this analysis, numerous treatment methods, both mechanical and chemical, can be considered to control water production. However, at times produced water management can be a huge challenge for operators. Their ability to reduce and recycle produced water can increase oil recovery, improve profitability, and minimize freshwater use, contributing to water conservation.

Water cut development in an oil reservoir is illustrated by the fractional flow curve (Figure 4.13). This curve explains the build-up of water cut with the growth in water saturation in the oil reservoir due to the

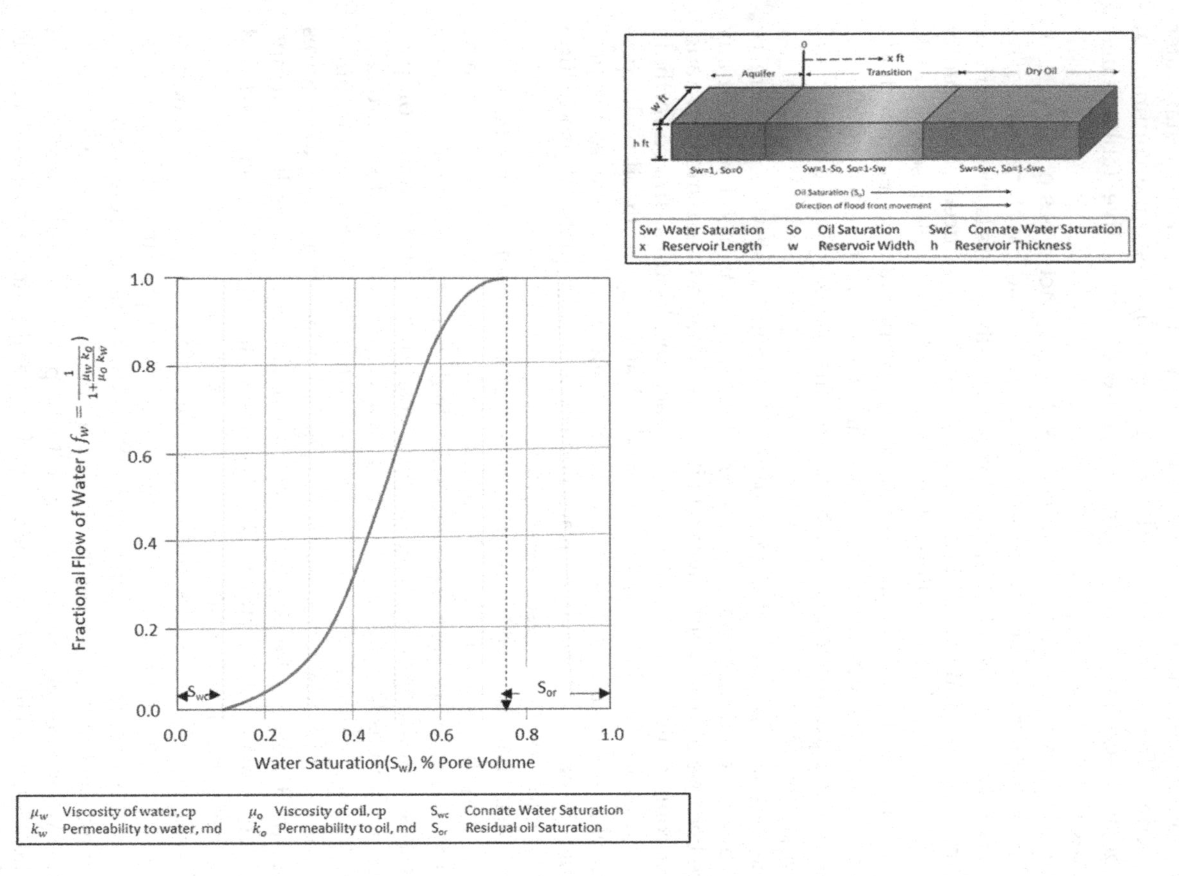

FIGURE 4.13
Fractional Flow Curve.

advancing waterfront. Many factors such as rock wettability, oil and water viscosities, production rate, formation dip, and initial gas saturation in a partially depleted reservoir can influence the fractional flow curve's shape.[7]

4.5.4 Reservoir Pressure Vis-à-Vis Bubble Point Pressure

The difference or "cushion" between reservoir pressure and bubble point pressure determines the magnitude of the undersaturated oil recovery; the higher this difference, the higher the potential for oil recovery without any pressure maintenance or secondary recovery methods. The performance between the initial reservoir pressure and bubble point is characterized by constant GOR and single-phase oil flow. The oil during this period is produced by rock and fluid expansion.

Reservoir pressure is a direct measure of reservoir energy. Therefore, it is a key diagnostic parameter to decipher the operating drive mechanism and its strength (Figure 4.14). The pressure varies with depth. Consequently, it must be referenced to a datum. It can be measured by placing a pressure recorder in the open as well as the cased hole. The most common practice of determining the reservoir pressure is to measure the shut-in pressures in various wells at different locations and take their average after converting those measurements to a common datum. These pressures are then mapped across the reservoir to generate isobar maps essential for explaining reservoir performance. When plotted against time or cumulative oil production, time-lapse pressure measurements can determine the rate of change in pressure and recovery efficiency. This information helps ascertain the time to reach bubble point pressure. Pressure measurements may be complicated if more than one zones are on commingled production.

Bubble point pressure is a pressure and temperature-dependent PVT property of oil. It is also dependent on oil and gas composition and may vary areally and vertically. Accurate knowledge of the bubble point pressure is critical to the timing of pressure maintenance/waterflood programs. It requires a meticulous PVT sampling campaign and analytics to understand the variation of bubble point pressure and other PVT parameters across the reservoir.

4.5.5 Fluid Saturations

Measurement of water or hydrocarbon saturation in prospective formations is part of a standard oilfield practice. These measurements are made in newly drilled/existing wells by recording a suite of logs used to identify the porous, permeable, and hydrocarbon-bearing intervals. This work is accomplished by a team headed by a petrophysicist, who decides which logs to be recorded. Petrophysics is a complex subject and an integral part of everyday decision-making. Fluid saturations provide useful information

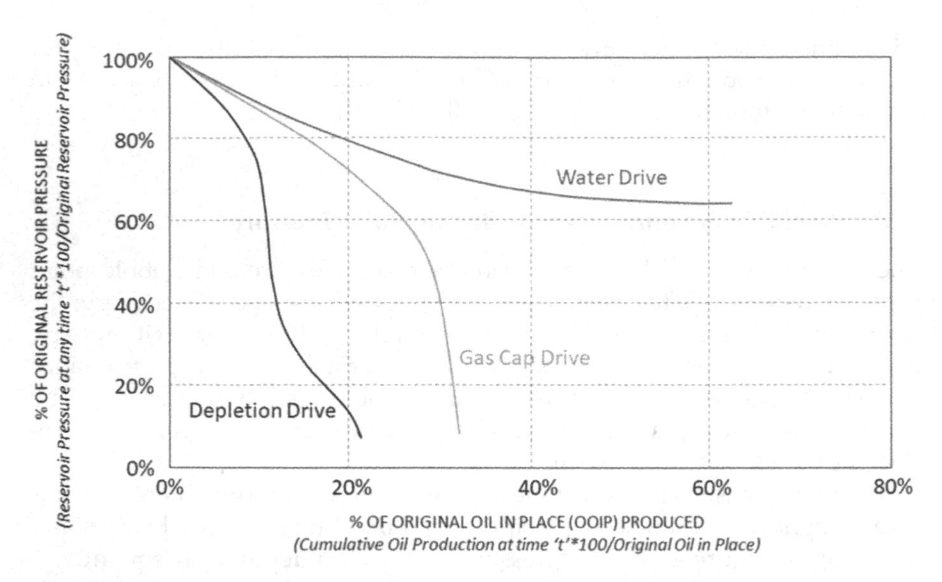

FIGURE 4.14
Reservoir Pressure and Operating Drive Mechanisms.

about the state of reservoir depletion. An oil reservoir would register a progressive reduction in oil saturation and increased gas and water saturations as oil is produced. Fluid saturations help plan well interventions, release new well locations, and diagnose several reservoir health issues such as unexpected excessive gas or water production, and so forth. Saturation measurements also form the basis for making crucial long-term reservoir management decisions for secondary and tertiary recovery planning. These measurements are often conducted in the open hole. However, tools and technologies are available to carry out the cased hole measurements as well.

Openhole resistivity logs can determine water and hydrocarbon saturations in a newly drilled well. Resistivity is a property that is the opposite of conductivity. It is expressed in ohm-meter (Ωm). For a complex material such as rock containing water and oil, the value of resistivity depends on the composition of the rock, formation temperature, salinity of water, porosity, pore geometry, and formation stress. Hydrocarbons do not conduct electricity, and therefore, the resistivity curve can indicate hydrocarbon presence in porous and permeable rocks. High resistivity values on resistivity logs demarcate hydrocarbon-bearing intervals. Porous and permeable intervals of the reservoir saturated with hydrocarbons being the real target, sonic and gamma logs are extensively used to support saturation interpretation. These interpretations are corroborated by actual testing of wells for validating the assumptions involved in the analysis.

Search for bypassed oil in mature fields requires knowledge of the current oil saturation behind the pipe. Pulsed neutron (PN) or thermal decay time (TDT) and Carbon/Oxygen (C/O) logging methods are used in a high-salinity (>20,000 ppm) environment to assess water saturation. Carbon Oxygen (C-O) logging was developed for low salinity fields where a Pulsed Neutron Lifetime (PNL) log is not useful. The reservoir saturation tool (RST) gained popularity because it can be run in an active well through the tubing, avoiding an expensive workover job. It can provide a record of oil cut or oil holdup both in vertical or horizontal wells. However, these tools have their strengths and weaknesses. A thorough understanding of the suitability of the logging tool vis-à-vis the measurement objectives is essential.

Core analysis is another technique that is applied for direct measurement of the fluid saturations in a rock. A core is a sample of rock used to examine the formation's lithology and contents. It can be recovered from an oil or gas well that is under drilling. It is then cut into multiple small cylindrical core plugs measuring about 0.75 to 1.25 inches in diameter and 3 to 4 inchesin length. Technology today provides an opportunity, though expensive, to extract "whole core" usually about 3 to 4 inches in diameter and up to 50 to 60 feet (15 m–18 m) long. Core offers an opportunity to physically see the formation and evaluate it in the laboratory to validate either log measurements or productivity tests. Visual inspection and laboratory studies on cores can provide useful information about the lithology, rock heterogeneity, capillary pressure, multiphase flow properties of the reservoir rock, anisotropy, etc.

Core analysis is subdivided into two categories: routine core analysis (RCA) and special core analysis (SCAL). RCA involves measuring basic rock properties such as porosity, permeability, fluid type, and saturation under atmospheric conditions. SCAL, on the other hand, is a set of more comprehensive tests. These tests include generating relative permeability versus fluid saturation curves fundamental to predict reservoir performance and EUR. SCAL also enables the correlation of the capillary pressure (P_c) with water saturation (S_w) in the formation and demarcation of dry oil, transition, and free water zones.

Knowledge of water or oil saturation is critical for identifying development locations and designing various fluid-injection schemes in the secondary or tertiary recovery phase. Additionally, a comparison of water or oil saturation over a period gives a good idea of reservoir depletion rate. It helps to decide the commercial viability of a wellbore or reservoir.

Due to the invasion of various drilling and completion fluids and changes in pressure and temperature conditions, the formation cores may experience significant changes in wettability and fluid contents as they are extracted from a depth and brought to the surface. New methods and technologies have been developed to reduce the impact of these limitations.

The health of the reservoirs operating under the secondary recovery scheme requires the following two additional parameters to be continuously monitored.

4.5.6 Sweep Efficiency

Water injection and waterflooding involve injecting water through water injection wells into an oil-bearing reservoir to displace oil from rock pore spaces. Injected water will preferentially move in the direction of the least resistance to flow. In 3-D porous space, the immiscible two-phase flow is governed by viscous, capillary, and gravity forces. Viscous forces are directly proportional to permeability and pressure gradient and control the velocity of displacing fluid. Gravity forces are proportional to the difference in the density of displacing and displaced fluids. The magnitude by which capillary forces dominate the viscous and gravity forces at the typical water injection rates in the field is reflected by the residual oil saturation (S_{orw}). The interplay of these three forces, that is, viscous forces, gravity forces, and capillary forces decides how the injected water will sweep the oil from the reservoir areally and vertically.

Areal sweep efficiency is defined as the fraction of the flood pattern's total area contracted by the displacing fluid. It is a two-dimensional (2-D) measure of the effectiveness of sweep in a flood pattern perpendicular to the flood direction. It increases steadily from the start of injection until the displacing fluid breaks through in producing wells. After that, it grows at a relatively slower rate.

Vertical sweep is evaluated by examining a vertical cross-section of the reservoir parallel to the flood direction.

Reservoir rock and fluid properties may vary areally and vertically. Consequently, the injected fluid moves as an irregular front. This front's shape is also controlled by the pressure gradients imposed by neighboring injection and production wells and their arrangement. The sweep efficiency that considers the influence of the reservoir's 3-D heterogeneities is known as volumetric sweep efficiency (Figure 4.15). Volumetric sweep efficiency is defined as the product of the areal and the vertical sweep in the waterflood pattern as shown by the following equation:

$$E_V = E_A \times E_I \tag{4.3}$$

where

E_v = Volumetric sweep efficiency, E_A = Areal sweep efficiency, and E_I = Vertical sweep efficiency.

These efficiencies are further defined as follows:

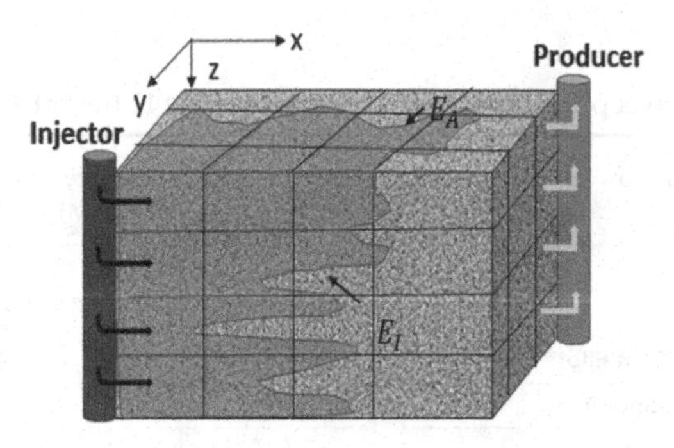

FIGURE 4.15
Volumetric Sweep Efficiency and its Components.

$$E_V = \frac{\textit{Volume contacted by injected water}}{\textit{Total reservoir volume}} \tag{4.4}$$

$$E_A = \frac{\textit{Area contacted by injected water}}{\textit{Total area of the reservoir}} \tag{4.5}$$

$$E_I = \frac{\textit{Vertical area contacted by injected water}}{\textit{Total vertical area between injector and producer}} \tag{4.6}$$

Quite clearly, volumetric sweep efficiency being the product of areal sweep and vertical sweep is a much smaller number than any of these efficiencies individually.

Before the evolution of 3-D reservoir modeling, estimation of volumetric sweep efficiency was based on analytical methods. These methods warranted that a complex 3-D problem was broken into two relatively simple 2-D problems that were easy to manage mathematically or in the laboratory. Physical models such as the electrolytic model, blotter-type model, potentiometric models, or analogs[8] paved the way for developing correlations to estimate sweep efficiency.

A summary of areal sweep efficiency values, thus determined for some selected flood pattern configurations and mobility ratio M = 1, is presented in Table 4.1. The reader is encouraged to refer to "The Reservoir Engineering Aspects of Waterflooding" by Forrest F. Craig Jr. for a more detailed discussion of these methods and ways to calculate the sweep efficiency post breakthrough as water injection proceeds further.

The factors that can influence areal sweep efficiency, in general, are the reservoir heterogeneity, mobility ratio, cumulative water injection, flood pattern, directional permeability, formation dip, and pressure distribution between injection and production wells, off-pattern wells, and presence of fractures. On the other hand, vertical displacement efficiency is controlled

TABLE 4.1

Estimated Areal Sweep Efficiency at Breakthrough for Different Flood Patterns[9]

Flood Pattern Type	Flood Pattern Configuration	Mobility Ratio (M)	Areal Sweep Efficiency (E_A) at breakthrough
1. Isolated 2-Spot		1	52.5% to 53.8%
2. Isolated 3-Spot		1	78.5%
3. Isolated Normal 5-Spot Pilot		1	80%
4. Isolated Inverted 5-Spot Pilot		1	105%
5. Developed 5-Spot Pattern		1	68%
6. Developed Normal 7-Spot Pattern		1	74%
7. Developed Inverted 7-Spot Pattern		1	73%
8. Direct Line Drive (d/a=1)	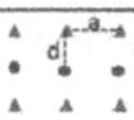	1	57%
9. Staggered Line Drive (d/a=1)	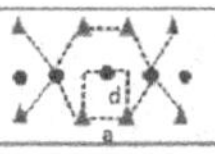	1	75%

by heterogeneity, mobility ratio, gravity effects due to the difference in densities of fluids, capillary forces, and vertical to horizontal permeability contrast.

Excessive water production due to early breakthroughs is a symptom of poor sweep efficiency. It leads to wasteful recycling of injection water without oil gain, increased operating costs, and weak economic returns.

4.5.7 Reservoir Souring

Reservoir souring is a phenomenon whereby Hydrogen Sulfide (H_2S) can cause sweet crude oil to turn sour because of water injection into hydrocarbon reservoirs over an extended period. The presence of H_2S in the production stream is indicated by increasing concentrations of foul-smelling gas with the odor of rotten eggs.

H_2S is a colorless gas heavier than air with the potential to settle in low spots unnoticed except for its smell. This foul-smelling "sour gas" is toxic and poses serious health hazards to both humans (Table 4.2) and the reservoir. Increased levels of H_2S in the production stream can increase the operational costs and cause a loss in production due to the possible shut-in of high H_2S content wells. In the absence of an effective mitigation plan, reservoir souring can seriously threaten material integrity, operations safety, and sales quality.

Injection of seawater, aquifer, or produced water for pressure maintenance or waterflooding is common for improving oil recovery. However,

TABLE 4.2

H_2S Warning Signs and Health Effects[10]

Concentration		Symptoms and Physiological Effects
Low Concentrations	0.01–1.5 ppm	Rotten eggs smell
	2–5 ppm	Unpleasant odor, nausea, burning of the eyes, and headaches with prolonged exposure
	20 ppm	Signs of fatigue, loss of appetite, headache, dizziness
Moderate Concentrations	50–100 ppm	Eye and respiratory tract irritation, loss of appetite, and stomach upset.
	100 ppm	Coughing, eye and throat irritation, loss of smell, and drowsiness. Worsening health conditions in general with extended exposure
High Concentrations	100–150 ppm	Loss of smell or paralysis
	200–300 ppm	Difficulty in breathing and irritation in eyes due to marked conjunctivitis from prolonged exposure of more than 1 hour
	500–700 ppm	Rapid unconsciousness, or the possibility of coma, severe damage to the eyes in 30 minutes, death after 30–60 minutes
	700–1000 ppm	Rapid unconsciousness, immediate collapse, or "knockdown."
	1000–2000 ppm	Nearly instant death

NB: 1 ppm = 1 part of H_2S gas per million parts of air (by volume)
H_2S concentration more than 100 ppm is considered immediately dangerous to life and health (IDLH)

sometimes this can promote the growth of anaerobic Sulfate Reducing Bacteria (SRB) and Sulfate Reducing Archaea (SRA), collectively called Sulfate Reducing Microbes (SRM), which is responsible for the generation of H_2S. These microorganisms breathe sulfate and reduce it to H_2S.

The souring brought about by living organisms (SRM) is known as biogenic souring. The favorable conditions for this kind of souring include the availability of nutrients besides temperature, pH, pressure, and salinity suitable for SRM to survive and grow. Waterflooding operations can provide this kind of environment. Injected water with sulfate ions and the reservoir hydrocarbons with carbon ions offer the necessary nutrients. Injected water can lower the reservoir temperature significantly, especially near the injector. The amount of H_2S generated by SRB is directly proportional to the amount of nutrients and sulfate ions present in the injection water. Studies indicate that the injection of produced water with Total Dissolved Solids (TDS) salinities lower than 100,000 mg/l can considerably increase souring potential.[11] A summary of the major factors controlling biogenic reservoir souring by SRBs is presented in Figure 4.16.

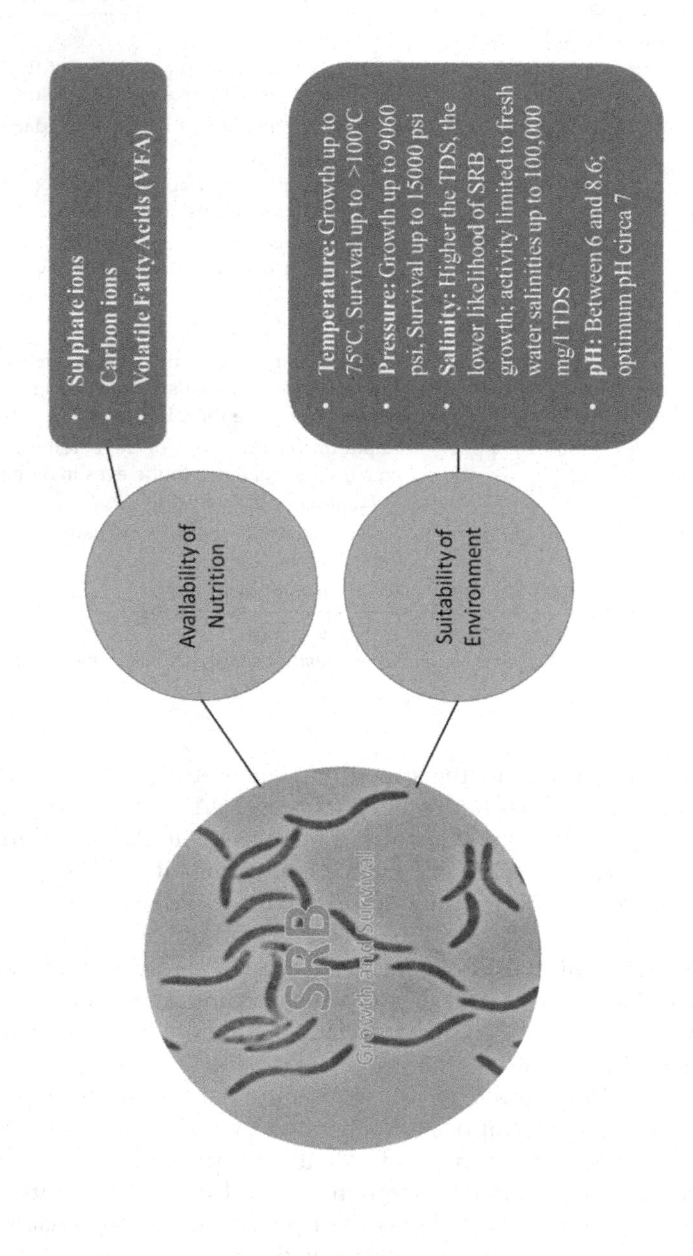

FIGURE 4.16
Major Factors Controlling Biogenic Souring by SRBs.

Reservoir souring has the potential to pose a severe challenge to the company's performance. Therefore, the operator must collect periodic samples from reservoirs under waterflood to monitor the H_2S content in produced fluids. Having ascertained the presence of H_2S, it is essential to determine if this is due to reservoir souring, if so, what proportion the problem might assume in the future. An in-depth predictive study may be necessary to evaluate the extent of the H_2S problem covering subsurface, surface, and sales aspects. Such an investigation helps design a mitigation plan consisting of the following methods:

- H_2S scavengers
- Biocide Treatments
- Nitrate Treatments
- Sulfate removal from injection water

Operators have used all these techniques to a varying degree of success. They may not completely cure the problem, yet they generally help avoid long-term reservoir shut-ins causing heavy production losses.

4.6 Factors Affecting Production Rate

Many chemical and mechanical processes can occur near or away from the wellbore and affect production rates in oil and gas wells. Near-wellbore restrictions are mostly one-time well issues or recurrent nature, impairing productivity and causing short-term production loss. Identification and mitigation of these problems are essential to restore production from the well.

A brief discussion of these problems is as follows:

a. Partial penetration
b. Formation damage
c. Formation collapse
d. Paraffin or Asphaltenes
e. Scales and precipitates
f. Emulsions
g. Mechanical failures
h. Corrosion

4.6.1 Partial Penetration

This terminology applies to when the well is not entirely drilled down or perforated to the formation's bottom. This can (i) limit the surface area available for the flow of fluids from the reservoir to the wellbore, and (ii) fluid flow in the reservoir may not be linear and reach turbulent velocity. These effects can combine to add a positive skin factor resulting in considerable pressure drops near the wellbore and resultant loss in well productivity (Figure 4.17).

The situation can be remedied, and actual well productivity can be achieved by deepening the well or adding perforations in the portion of the well hitherto uncovered with perforations.

4.6.2 Formation Damage

The term formation damage refers to the loss of a formation's production capacity by various field operations such as drilling, workovers, and production. Overweight drilling and workover fluids can invade the prospective hydrocarbon zones and damage the formation permeability. During production, the displacing phase may bypass the oil and accumulate in the vicinity of perforations. The build-up of the displacing phase saturation near the wellbore reduces the formation permeability to oil flow, adversely affecting the oil rate. A variety of reservoir phenomena are responsible for formation damage. Some of these are changes

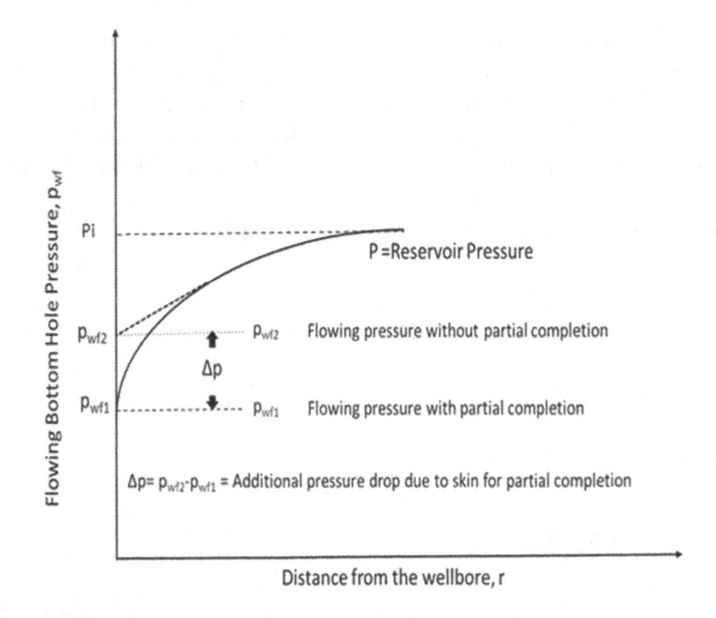

FIGURE 4.17
Additional Pressure Drop due to Partial Completion.

in the rock's wettability, alteration in lithology, variation in the fluid viscosity, and migration of fines within the reservoir. The extent of this damage can be measured by a positive skin factor calculated from a pressure build-up or drawdown test data. The wells that suffer from formation damage require stimulation treatment to restore/enhance the product rates.

4.6.3 Formation Collapse

Unconsolidated sands or weakly cemented formations subject to occasional collapse/failure because of severe pressure drop around a wellbore. Besides, there may be instances when a reservoir is so acutely depleted that the overburden pressure far exceeds the formation pressure lowering the formation permeability.

A small hydraulic fracture treatment can sometimes help reduce the chance of formation collapse.

4.6.4 Paraffin or Asphaltenes

Paraffin or wax formation and deposition is one of the most common issues in oil-producing wells. This problem occurs if the produced crude oil temperature falls below its pour point temperature (42°C and 44°C). Paraffin can deposit in the wellbore and flowlines forming soft to hard, brittle accumulations. The problems sometimes can be very acute and, if left untreated, can have a crippling effect on oil well production. Paraffin wax build-ups in oil wells are managed using mechanical scraping, hot oil or hot water circulation, coil tubing, chemical/wax solvents, and sometimes combining them.

Asphaltenes derive their name from "asphalt" obtained as a residue after the distillation of bitumen. They have a complex molecular structure with high molecular weight constituents consisting primarily of carbon, hydrogen, nitrogen, oxygen, and sulfur. Asphaltenes precipitation may be triggered by variations in temperature, pressure, and chemical composition of crude oil as experienced by reservoirs under depletion. Depending on the degree of severity, it can choke perforations, tubing, or wellhead to cause a significant flow impediment. Studies have indicated a direct correlation between bubble point pressure and asphaltene precipitation. This finding has been used to determine the most probable location of asphaltene formation between the wellhead and the reservoir. Significant and sudden changes in temperature can prompt twin deposits of asphaltene and paraffin together.

Like paraffin, asphaltene deposits can also be handled using mechanical devices or aromatic solvents such as xylene. Some chemicals can be used as asphaltene precipitation inhibitors.

4.6.5 Scales and Precipitates

Scales and precipitates can form because of the following:

- Sustained exposure of wellbore to fluids that tend to form scales
- Mixing of the foreign and native fluids,
- Use of incompatible workover or drilling fluids and
- Change in reservoir pressure and temperature conditions
- Reaction of chemicals (acid stimulations, etc.) with formation fluids and rock

Chemical inhibitors are found to be effective in dealing with scales and precipitates. If the problem persists, the well may be cleaned out by workover operation.

4.6.6 Emulsions

An emulsion is a mix of two or more liquids, which are not miscible with each other. Since oil is typically co-produced with water, emulsions can be encountered in almost all oil production phases inside the wellbores, wellheads, and wet crude-handling facilities. In crude oil production, water-in-oil emulsions are typically produced where water as the discontinuous phase is dispersed into oil that forms the continuous phase. Oilfield operations have everything necessary to create emulsions; crude oil, water, an emulsifying agent, and agitation. Organic acids, asphaltenes, resins, drilling fluids, stimulation chemicals, scale, waxes, fine solids, and so forth can work as emulsifying agents or stabilizers. Opportunities for agitation exist starting when the crude oil flows from the reservoir into the production tubing. The substantial pressure drop from the wellhead to the manifolds, turbulence near the perforations, chokes, and valves can cause intense mixing of oil and water. The use of subsurface pumps for artificial lift in oil wells can further create opportunities for emulsion formation.

Removing any of the four components, crude oil, water, an emulsifying agent, and agitation, will break the emulsion or prevent its formation. Emulsions must be broken or demulsified into oil and water almost entirely before the oil can be transported further to meet sales specifications. The application of heat and the use of an appropriate chemical demulsifier promote demulsification. It is generally followed by a settling time with electrostatic grids to promote gravitational separation.

4.6.7 Mechanical Failures

Fluid flow from the reservoir to tanks occurs in a closed system. Reservoir fluids at high pressure and temperature must travel a long way en route from their subsurface abode to storage tanks via the wellbore, flow lines, and processing facilities. Mechanical failures can occur anywhere in a well, pipeline, or facility due to inadequacy of material, off-design or unintended service conditions, maintenance deficiencies, improper operation, or aging infrastructure.

Unfortunately, it is impossible to eliminate equipment failures, no matter how effective the maintenance program is. Sustained long-term exposure of downhole equipment or flowlines to formation fluids makes them susceptible to breakage due to corrosion. Similarly, the material with rotary parts or rubber seals has a design life that can be severely affected by harsh subsurface or surface environment. Even small failures due to wear or general corrosion can account for substantial material losses and downtime every year. Catastrophic malfunctions can cause varying degrees of damages to health, safety, and the environment.

Companies usually have a rigorous maintenance schedule for servicing and repair of critical surface equipment. The dedicated well maintenance team is responsible for the regular maintenance of wells. Similarly, corrosion protection and inspection teams must ensure proper maintenance of trunk pipelines and other surface equipment. Likewise, the Field Development teams' role is to collect data about well integrity and take preventive action.

4.6.8 Corrosion

Corrosion is a common serious problem encountered in oilfields. Oil companies have to divert their time, effort, and capital in corrosion or integrity management. If they do not, they must brace themselves to bear downtime, economic losses, and even accidents.

In a wellbore, corrosion is generally caused by water, carbon dioxide (CO_2), and hydrogen sulfide (H_2S), which, along with the scale, can weaken the metal or deteriorate the properties of every component made of metal or rubber. Corrosion in pipelines is generally related to the amount and nature of the sediments. High flow rates tend to cause turbulence as they blow away the pipeline's deposits at high velocity to promote corrosion. Low flow rate regimes allow sediments to settle at the bottom, inducing pitting corrosion. The rate of corrosion can be aggravated by erosion, microbiological activity, and multiphase fluid flow.

Corrosion inhibitors are quite popular in the industry for managing corrosion. They are broadly classified into organic or inorganic inhibitors or, more specifically, into anodic, cathodic, or mixed corrosion inhibitors.

4.7 Significance of Pressure Maintenance to Reservoir Health

Good reservoir health is essential to achieve goals of high oil recovery and sound economics. Water injection is a key recovery program to maintain reservoir pressure near the bubble point pressure. Loss in reservoir pressure is the consequence of three factors: first, not enough water being injected into the reservoir; second, water not being injected at the right place; and the third, poor water quality. Experience has shown that water must be injected in a place where it is most needed. Injecting smaller volumes of poor quality water far away from the pressure sink will exacerbate the situation. For successful pressure maintenance, the water needs to go where the oil is. Operators wary of injecting water in their best-producing area end up having poor water injection projects.

Poor pressure maintenance may also result from poor planning, midcourse changes in project definition and scope, delays in project execution, and extended project cycle time.

4.7.1 Timing of Pressure Maintenance

The timing of pressure maintenance is a critical factor in maintaining a reservoir in good health. Proper implementation of pressure maintenance project(s) helps to conserve reservoir energy and restore the declining reservoir pressure to a level where wells can maintain a high-enough production rate to make economic sense.

Water injection needs to be initiated when the reservoir pressure approaches the bubble point pressure. Maintaining reservoir pressure at bubble point pressure helps maximize the oil recovery because each barrel of water can displace the maximum oil volume of reservoir barrels, and oil viscosity is most favorable.

However, maintaining reservoir pressure by injecting water at or near the bubble point may not always be possible for several reasons. In heterogeneous reservoirs, the bubble point pressure may vary areally and vertically with depth. The surface system may not make such fine adjustments as may be necessary to meet this requirement.

Pressure maintenance projects are expensive. Therefore, companies like to be sure of the need, timing, and size of the project. Unfortunately, this information can only be obtained from reservoir performance data (time-lapse plots of reservoir pressure, oil rate, GOR, and water cut) for several years. A rising, stabilizing, or declining trend of these parameters is used to decipher the reservoir drive mechanism and its strength. The modern reservoir modeling approach is very beneficial to run the sensitivities about the timing of pressure maintenance and injection water requirement. This approach helps to support the decision-making process over a range of uncertainties.

High-value capital projects for the design and construction of injection facilities usually have a project cycle of 5 to 7 years, sometimes even more. Such projects typically include creating an injection facility, drilling injection wells, and the pipeline network connecting these wells with the facility. Many other time-consuming activities must precede these tasks.

If a pressure maintenance project were to be designed based on the assurance provided by reservoir performance data, the project would be late by a few years. In this period of delay, two things happen. A lot of reservoir energy is irretrievably exhausted in the form of produced solution gas. In this period, the need for water injection grows further than the initial requirement. Therefore, when commissioned, the pressure maintenance project, is in effect, undersized to meet the new reservoir needs. As a result, the operating drive mechanism mostly lacks the energy required to deliver planned production rates.

4.7.2 Water Injection Requirement

At any instant, the water injection requirement may be calculated based on the cumulative reservoir voidage created by oil, gas, and water production from the reservoir. A reservoir will regain its initial pressure if all the produced fluids from the reservoir, including the gas, can be returned to the reservoir. For pressure maintenance at bubble point pressure, the water injection requirement can be estimated by replacing the cumulative voidage created by oil, gas, and water production since the bubble point pressure.

The water injection requirement is maximum during the fill-up phase since there is no return water. To achieve early fill-up, it is a general practice to maintain a high injection rate between 1 and 2 b/d/acre-ft during this period. However, there is a need for caution because fractures or high permeability streaks between injectors and producers can transport injected water to producing wells, causing premature water breakthroughs. After the fill-up period, the injection rate is nearly halved between 0.5 and 1 b/d/acre-ft. These numbers are only indicative and must be checked by a detailed study. Injection planning, including pattern design and facility design pressure, should be based on these considerations.

4.7.3 Source of Injection Water

Oilfields need large water volumes to inject over nearly two-thirds of their producing life for secondary and tertiary recovery operations. This water may be available in the form of produced water from the same or other oilfields, seawater, or subsurface fresh/brackish water. It is important to survey all possible water sources and select one or more sources that assure meeting quantitative requirements and lower treatment costs before injection. The use of brackish water sources for secondary or tertiary recovery projects is common due to the current need for freshwater for living beings.

Injection of water also results in water production, though not necessarily immediately. Since strict modern environmental regulations require the safe disposal of produced water, reinjection of the produced water to the oil reservoir for additional oil recovery makes technical and economic sense.

The volume of produced water that is returned to the reservoir gradually increases as the waterflood progresses.

4.7.4 Water Quality Needed to Maintain Injectivity

The water quality directly affects the rate of water injection into the subsurface formation and injection pressures. The compatibility of injection water with native rock and fluids is essential. Incompatibilities between different waters (native and foreign) and the reservoir rock must be considered while setting water quality specifications. While recognizing that poor injection water quality will result in lost oil production, a proper balance between water quality and cost must be achieved. Water quality specifications such as 98% removal of particles above 2 microns, oil < 5 ppm, and oxygen < 50 ppb can usually provide sufficient water quality. However, some companies may circumvent these specifications because of good reservoir quality and high cost.

Injection water must be free from substances that can cause corrosion, scale, formation damage, or sour. Levels of dissolved acidic gases besides dissolved and suspended solids, bacteria, and dispersed oil in injection water should be established. Any concerns of precipitation resulting from the mixing of incompatible ions must be addressed in advance.

Oxygen, carbon dioxide, and hydrogen sulfide gases dissolved in water may cause corrosion to subsurface and surface material and equipment. Water quality must be continuously monitored and controlled to neutralize its impact.

Oil field waters contain cations such as sodium, calcium, magnesium, barium, strontium, ferrous, and ferric and anions such as carbonate, bicarbonate, sulfate, and sulfide chloride, bromide, iodide, and silicate in the form of dissolved solids that tend to form scales. A proper scale inhibition program is required to mitigate the problem.

Pore-size distribution, composition, and distribution of clays are essential considerations from an injection point of view. Complete water analysis of both the injection source water and connate water enables the evaluation of scaling, clay swelling, and brine incompatibilities. Injection water containing tiny suspended solid particles of sand or clay or precipitates of iron sulfides, calcium carbonate, and other scales may plug the formation pores and reduce injectivity. Settling tanks and guard filters can help contain this problem to some extent.

Microbiological or bacterial growth can occur in water injection and oil production systems where minor amounts of water may be present. This growth can cause serious problems such as reservoir souring due to the formation of H_2S that, in turn, can pose severe HSE risks to company assets and personnel.

Reinjection of produced water may encounter a common problem of dispersed oil in water. Continued injection of this kind of water in an injection well may require higher injection pressure because of reduced relative permeability to water.

Inappropriate water quality can lead to partial or complete plugging of the reservoir. Restoring the injectivity may involve expensive stimulation jobs (acid treatment or hydraulic fracturing) or expensive well-workovers. In the extreme case, the injection well facing this problem may have to be sidetracked.

4.7.5 Injection Pressure Requirement

Determining the water injection rate and injection pressure is critical to waterflood development and the water injection facility's design. These are two crucial parameters that decide the size and cost of the injection facility. Engineering design in terms of facility capacity needs to be flexible, as initial plans do most likely change. The design pressure of the facility, on the other hand, must be high enough to pump water through injection wells into the worst reservoir rock successfully.

Both of these issues are addressed based on actual well tests in the field. These tests, commonly known as Injectivity Test and Pressure Fall-off Test, are conducted on injector wells (Figure 4.18).

For Injectivity Test, the injection well is initially shut-in to allow the reservoir pressure to stabilize. Subsequently, water at a constant rate is injected into the well with a continuous pressure response record. This test

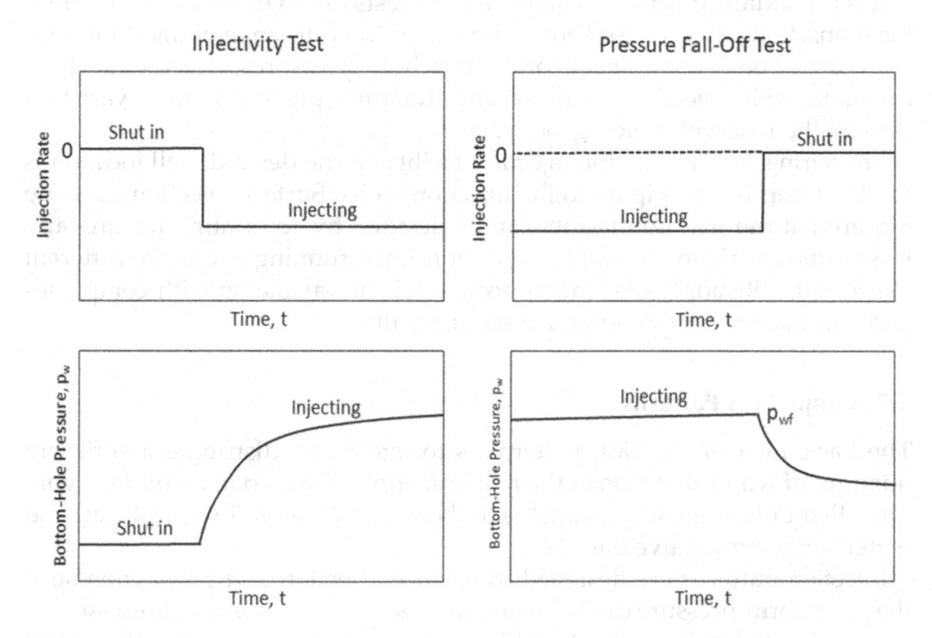

FIGURE 4.18
Rate Sequence and Pressure Response for Injection Well Testing.

procedure and its interpretation are quite analogous to pressure build-up tests conducted on oil and gas production wells.

A pressure fall-off test for a target reservoir is conducted by maintaining a constant water injection rate into the injection well until a stable pressure is reached. After that, the pump is turned off, and the rate at which pressure decreases is measured. The test's injection rate and pressure measurement data are used to calculate formation permeability, skin factor, reservoir pressure, and injectivity index that are usually a reliable indicator of injection well performance. This test procedure and its interpretation are analogous to pressure drawdown tests conducted on oil and gas production wells.

The majority of the injection programs are planned to inject water below fracturing pressure. However, poor reservoir rock, coupled with stringent water quality specifications, may force the operating design pressure to exceed the fracture pressure. In both situations, knowledge of the fracturing pressure (formation breakdown) is essential. A step-rate injectivity test is typically used to estimate the fracture pressure in an injection well. This test is conducted by injecting water in a series of increasing rates lasting over the same duration and measuring the corresponding tubing head pressures. Four to six rates may be used. The change in the slope would indicate fracture pressure (Figure 4.19). This test procedure is analogous to the flow-after-flow test that is more popular with gas production wells for estimating their absolute open flow potential.

It is a standard practice to carry out such tests on several wells at different locations. Such an exercise allows capture of the full range of injection rates and corresponding wellhead or bottom-hole pressures. These data help calculate well injectivity indices and fracture pressure; their variation reflects the reservoir heterogeneity.

Pressuring water from the injection facility to the desired well location is the final step before piping to the injection wells. Surface injection pressure required at the injection facility can be decided by accounting for pressure losses due to friction across the injection lines running up to the different injection wellheads. Detailed surface models are available with companies and contractors for the optimization of results.

4.7.6 Injection Patterns

The basic idea of injection patterns is to inject and distribute a sufficient quantity of water that aids in the maximization of sweep in an oil reservoir. This distribution must be equitable to the withdrawal of fluids (oil, gas, and water) on a cumulative basis.

Injection patterns are designed to balance injection and production such that a uniform pressure can be maintained across the reservoir. It is essential to prevent the migration of fluids across pattern boundaries. For these reasons, four key factors that are kept in mind for designing injection

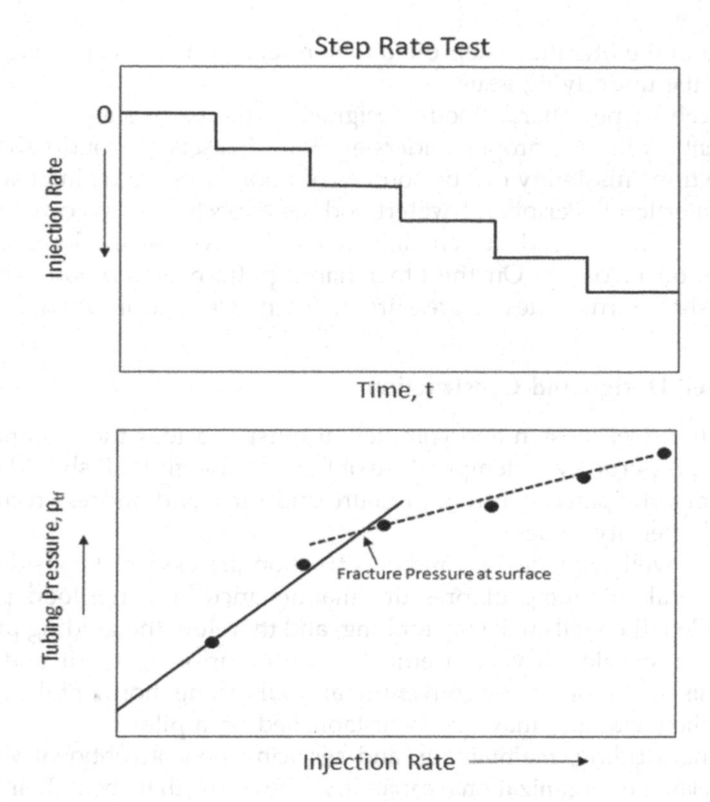

FIGURE 4.19
Injection Rate Sequence and Pressure Response for Step Rate Test.

patterns are well spacing, injector to producer ratio, the productivity of producing wells, and injectivity of injection wells. The arrangement and design of injectors also consider any permeability anisotropy and orientation of natural fractures, if any.

The utopian wish of having perfectly regular injection patterns is most often compromised by the compulsion to utilize existing well locations. In reality, overriding considerations of well-cost and surface management usually take precedence and might lead to irregular or nonsymmetrical patterns.

The concept of regular and symmetrical patterns originates from the assumption that the reservoir is homogeneous. However, this is seldom the case. Theoretically speaking, a heterogeneous reservoir with varying rock and fluid properties will most likely require irregular and nonsymmetrical patterns. Nevertheless, regular and symmetrical patterns are preferred for waterflood management because they make the difficult injection and production allocation task relatively easy.

There is always a debate about whether one should deploy a peripheral or pattern mode of injection. The clear merits of both these methods are provided in the literature. Representative reservoir models are a big help to examine the underlying issues.

Ill-conceived peripheral floods designed in the early stages of field development without a proper understanding of reservoir conductivity and pressure transmissibility can be sources of poor pressure maintenance and failed waterflood. Peripheral waterflood, as expected, is a slower process than pattern waterflood. It generally works in favor of the long-term objective of oil recovery. On the other hand, pattern waterflood can better address short-term issues of pressure maintenance and production targets.

4.7.7 Well Design and Construction

An injection well design and completion must consider the volume, composition, properties, and temperature of the injection fluid. It should cater to maximum anticipated injection pressure and rates and address recognized potential integrity issues.

Injection well type, design, and construction are essential considerations in the overall planning of pressure maintenance or waterflood projects. Once drilled, the wells will stay for long, and therefore the guiding principle to "keep it simple" is very useful. In case of pressing merit and strong sentiments in favor of nonconventional wells (long horizontal or multilateral), their viability may first be established on a pilot basis.

Planning, drilling, maintaining, and servicing nonconventional wells are closely related to organizational capability. The teams that spent their lifetime acquiring and analyzing data from conventional wells may be overwhelmed by the challenges posed by nonconventional wells. The operations of drilling, coring, logging, bottom hole sampling, and surveillance in non-conventional wells can challenge the teams' expertise. Besides, the cost of waterflood operations and well surveillance programs increases exponentially.

Well-design must not only be based on the maximization of techno-economic benefits from the secondary recovery phase alone. Its viability must consider the cost of drilling and operations beyond secondary recovery operations as well. Implementation of EOR in a reservoir, starting from pilot to full scale, usually assumes patterns consisting of conventional wells. Transitioning a waterflood with nonconventional wells to EOR with conventional wells would be contentious from technical and economic points.

4.8 Corrective Measures for Restoration of Reservoir Health

Fluctuations in reservoir health are common and frequent. A reservoir suffers from poor health, mainly due to poor planning or negligence. Incorrect understanding of the reservoir and its behavior, excessive withdrawals due to high production targets, and water injection delays are common reasons for the deterioration in reservoir health.

Unforeseen circumstances of blowouts due to war and natural disasters such as earthquakes can also cause conditions that result in poor reservoir health. While conflict and tremors are beyond oil companies' control, most performance-related health issues can be managed by proper planning and proactive management.

A proper surveillance plan is essential for systematically monitoring reservoir and well health. Analysis of routine time-lapse data of pressure, oil/gas and water rates, GOR, and water-cut collected from flowing and shut-in wells aids in identifying and even predisposing reservoir/well health issues.

4.9 Assessment of Health by Reservoir/Well Health Indicators

Good reservoir and well health is a prerequisite for efficient performance and successful production and recovery targets. However, many sources, such as those discussed earlier, can trigger various reservoir and well-health issues. These problems can grow if not attended to in time and result in loss of well productivity due to increased water cut, high GOR, and a sharp decline in reservoir pressure. The question becomes acute as this phenomenon envelopes majority of wells in a reservoir over time.

Health checks are performed at the well and reservoir levels. Regular monitoring of production from individual wells provides initial clues for fluctuations in oil, gas, and water production or injection rates. Wells are tested by diverting the flow to the test separator located at the production gathering centers (GCs) or using portable test separators near the wellhead. The number of wells in giant oilfields may be staggering. It affects the quality and frequency of production tests. A workaround to this problem is to test only the key wells in the reservoir regularly. Some differences are easy to explain because they can be directly related to the known field events. The other differences have to be investigated. The diagnosis may reveal that reservoir/wells are not performing as expected because of the following:

- The planned inputs (wells, workovers, injection, etc.) were not provided on schedule
- The inputs did not match the reservoir needs; they were inadequate
- Production was not appropriately scaled up or down in proportion to the inputs
- The assessment of the size and behavior of the reservoir was not correct

These situations harm the health of the reservoir and wells. If allowed to persist for long, they can severely affect project economics. IOCs and NOCs who take pride in using best practices have to demonstrate that they take special care to maintain their reservoirs in good health.

Table 4.3 lists the essential reservoir health criteria and reservoir health indicators (RHIs) for the reservoir health check. Similarly, Table 4.4 contains the health criteria and well health indicators (WHIs) for well-health check.

TABLE 4.3

Reservoir Health Criteria and Reservoir Health Indicators (RHIs)

Theme(1)	Criterion(2)	Performance Measure(3)	Target(4)	Tolerance Rule(5)
Drilling & Completion	Drilling Performance	Number of wells drilled in a year	Set the estimated number of wells (N)	±10%
	Drilling Quality	Oil Rate Delivery, BOPD	Sum of actual initial oil rates of production wells (ORD)	±10%
Production	Oil rate per reservoir	Daily Oil Production Rate, BOPD	Sum of allowable oil rate of all production wells (AOR)	±5%
	Decline Rate	The annual decline in production rate, % per annum	Set an estimated number (D)	±1%
	GOR per reservoir	Weekly cumulative GOR, ft^3/STB	Set an estimated number corresponding to the allowable oil rates (GO)	±25%
	Water Cut per reservoir	Weekly water cut, %	Set an estimated number (WC)	±2%
	Oil well inventory utilization	Number of wells online (with 90% uptime)	Set a number (W)	−5%
Reservoir Pressure	Current reservoir pressure	Average reservoir pressure referenced to a datum, psi	Set a number based on predicted reservoir pressure (P)	±50 psi
	Need for pressure maintenance	The difference between the current reservoir pressure and bubble point pressure, psi	$\Delta p = (P - P_b)$	±50 psi

(Continued)

TABLE 4.3 (Continued)

Reservoir Health Criteria and Reservoir Health Indicators (RHIs)

Theme(1)	Criterion(2)	Performance Measure(3)	Target(4)	Tolerance Rule(5)
Pressure Maintenance	Injection rate per reservoir	Sum of injection rates of all injection wells, bwpd	Set a number based on the plan (I)	±10%
	Cumulative VRR since the BPP	CVRR = Cumulative fluid withdrawal/ Cumulative injection, %	Cumulative VRR = 100%	±5%
	Effectiveness of injection	Areal sweep efficiency, %	Set an estimated number (SE)	±2%
Recovery Outlook	Current recovery	Cumulative oil produced as % of OOIP	Set a number (R) (based on the P50 value of OOIP)	±1%
		Cumulative oil produced as % of mobile oil	Set a number (M) (based on the P50 value of mobile oil)	±1%
	Residual Oil Saturation	S_{orw} for water, %	Set a number for S_{orw} based on well logs, core analysis, SWCTT, etc.	±2%

The performance measures for RHIs and WHIs must include critical management programs/assets. The targets and tolerance rules listed in columns (4) and (5) may be set based on the industry benchmarks or standards in line with the business guidelines, performance forecasts, reservoir studies, field and laboratory tests, and world analog databases. The reservoir's actual performance and health check can be accomplished by comparing the year-end numbers against the targets and rules.

The reservoirs or wells, whose performance/health measures are outside the prescribed range, are expected to have performance or health issues. The engineers then have to dive deeper into well data to pinpoint the source of the problem.

It is necessary to include a caveat at this point. The criteria for RHIs and WHIs are not cast in stone. They are subject to change depending on the fluid quality, reservoir heterogeneity, and reservoir management focus. They can be expanded or narrowed depending on experience with reservoirs. The range of reservoir/well health measures, targets, and tolerance rules provided in Tables 4.3 and 4.4 can also be finetuned based on the specific objectives of performance improvements.

TABLE 4.4

Well Health Criteria and Well Health Indicators (WHIs)

Theme(1)	Criterion(2)	Performance Measure(3)	Target(4)	Tolerance Rule(5)
Drilling & Completion	Well Integrity	No leaks in the casing, cement, downhole packers, and plugs	Integrity issues	Zero (no issues)
	Drilling Efficiency	Non-productive time (NPT), % of plan	Non-productive time allowed	Max 7%
	Well quality	Oil rate, BOPD	Initial oil rate expected from the newly drilled well after clean up	±10%
Production Performance	Oil Rate	Daily Oil Production Rate, BOPD	Allowable oil rate of the production well	±10%
	Decline Rate	Decline in production rate, %	Set an estimated number for annual decline rate (D)	±3%
	GOR	Daily cumulative GOR, ft^3/STB	Set an estimated number for GOR corresponding to the allowable oil rate	±25%
	Water Cut	Daily water cut, %	Set an estimated number for WC	±2%
	Shut-in BHP	Stabilized bottom hole pressure referenced to a datum, psi	Set an estimated number for P_s corresponding to the allowable oil rate	±100 psi
Injection Performance	Injection Rate	Daily water injection rate by well, BWPD	Daily allowable water injection rate	95% of AWIR
	Limiting Injection Pressure	Maximum bottom hole injection pressure (MIP), psi	Set MIP < Facture Closure Pressure (FCP)	MIP to be 2%–3% lower than the FCP
	Shut-in BHP vis-à-vis BPP	Operating pressure for waterflood, psi	$P_s = P_b$	±50 psi

The health check for oil and gas reservoir/wells performed on human healthcare analogy can serve very well to assess the state of their fitness based on predicted and actual performance. Field development teams can suitably modify these criteria to include more or fewer parameters. Unique problems such as asphaltene precipitation, the formation of H_2S because of souring, corrosion, and sand production can also affect a well's health. If widespread, these problems can be treated as a reservoir health issue and included in Table 4.3. If the problem is limited to a few wells, it can be included in Table 4.4. The method of setting new performance measures remains the same. The values and range of RHIs/WHIs can also be amended as deemed fit in the light of the experience with reservoirs.

There are two clear advantages to this method. First, the petroleum reservoir and well health measurements move from a qualitative to a quantitative domain, enabling a quick assessment of the problem's gravity and associated potential risk. Second, this methodology allows the ranking of reservoirs/wells based on their health ratings, which helps decide the intervention's priority.

4.10 Impact of Poor Health on Reservoir Management

Poor reservoir health defeats the very objectives of reservoir management. It reflects poor reservoir performance and low recovery efficiencies. Production and injection wells do not meet their goals because they cannot produce or inject at target rates. In addition to this, the productive life of these wells is significantly shortened with frequent interventions. Drilling and workover of additional wells at extra cost becomes necessary for the company to achieve its target. Sometimes this may require the acquisition of further drilling and workover rigs. Poor reservoir health may also affect the water injection facility's timing and capacity, resulting in the revision of the development plan. To summarize, poor reservoir health can trigger many short- and long-term reservoir management issues, which can upset the operating company's original production and budget.

Notes

1 https://www.nationalgeographic.org/encyclopedia/petroleum/
2 Mobility Ratio = $(k_{rg}/\mu_g)/(k_{ro}/\mu_o)$
3 Emil J. Burcik; Properties of Petroleum Reservoir Fluids, pp. 47–75
4 https://www.who.int/about/mission/en
5 Crain's Petrophysical Handbook; https://spec2000.net/16-decline.htm
6 Ahmad, Mushtaq (2016) Re: What is the difference between GOR (gas oil ratio) and Rs (gas solubility)?. Retrieved from: https://www.researchgate.net/post/what_is_the_difference_between_GORgas_oil_ratio_and_Rs_gas_solubility/573574baf7b67e19125215b1/citation/download.

7 Forrest F. Craig, Jr.; The Reservoir Engineering Aspects of Waterflooding, Monograph Vol. 3, Henry L. Doherty Series, SPE, 1971, pp. 38–43
8 Forrest F. Craig, Jr.; The Reservoir Engineering Aspects of Waterflooding, Monograph Vol. 3, Henry L. Doherty Series, SPE, 1971, p. 52
9 Forrest F. Craig, Jr.; The Reservoir Engineering Aspects of Waterflooding, Monograph Vol. 3, Henry L. Doherty Series, SPE, 1971, pp. 48–59
10 United States Department of Labor OSHA; https://www.osha.gov/SLTC/hydrogensulfide/hazards.html
11 Oilfield WIKI, Reservoir Souring; http://www.oilfieldwiki.com/wiki/Reservoir_souring

5

Reservoir Management Policy Framework

5.1 Introduction

Reservoir management is a system that consists of several processes and sub-processes aimed at maximizing the oil/gas rate, recovery, and profits. It uses human capabilities, appropriate technologies, and financial resources to safely and economically recover oil and gas. Operating companies capitalize on their experience and the use of best practices to increase asset value. However, this journey is quite adventurous as there are operational, economic, and strategic risks on the way. Despite these risks, there are opportunities to make a profit from oil and gas investments. Therefore, reservoir management focuses on maximizing the benefits and reducing risks within and outside its purview.

5.2 Elements of Reservoir Management

A robust reservoir management system provides for a sustained, relatively risk-free operational and financial performance. It embraces a company's operating philosophy and critical business requirements with strategic, commercial, and social obligations and lays out a well-rounded template for decision-making. The system is supported by four pillars: Organization, Capability, Policy, and Governance (Figure 5.1).

5.2.1 Organization

For a business, the organization is a necessary means to allow the combination of human, physical, and financial resources to achieve its commercial end. An individual, no matter how talented, capable, or hard-working, has limits to his achievements. To quote Helen Keller,

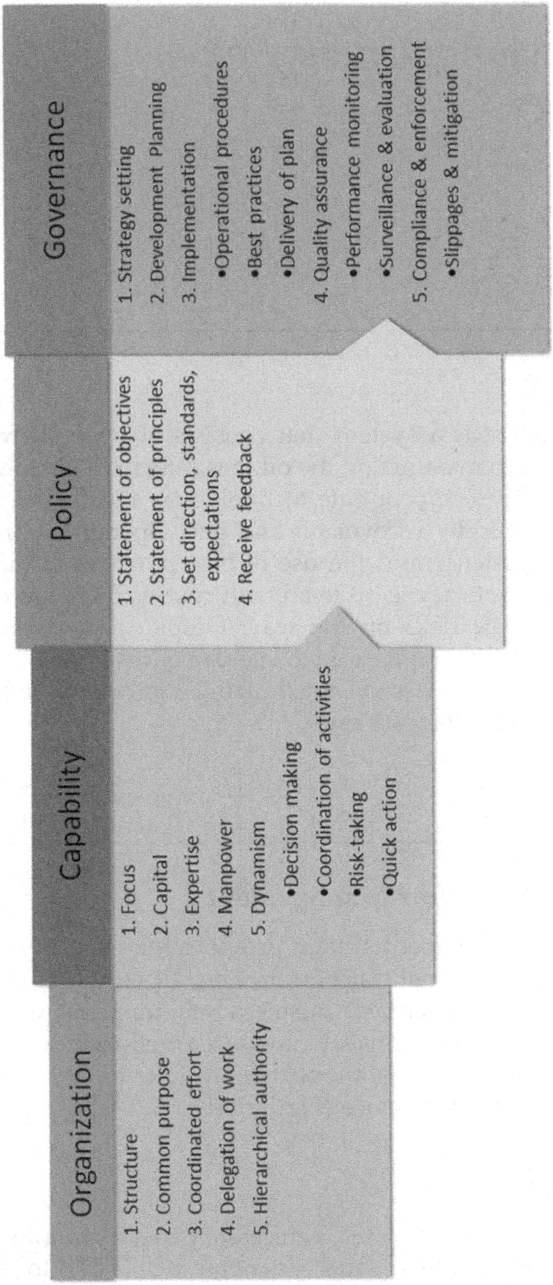

Organization
1. Structure
2. Common purpose
3. Coordinated effort
4. Delegation of work
5. Hierarchical authority

Capability
1. Focus
2. Capital
3. Expertise
4. Manpower
5. Dynamism
   •Decision making
   •Coordination of activities
   •Risk-taking
   •Quick action

Policy
1. Statement of objectives
2. Statement of principles
3. Set direction, standards, expectations
4. Receive feedback

Governance
1. Strategy setting
2. Development Planning
3. Implementation
   •Operational procedures
   •Best practices
   •Delivery of plan
4. Quality assurance
   •Performance monitoring
   •Surveillance & evaluation
5. Compliance & enforcement
   •Slippages & mitigation

FIGURE 5.1
Four Pillars of Reservoir Management System.

"Alone we can do so little; together we can do so much." A business organization consisting of many people working together, having varied talents, ways of thinking, skillsets, and experience, can create boundless opportunities for commercial success.

American organizational psychologist Edgar Schein identified an organization with four essentials-common purposes, coordinated effort, the delegation of work, and hierarchical authority. Generally, all businesses are managed by a hierarchical structure that establishes roles and positions within the company. An effective organizational structure complements interdependence with independence. The company management decides goals, objectives, and business strategies to achieve the goal. It also organizes the necessary finances, tools, technologies, and workforce. The managers and employees work together within the area of their expertise to accomplish the goal.

Organizational structure is a setup that determines the hierarchy of people, their function, the workflow, and the reporting system. Various units and departments of an organization must work together to deliver results according to accepted standards. A properly structured organization with a clear and common goal is the first essential requirement for implementing the reservoir management process.

5.2.2 Capability

Organizational capability is the term used to define a company's ability to manage resources such as capital, employees, expertise, and time to produce results in terms of quality, quantity, and cost. A company's organizational capabilities must focus and align with its business goals.

For an oil company to survive today, it should have the added qualification of agility. Agility means an individual or organization's ability to be flexible and move quickly in the face of unexpected changes. An organization that is adept in decision-making, skillful in calculated risk-taking, competent in handling cross-team activities, and capable of taking quick actions can stay on top in any situation.

An organization's success depends on the competence of its management and employees. Personnel training and development are the crucial functions that an organization must invest in to build the necessary skill, capability, and knowledge base to ensure success. This process of learning and improvement applies to both management and employees, whether new or old. New employees must be trained to understand the role, responsibility, and requirements of their job. Other employees need the training to refresh the theory and practice in their service discipline and apply new tools and technologies to improve their performance. It is a continuous and never-ending process that brings about quality improvement in thinking, working, and delivery.

Training and development of employees to recover more difficult oil and gas in the face of increasing competition from alternative energy sources require an oil company to be continually readjusting its tactics and strategies. Despite the advent of new technologies and best practices, oil and gas companies find it difficult to replenish the produced reserves, deal with operational and environmental hazards, and retain trained human resources. On top of this, volatile oil prices increase production costs, and an oil company must react quickly to market conditions.

5.2.3 Policy

A policy is a broad statement of principles and guidelines, not a directive that an organization lays down to achieve its business goals. Policies assist in decision-making. Factors such as advanced knowledge, innovations, and changes in the social, political, economic, and legal environments at the national and international levels influence policymaking. A policy should be broad and visionary. It should be unambiguous, fair, and firm. It must be based on a long-term or strategic view of the business environment. Periodic policy reviews, updates, and modifications based on results and feedback are essential to keep it current with the social, economic, and legal settings.

It is the management's responsibility to circulate company policies to all the stakeholders, including employees and clients of the company. The companies usually publish their employee welfare plans and procedures in the form of a booklet or on the company website that is widely accessible to everybody. Despite this, many people in the company may not know enough about the reservoir management policies, plans, and strategies, limiting the opportunity to receive some brilliant ideas and feedback. Improved communication in this respect within and outside the company is always beneficial.

5.2.4 Governance

In the context of applying reservoir management principles to oilfield development, the governance is to "enable, observe, and regulate" rather than its customary practice to "control, dominate, and command."

The governance framework includes input on stake holdings, strategy setting, joint venture formation, conflict resolution, petroleum legislation, tax regulations, incentives, and so forth. The company management is responsible for setting up enabling systems and processes that facilitate proper supervision, control, and information flows. It also ensures that company structure, capabilities, and policies are conducive to achieve reservoir management objectives set by the government and the company.

The company management recognizes that it is impossible to keep a 24 x 7 watch on every important activity. As a result, it establishes a process of governance of reservoir management activities. It delegates authority at various levels in the organization to properly manage a variety of activities.

The governance process has an extensive scope and many stakeholders. It consists of many activities representing a relay race where one team referred to as Service Provider forms the input for another team acting as a Customer. The final delivery of the plan depends on the successful completion of all activities that are part of the process. Take, for example, the most common task of producing oil from a well. This task has several activities, processes, and sub-processes that must be completed before oil goes into the oil tank. The Field Development team must release a location; it must be checked by a group consisting of representatives from various asset teams and sometimes even by government officials before the drilling department drills it. After successful drilling, it is tested and connected to the GC, where oil, gas, and water are separated and sent for their respective end-use (Figure 5.2).

This example highlights the importance of governance in a multiple-team activity. Such activities where a team builds on the input of other groups are the essence of an organization. Team Leaders and Managers provide decision-support and additional resources if required. Excellent teamwork and effective communication are the basic requirements to accomplish such tasks in time and according to specifications. The failure of any team not completing its work in time and according to standards has a domino effect causing the program's collapse.

5.3 Typical Governance Model of an E&P Company

As discussed in Chapter 3, a few individuals or shareholders hold ownership and control of a private limited company. Such companies do not offer their stock to the general public.

A public limited company has the advantage of raising capital from the public by selling its stock. The word "limited" implies that shareholders' liability is limited to their share of holding in the stake in the case of a company running into a loss.

Company affairs are controlled by a governing body or Board of Directors (BOD). BOD is a group of individuals authorized to establish policies, monitor company performance, and make decisions. BOD sets up a framework of policies, value systems, procedures, and protocols for management and employees. It creates a governance structure to deal with decision-making, involvement in operations, and the reporting relationships between the BOD and the staff. It also includes the mechanisms to balance the directors' powers to ensure accountability, fairness, and transparency in business dealings with all stakeholders.

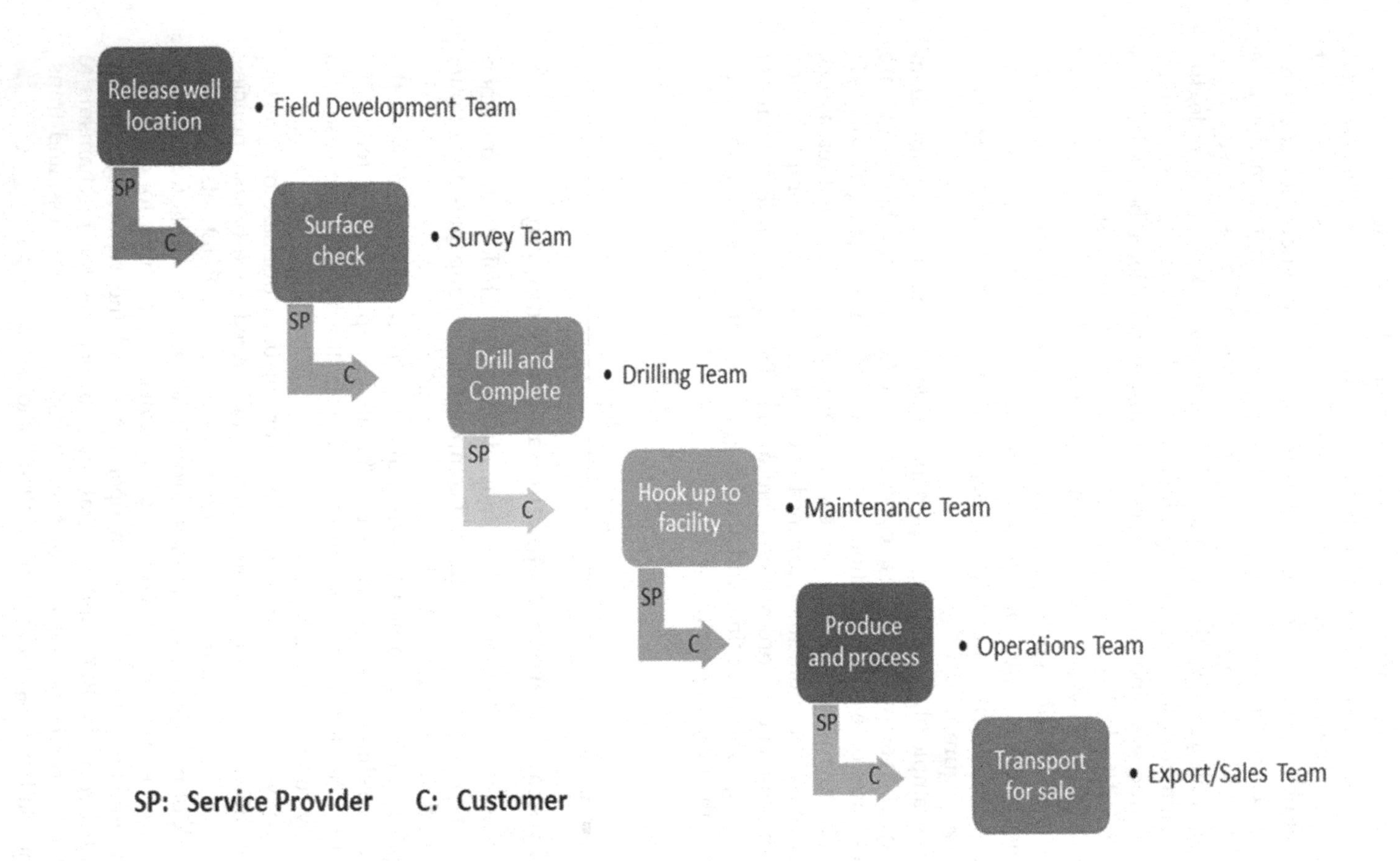

FIGURE 5.2
A Summary View of the Process for Delivery of a Production Well.

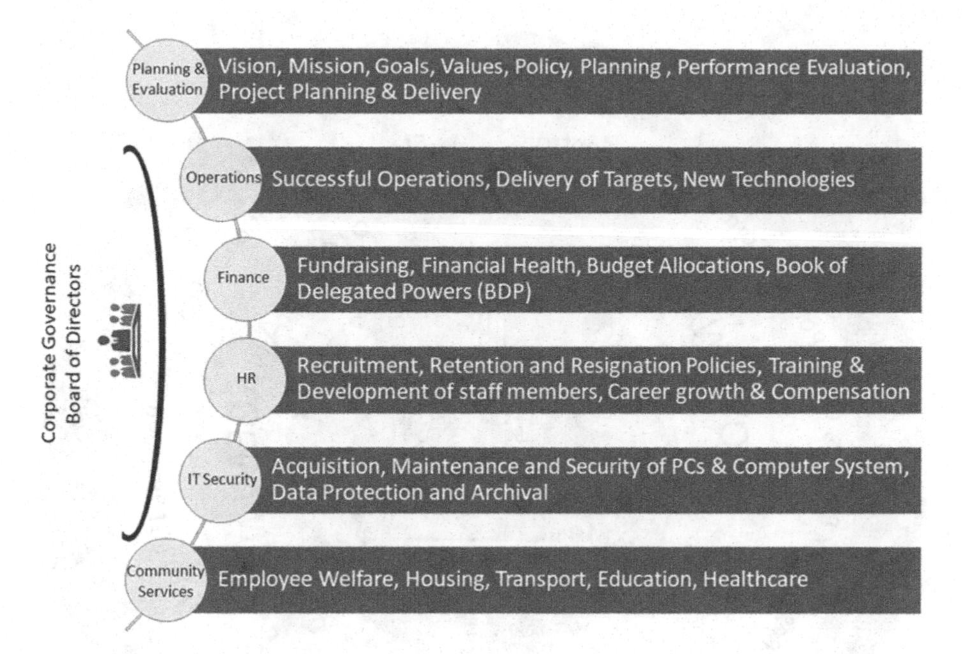

FIGURE 5.3
Basic Functions of the Board of Directors (BOD) of an E&P Company.

All oil and gas companies, irrespective of their size, will have the following vital functions to be managed by Directors at the board level (Figure 5.3).

- **Planning & Evaluation:** Setting the vision, mission, organizational goals & objectives, values, planning process, performance evaluation, policy-making, setting procedures and protocols, project approvals.
- **Corporate Operations:** Ensuring successful continuance of organizational operations to achieve results, setting performance targets, use of appropriate technology
- **Finance:** Designing Book of Delegated Powers (BDP), Options for Fundraising, Management of company finances, budget allocations for operations, and other relevant activities
- **Human Resource:** Recruitment, Retention and Resignation Policies, Training & Development of Staff Members, Career Growth & Compensation
- **IT Security:** Acquisition, Maintenance and Security of PCs and Computer Centers, Data Protection and Archival
- **Community Relations:** Employee Welfare, Housing, Transport, Education, Healthcare.

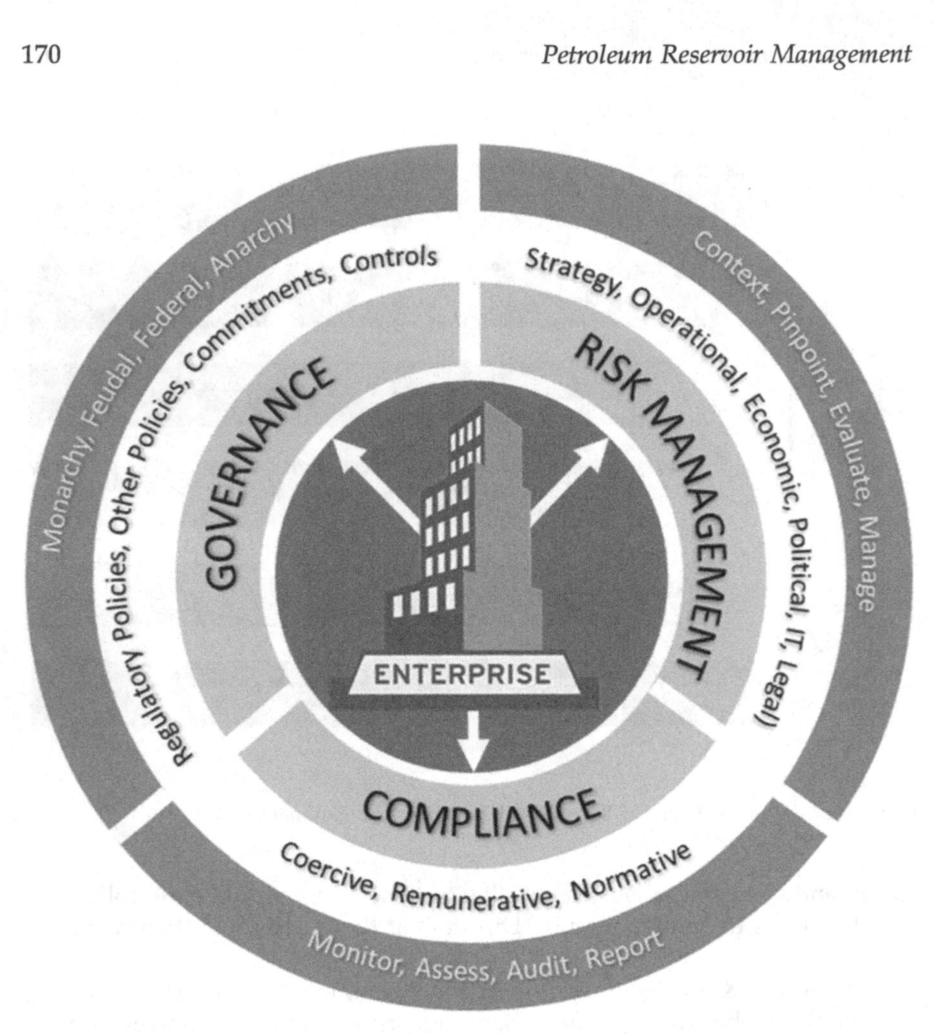

FIGURE 5.4
Relevance of GRC to Reservoir Management.

GRC is a technical acronym used to express the symbiotic relationship between governance, risk, and compliance (Figure 5.4). Petroleum E&P programs are full of uncertainties and risks. Management of threat at the corporate level has a much broader scope than operational or financial risks. The scope and scale of a GRC program vary with each organization. As stated, governance is the combination of policies and procedures introduced by the BOD. Proper management facilitates the quick translation of critical strategic and operational decisions into action. It ensures that the systems and processes prescribed by the organization are effectively followed to manage the combined efforts of working teams towards a shared objective. The governing body does not have to indulge in micro-management of operations and company affairs. It can delegate its power to one of the directors or even further down the line if required.

The style of governance depends on the type of personality dominating the BOD and the hour's need. The same governing body known to be stubborn can be flexible in the face of an emergency or disaster. Here are four basic styles of governance that compare the system of erstwhile political governance with that of modern corporations[1]:

- **Monarchy (Enterprise):** This style is associated with a monarchial type of decision-making and governing set up where the CEO and his selected few exercise all powers. If required, they may seek input from others in the interest of the betterment of the enterprise.
- **Feudal (Local)**: This style refers to a governing style where priority is accorded to a particular function or region over the enterprise's interests.
- **Federal (Balanced):** The federal form refers to shared decision-making based on usual "give and take" that strengthens the importance of functional groups/regions and those of the enterprise.
- **Anarchy (Personal)**: In this form of governance, the decisions are made to secure personal benefits based on the importance of certain individuals rather than the enterprise's interest.

Petroleum E&P programs are full of uncertainties and risks. An enterprise or business risk is a possibility that prevents the organization from achieving its business goal. Enterprise Risk Management (ERM) includes periodic assessment of possible risks such as competitive, economic, strategy, reputational, environmental, legal, political, and compliance risks. ERM is, therefore, a prerequisite for the company management to prevent or control damage to company assets, finances, and reputation by implementing risk mitigation plans/actions.

Compliance and enforcement are essential aspects of governance. Compliance focuses on adhering to corporate policy standards and value systems for achieving organizational goals and objectives. These elements have added significance for countries that have established a national authority to regulate E&P activities. Permit and license holders are allowed by the local government or its nominee to perform oil and gas activities in the designated areas. These permissions or licenses have a detailed account of provisions that are necessary to carry out E&P activities in the region. The operator's failure to act by industry laws and regulations, internal policies, or prescribed best practices can expose it to legal penalties and material loss. Compliance is accomplished through proactive monitoring of events, investigation of alleged non-compliances, and enforcement of sanctions if required. It can be exercised through the following means-

- **Coercive power:** Application or threat of use of physical sanctions, social, emotional, or financial power

- **Remunerative power:** Monetary payments (and punishment), social awards and recognitions
- **Normative power:** Symbolic rewards and deprivations, promotions or demotions in a professional career, etc.

These powers are administered based on authority, social status, financial standing, expertise in a field, physical strength, and information sought by others.

An oil and gas field is developed according to the objectives of the business strategy. Extensive surveillance is carried out at the individual well level to identify health issues and plan any remedial measures. Well and reservoir performance is continuously monitored to determine if the development plan works according to the design. In the case of gaps or slippages, mitigating options are considered and executed.

5.4 Reservoir Management Policy Issues

Policies are a set of principles applied to create order and consistent decision-making in organizations. They establish the code for conducting business and organizational performance. People observe and act according to what is noticed, rewarded, or criticized. With proper governance mechanisms, risk management, and compliance policies can be the foundation of a distinctive corporate culture characterized by consistent and reliable performance.

Several reservoir management policy issues are critical for both the national government and operating E&P Companies. Table 5.1 lists some of the top policy issues that shape the operating philosophy of E&P Companies working in an area. Table 5.1 presents a brief discussion of the main themes.

5.4.1 Petroleum Exploration & Production

A country or company aspiring to venture into petroleum exploration and production must have access to oil and gas resources, capital, and an experienced workforce. If it falls short in any of these, it may consider seeking participating interests in E&P acreages abroad or invite foreign capital and expertise to bridge the gap. Different options are considered and strategies adopted to implement the following policy themes-

5.4.1.1 Award of E&P Rights

To ensure energy security, sovereign countries with access to petroleum resources may call for licensing rounds. Licensing is a process under which

TABLE 5.1

Reservoir Management Policy Issues Associated with E&P Companies

Theme		Options	Explanation
1	Petroleum Exploration & Production	Who are the players for E&P operations? Need for NOC or its involvement?	Better control on strategy and speed of operations, value and job creation
		Participation of foreign companies through Concession/ Production Sharing Contracts/ Service Agreements	Access to expertise, technology & capital improves chances of success, risk and reward sharing
		Accessing oil and gas resources in foreign countries	Security of energy supply and revenue earnings for the guest and host country
	a. Award of E&P Rights	Licensing Rounds	Acreage, time frame, and bidding criteria decided by the national governments
		Open Door Policy	Proposals received from interested parties for specific areas at any time
		Auctions	Award to the highest bidder
	b. Prioritization of Areas	Develop what is known first	Start developing with what is available and known, reduce uncertainty and risk
		Develop by the size of the prize	Assess the potential of areas before prioritizing development, make the best use of available resources.
	c. Emphasis on Exploration or Development	Exploration grows new reserves	Risks to investments, no revenue generation
		Development exhausts reserves	Not much risk generates revenue. No reserve growth-a concern for future
		The mix of Exploration and Development	What mix is appropriate?
2	Outsourcing versus In-House	Geophysical surveys	Conduct surveys, process & interpret data
		Drilling, Completion, and Workover Services	Drill, complete wells, and do rig workovers
		Logging & Perforation Services	Well logging and perforation services

(Continued)

TABLE 5.1 (Continued)

Reservoir Management Policy Issues Associated with E&P Companies

Theme		Options	Explanation
		Well Testing Services	Repeat formation test (RFT), modular dynamic test (MDT), Pressure Build Up, Fluid Sampling, Portable Tests
		Well Hook-Up & Maintenance	Connect drilled wells to the production/ injection facility, maintain X-mas trees, flowlines & valves
		Laboratory Services	Core Analysis, PVT, Geo-chemical, Geo-mechanical services
		Special Studies	G&G, Reservoir Engineering, Reservoir Simulation & Petroleum Engineering Studies
	a. Role of Service Contracts	Specific Criteria (service, quality, efficiency, experience, track record)	Minimize discretion, identify technically and commercially suitable service provider
		Competitive bidding	Promotes transparency, competition
		Preference to local contractors or level playing field	Support to the local economy or quality and fairness
3	Production from Border Fields	Maximize the rate of oil/gas exploitation from border fields without much investment	Quick revenue generation if no friendly relations with border country
		Maximize oil recovery with significant investments	Maximize the benefits of friendly relations with border country, establish a protocol
4	Depletion Policy	Produce fast and relinquish acreage	Quick revenue generation without significant investments, no concern to reservoir health
		Maximize NPV	NPV maximized, not necessarily oil recovery
		Maximize oil recovery	Oil recovery maximized, not necessarily NPV
		Optimize techno-economics	

(*Continued*)

TABLE 5.1 (Continued)

Reservoir Management Policy Issues Associated with E&P Companies

Theme		Options	Explanation
			Optimization for oil recovery and NPV achieved
		International commitments	The host government may have a commitment with other countries to export/import oil and gas
		Expectations of oil price	Oil price forecast may force the company to produce it or hold it back in the reservoir
5	Produced Water Management	Surface disposal after treatment	Reuse after necessary treatment
		Dispose into shallow disposal wells	Dispose into shallow reservoirs. Ensure it does not contaminate the land and aquatic environment in the neighborhood
		Use to maintain reservoir pressure	Produced water can be reinjected into an oil-producing reservoir for pressure maintenance
6	Associated Gas Management	Sell to users	May require initial processing and compression for supply
		Flare into atmosphere	If no market for gas, can it be flared (loss of energy resource and revenue)?
		Inject back into the reservoir	Like produced water, it can be reinjected into the reservoir for pressure maintenance
7	Fiscal Policy	Royalties, Taxation, tax concessions	Government's take for granted the opportunity to exploit hydrocarbons
		Compensation for Risk-Taking	Grant of incentives, discounts
8	Employment Opportunities & Compensation	Blue and white-collar jobs	What mix of local and foreign content?
		Compensation packages for locals and outsiders	The difference in compensation packages?

(Continued)

TABLE 5.1 (Continued)

Reservoir Management Policy Issues Associated with E&P Companies

Theme		Options	Explanation
9	Transfer of Technology	Competence building	Attain self-sufficiency in a given time frame
		Adopting best practices	Good reservoir health
			Modernization of industry/operations
10	Regional or Local Development	Local content policies	Sourcing a certain percentage of intermediate goods or inputs from local producers
		In-country value or franchising	Right to price and quality competitiveness
11	Data Confidentiality & Security	Data is an important corporate asset	Digital and analog data: storage and security vital for company business
		A large volume of the database (G&G, Engineering, Drilling, Financial, HR data, etc.)	Data gathered from acquisition, interpretation, studies, and operations required for future work programs, reference, audits, and so forth
12	HSSE (Health, Safety, Security & Environment) Compliance	Health, safety & security of company employees and assets	Strive for improvement in the health, safety, and security of our people at all times
		Environment protection	Protect air, soil, sea, and aquifers from pollution due to oilfield operations

the petroleum resource ownership can authorize a national or international company to explore or commercially develop a specific geographical hydrocarbon-bearing area. The procedure for licensing can be based on a bid system or a grant system. Companies or groups of companies submit their bid with commercial and technical components to explore or develop an area over a specified period in the bidding system. The commercial element in the proposal contains the fees, rent, royalty, and taxes as applicable. The technical component of the plan includes the biddable work program. The project is awarded to the bidder who offers the highest biddable government share of revenue and the biddable work program.

Under the open-door policy or grant system, the license is granted to a company or consortium of companies that can demonstrate the maximum interest and capability to achieve the government objective.

Auction is one of the standard methods to award petroleum licenses and contracts to the highest bidder. Four primary forms of auctions[2] are reported in the industry: ascending bid, descending bid, first-price sealed-bid, and second-price sealed bid. In an ascending bid system, the auction price is raised gradually, and the bid is finally awarded to the highest bidder. The initial price is set at a high value in a descending bid, which is progressively lowered until accepted by one of the bidders. First-price sealed bids are awarded to the highest bidder based on the bids received in the sealed envelope. Each bidder can bid only once. Second-price sealed bid system allows the highest bidder to equal the price quoted by the second-highest bidder. Bids are invited and received in the sealed envelopes, as in the case of first-price sealed bids. However, first-price sealed-bid appears to be a more popular auction method than others.

5.4.1.2 Prioritization of Areas

The countries or companies with limited funds or an extensive inventory of petroleum acreages must prioritize focus areas. It is an excellent approach to assess the risk and reward situation of all the petroleum properties that are a candidate for E&P activities. Investing in high-return projects results in fast economic growth, which leverages less attractive assets later on. The opposite of this may prove financially unexciting.

5.4.1.3 Emphasis on Exploration or Development

It is a common dilemma whether a petroleum company must pay its undivided attention to exploration or development. Exploration is the key to reserves growth but has high investment costs and uncertainty about success. It implies more significant risks to financial investments. In contrast, a field development program is the source of reserve depletion. Still, it generates revenue essential for the company's growth and support to its exploration projects.

While some oil companies with excellent experience and financial strength may consider exploration to be their forte majority of companies, particularly the NOCs, try to balance their portfolio by mixing exploration and development projects. The critical policy question, how much emphasis be placed on Exploration or Development, is answered by setting the target value for Reserve Replacement Ratio (RRR).

$$RRR = \frac{\textit{Volume of proved reserves added to the reserve base during the year}}{\textit{Volume of reserves extracted during the year}} \tag{5.1}$$

The value of RRR greater than 100% means current production can grow beyond the present level. RRR equal to 100% means current production can sustain at the present level, and RRR smaller than 100% implies production level is likely to decline.

5.4.2 Outsourcing versus In-house

It is a significant policy decision to be made. Should an oil company engage itself primarily in the core exploration and production activities or submerge itself in performing support activities? The volume of work related to these support activities and their cost component are formidable. Some of these services/operations that are performed on a recurrent basis are as follows:

- Conduct geophysical surveys and process data
- Drilling, completion, and workover of wells
- Well logging and perforation services
- Well sampling and testing services (RFT, MDT, Pressure Build Up, Fluid Sampling, Portable Tests)
- Well hook-up and maintenance (Connect drilled wells to the production facility, maintain X-mas trees, flowlines & valves, service oil wells)
- Laboratory Services (core and PVT analysis, geo-chemical, geomechanical services)
- Special Studies (G&G, reservoir engineering, reservoir simulation & petroleum engineering Studies)
- Design and construction of oil/gas processing and transportation facilities, and so forth

The petroleum industry has several competent and qualified service providers and consulting companies who can share this workload and deliver value for money on a contract or fee basis. Many oil companies use their services to remain lean. However, some NOCs and host governments look at this as an opportunity for employment generation, training and development of people, and, most importantly, cost savings. They prefer to perform these functions indigenously.

5.4.2.1 Role of Service Contracts

Many oil companies follow the policy to award service contracts to local or international service providers. The benefits of this policy are manifold. A service contract saves the operating company many hassles of equipment procurement, maintenance, and repairs. The operating company can relax in knowing that the service provider being an expert in the field will deliver an excellent service at a competitive price. Some large operating companies can use many service providers to generate competition and distribute the workload amongst them.

Operating companies follow a competitive bidding process to minimize personal discretion. Their bid evaluation criteria include several performance measures such as quality of service, efficiency, experience, track record, price, and so forth. They may also prefer to promote local service providers as a matter of policy or cost considerations.

5.4.3 Production from Border Fields

Investments in the border oilfields of a country are guided by the relationship between the host and the border nation. Friendly relations promote investments and "business as usual" to achieve maximum economic gains. Tensions or hostilities between the border countries cause an atmosphere of mistrust. Consequently, a smaller host nation may be inclined to step up its efforts to recover as much oil as possible in a short time without additional investments.

5.4.4 Depletion Policy

Depletion policy is the centerpiece of development planning. It allows the transformation of a natural resource into revenue. Based on exploratory and appraisal efforts, the host nation has a fair idea of the petroleum resource size. It must then decide how to exploit this resource and, at what pace, based on the medium- and long-term forecast of national energy balance, that is, requirements and availability from various energy sources such as coal, hydropower, nuclear, crude oil, and natural gas. In case the host nation does not have any experience in petroleum E&P operations, it may consider inviting IOCs or POCs to extract oil/gas along with investments or for a fee. The problem with IOCs or POCs is that their operating philosophy and commercial interests are usually at variance with those of the host government. The primary business goal of a POC/IOC is economic. However, the government's objective of developing an oil and gas resource is not pure revenue generation; it is also fulfilling its socio-economic goals. This philosophical difference in approach can result in different designs of the depletion policy.

It is recognized that theoretically, a homogeneous reservoir with excellent vertical and lateral connectivity can be drained by a single well. However, oil recovery at this pace will take too long or infinite time. The other extreme is to achieve the same magnitude of oil recovery in a very short or infinitesimal time through an infinite number of wells. Though hypothetical, these cases illustrate that resource allocation, investments, oil recovery, and time are closely interrelated. The oil recovery increases with time and investments up to a specific limit. Providing excessive inputs such as wells beyond a particular limit may not yield additional oil. Extra wells add to cost and interfere with each other's drainage areas, reducing wells' cost-effectiveness.

The value of investments is ascertained based on the concept of "time value of money" (TVM). This concept articulates that money available at present is worth more than the same amount in the future because of its potential to earn interest. An operating company with a desire to invest in the oil business seeks to maximize its profit in terms of Net Present Value (NPV), which is the difference between the sum of discounted cash inflow and cash outflow. A positive NPV results in profit, while a negative NPV results in a loss.

Based on the principle of TVM, a petroleum resource must be extracted at the earliest date to gain from potential value creation. However, it does not mean that a company can profit from huge upfront investments in drilling an infinite number of wells for oil recovery. Excessive investments in drilling a large number of wells reduce the project duration to such a short time that the project results in a net loss or negative NPV before the pay-out period.

Operating oil companies use this knowledge to design a depletion plan, which helps them maximize economic benefits in NPV with judicious use of investments and other physical resources. Figure 5.5 shows the results of the economic evaluation of a hypothetical reservoir, which can hold an assumed plateau rate of 10,000 BOPD for 4 years. Project life is characterized by significant investments in the initial phase for building infrastructure and facilities. Production starts in the 7th year. Phasing and magnitude of capital investments have a substantial impact on the project NPV. Based on discounted cash flow (DCF) in this example, production becomes uneconomic after the 27th year at the prevailing crude oil price.

The cumulative NPV curve starts flattening after 18 years and presents a decision point for the operator. Smaller private oil companies with limited financial capacity and resilience may have to consider "go or no go" with the project further. Larger POCs, IOCs, and NOCs must find if and how the new investments can help improve oil production, reserves, and project economics.

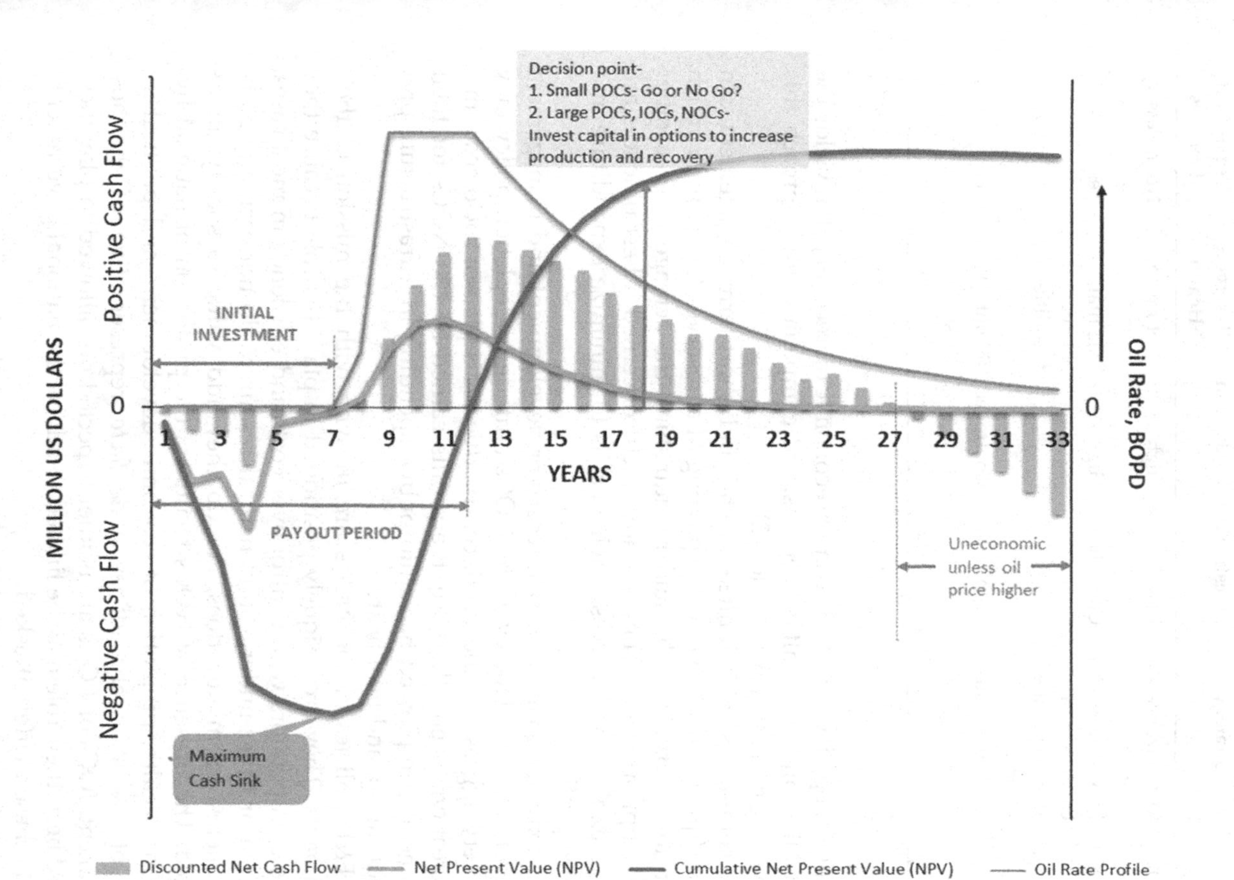

FIGURE 5.5
Impact of Oilfield Economics on Depletion Policy and Planning.

TABLE 5.2

Development Objectives of the Operating Companies

Type of Company	Strategy	Investments	Production	Reservoir Health	Use of Best Practices
Small POCs	Maximize wealth creation	Low	High	Not much concern	Not necessary
Large POCs or IOCs	Maximize wealth creation with attention to the value	Controlled	High	Maintain in good condition	Yes
NOCs	Maximize value creation	High	Controlled	Maintain in good condition	Yes

The depletion policy aims to maximize economic oil recovery that yields the highest NPV and other allied benefits. Oil companies can approach this objective in different ways (Table 5.2).

Some companies with limited capital and short-term engagement can make a "quick buck" by overexploiting the reservoir(s) without much concern for reservoir health and relinquishing the acreage to grab other attractive opportunities. They are not much concerned if best practices are in use to exploit the reserves. Their goal is to minimize expenditure and maximize profit.

Other companies can approach reservoir development and management more professionally. However, the NOCs attitude to depletion policy may not be very aggressive and profit-centric due to its plan of socio-economic value creation. Even otherwise, the profits earned by the NOCs for their governments are plowed back into public spending in infrastructure projects, education, and healthcare.

The E&P policies of the NOCs are in sync with the mission of their governments. They are uniquely positioned to play this role because they are also the operators with control over costs and efficiency in most cases. Petroleum or other underground natural resources do not earn any automatic interest as money does, nor do they add value to society unless exploited. The presence of NOCs facilitates the protection of national hydrocarbon wealth, promotion of economic development, and political interests of the state abroad[3] as a *de facto* representative of the host government. IOCs or POCs are neither expected nor allowed to play these roles. At best, their role may be limited to advise the national government/NOC on these matters if asked.

The next important question is at what pace the host nation wants to extract petroleum resources. The host nation is often willing to deploy what it takes to maximize its economic gains in a short time for converting its petroleum wealth into monetary value. Depletion policy can translate this aspiration into the development plan through quick production "build-up" by drilling more wells to achieve the "plateau" rate in a relatively short time. The pace of exploitation can be set based on Exploitation Index (EI) or Reserves to Production Ratio (R-P Ratio). EI describes the pattern of using up existing reserves and is the reciprocal of the R-P Ratio. R-P Ratio represents the life of reserves for the current trend of production.

$$El = \frac{Annual\ Production}{Proven\ Reserves} \tag{5.2}$$

$$R - P\ Ratio = \frac{Proven\ Reserves}{Annual\ Production} \tag{5.3}$$

These ratios are considered with organizational capability, resource availability, and timelines to create a framework for reservoir management and development plans. The picture on the left-hand side in Figure 5.6 presents a hypothetical example for a high, mid, and low production rate paradigm with an assumed fixed reserve base of an oilfield. R-P Ratio and Exploitation Index graphs on the right-hand side present the pattern of used up reserves for low, mid, and high rate cases during the build-up, plateau, and decline phases. High rate case assumes drilling, completion, and connection of more wells to the production facility. It calls for a higher capital budget, drilling rig equipment, workforce, and tighter time schedules. Mid and low-rate cases have a less aggressive drilling schedule and lower budget, equipment, and material requirements.

The depletion policy for net oil-importing countries aims to maximize their indigenous production. They sign long-term commercial agreements with oil-exporting countries to ensure an uninterrupted supply of crude oil and its products to meet their energy needs.

Price shocks caused by political upheavals or instabilities force exporting countries to reconsider their oil production at a reasonable time.

5.4.5 Produced Water Management Policy

Water may be a byproduct of the oil and gas production stream processed at the gathering centers. Almost all companies use the strategy to reduce, release, reinject, and reuse the produced water (Ref Chapter 6) depending on the volume and production timing. However, all these options have high cost-components attached to it.

Produced water cannot be directly reused for agriculture and other purposes. It can also not be released onto the land or sea without proper treatment due to strict environmental regulations. Salts limit crops and plants' ability to grow, and in excessive amounts can be toxic too. Free or dispersed oil present in the produced water is toxic and harms the

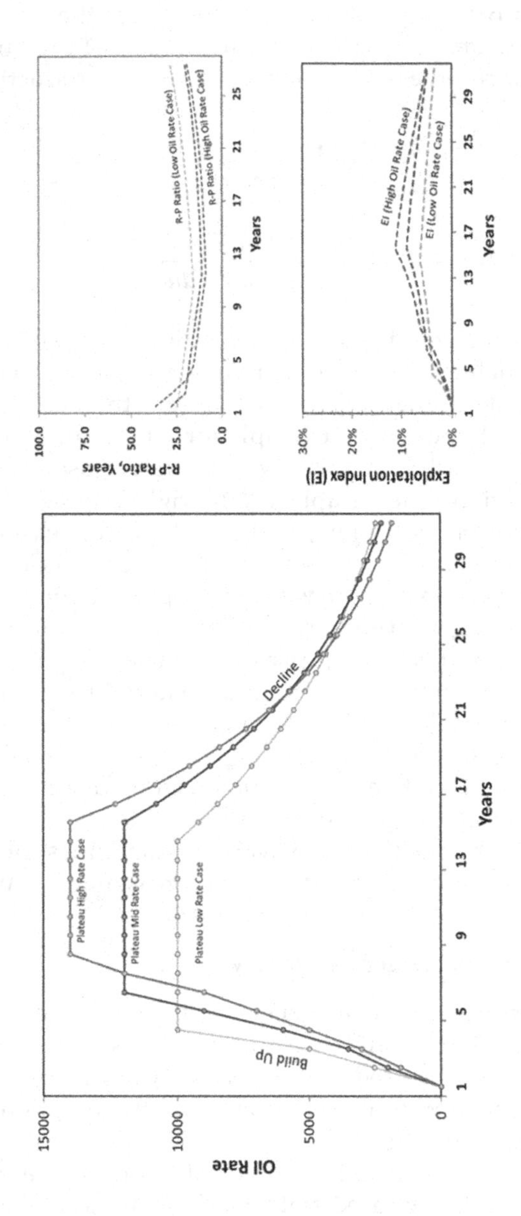

FIGURE 5.6
Depletion Policy Determinants.

ecosystem. For the same reasons, the produced water is also unsuitable for surface disposal, especially if the water volume is significant.

Disposal of the produced water into one or more shallow subsurface reservoirs is an option if it cannot be reinjected into an oil-producing reservoir for pressure maintenance. Operating companies need to find ways to balance the cost of their choice they make with the attendant benefits. This issue may also require consultations with the local government, as many states/countries may have concerns about water scarcity.

5.4.6 Associated Gas Production Management Policy

Crude oil production is accompanied by associated natural gas. It is used as a conventional heat source for manufacturing cement, glass, steel, paper, and many food products. It is also supplied to households for cooking and heating the homes in cold countries.

An oil company must create the needed infrastructure to supply natural gas to the consumer industries in its neighborhood. In the absence of the consumer industry in the area, the operator is sometimes forced to flare this gas, negatively impacting the environment and revenue.

Like produced water, the associated gas can also be injected back into the reservoir for pressure maintenance. Some companies may also like to store it in suitable underground reservoirs for use in the future. However, high upfront investments for drilling wells and building gas handling facilities are necessary to exercise these options.

According to the World Bank estimates, approximately 140 billion m^3 of natural gas is annually flared worldwide. As a result of this practice, not only a valuable hydrocarbon energy resource is wasted, but about 300 million tons of CO_2 is emitted[4] into the atmosphere contributing to global climate change. The concerned governments, oil companies, and development institutions are expected to respond to the World Bank initiative of eliminating gas flaring by 2030.

5.4.7 Fiscal Policy

Fiscal policies are critical to translating the national natural resources into economic benefits for the host country. These policies are designed to incorporate all legislative, taxation, contractual, and fiscal regimes that govern petroleum operations, including E&P activities.

One must appreciate that the host government is not interested in gaining control of E&P operations and investments planned by the operating company. Its main interest is to use the operating company's strengths to secure its own economic growth, access to technology, and investments to support national E&P programs. it may pursue these objectives by introducing lucrative provisions such as risk-sharing, cost recovery, incentives, discounts, tax holidays for a limited period, and in deserving cases.

The prospective operating companies, on the other hand, look for fair returns on their investment in exchange for their E&P efforts in a specific timeframe. The fiscal policies aim to provide a deal that is acceptable to both parties.

In a typical petroleum E&P environment, the host government maintains the following fiscal arrangement-

Pre-discovery provisions: Signature Bonus and Rentals, Discovery or Prospectivity Bonus to be paid to the host government

Post-discovery contract terms: Royalty, Production Bonus, and Crypto Fees (levies, duties, and other financial obligations of oil and gas producing companies)

Profit-based stipulations: These stipulations include corporate income tax, education tax, and profit oil shared by the government and the investors in a negotiated measure. Profit Oil represents the revenue after the deduction of royalty and recoverable costs.

5.4.8 Employment Opportunities and Compensation

The petroleum sector employs a large number of people across the globe. This sector provides direct and indirect employment to people supporting oilfield operations and a wide range of goods and services. Public sector undertakings, in particular NOCs, play a critical role in the industrial and financial growth of a country. By sheer virtue of their size and mandate, NOCs generate various employment opportunities at various levels: menial to high tech. Courses are designed and taught at universities to prepare students aspiring to join the petroleum industry. For sophisticated oilfield operations and activities, most of the companies run skill development programs. These programs allow employees to upgrade their skills and compete for higher positions in companies. The evolution of IT has revolutionized the way people work now. Its integration with disciplinary core skills has become a primary requirement.

Compensation is the combination of salaries, wages, and benefits that an employee receives for doing a particular job. Benefits may include housing, transport, healthcare, insurance, training, and education, such as conferences and short courses, superannuation, and so forth. The issue of compensation is closely linked with an employee's role in the organization, nature of employment, and level of skills required to perform a job. Petroleum companies want the best employees who can create value for the company. They want the talent with fresh ideas for the successful extraction of oil from inland or offshore locations. A compensation package not in conformance with the employee's role and contribution can hurt both the company and the employee.

5.4.9 Transfer of Technology

Technology is about using human intelligence to solve problems with tools, materials, and systems-application of technology results in savings of time, cost, and effort. Many countries around the world have significant quantities of petroleum resources. But all of them do not have the know-how to find and extract it. These nations' economic development and growth depend on oil/gas extraction with external support. This dependency costs these nations a fortune. They realize that by mastering the technology, they can be self-sufficient in managing their petroleum operations.

IOC and NOC alliances in the E&P sector have experienced mixed results for the transfer of technology. Some NOCs have done well in absorbing the technology, while others have not succeeded much. Some host nations cite conflict of interest on the part of IOCs as the main reason for the sluggish transfer of technology, while the IOCs point to a lack of seriousness on the host nation's part.

5.4.10 Local Development

Infrastructure development is closely related to the economic growth achieved by various sectors. However, the impetus the petroleum sector has provided in transforming society is evident in Middle East countries. Infrastructure and economic growth feed on each other. Lack of infrastructure leads to reduced economic growth. Growth, in turn, makes demands on infrastructure.

Oilfield activities and operations invigorate local and regional markets based on the requirements for goods and services. For example, the procurement of heavy equipment on the "rent or lease" basis or sub-contracting of low-tech activities such as laying of flowlines or routine servicing of heavy transport and machinery may be quite common in the new order. Such requisitions can fuel the market to a higher level of expectancy and growth. Protectionist measures by host governments in local content policies and franchising can also allow the local or regional market to compete, innovate, and expand their businesses. These methods act as a catalyst for the growth of the local economy.

5.4.11 Data Confidentiality and Security

Data is a collection of facts and information in words, numbers, observations, or images. Oil business uses, generates, acquires, processes, and interprets the huge amount of data in an electronic or physical form about their employees, customers, operations, activities, and financial status. Consequently, it has to have a full department and computer set up to store and protect data properly. Typical data inventory for a petroleum company engaged in E&P activities includes raw data such as seismic records, well logs, conventional and special core analyses, fluid analyses, static pressures, pressure-transient tests, flowing pressures, periodic well production tests, and monthly produced volumes of oil, gas, and water. Likewise, interpreted data could include a variety of maps, analyses, records, and reports, for example, the seismic conversion of time-to-depth maps, seismic attribute maps, log analysis, formation tops, structure and isopach maps, cross-sections, static and dynamic models, the layer-wise split of original oil and gas-in-place and initial recoverable reserves, crude oil and gas sales revenue, details of various contracts and payments, CAPEX-OPEX, budget availability and utilization, records of meetings, company targets, and performance, staff appraisal, salary, and so forth.

Data can be human or machine-readable. Human-readable data is unstructured and can be deciphered only by humans. Only humans can understand the meaning of an image or block of text, not the machines yet. On the other hand, machine-readable data refer to structured data that can be processed by computer programs or software.

Data confidentiality is a crucial business requirement in today's highly competitive commercial environment. Data or information are recognized as an essential corporate asset that must be secured in line with company policies. Issues concerning confidentiality arise when data is disclosed to unauthorized elements for fear or favor. This confidential information can be misused in the wrong hands to commit illegal activity (e.g., fraud, terrorism, discrimination, corruption). Failure to protect sensitive business information can lead to an irreversible loss of business.

Data security is the protocol to ensure that:

- Data is accessible to authorized users with appropriate rights or privileges to populate, view or edit in the company database for the efficient transaction of business and
- Data is adequately protected from unauthorized users/unscrupulous elements inside and outside the company to safeguard its business interests.

The primary aim of data security is to protect the confidentiality, integrity, and availability of accurate data. Reliable and good quality data is an asset for decision-making at various levels and departments across the company.

5.4.12 HSSE Compliance

HSSE is the top priority for any oil and gas company. Oil and gas occur in deep underground reservoirs at high pressures and temperatures. During drilling, production, and transportation, oil and gas can pose the risks of blowouts, fire, and explosion, causing harm to people and assets. Oil and gas well drilling and service activities, both onshore and offshore, use heavy, sophisticated equipment, machinery, and tools. Even with safety training, the equipment, material, fuel, and chemicals stored at well sites/facilities for operational use can create serious hazards for workers. Apart from non-compliance with safety policies and prolonged work schedules, lack of training, supervision, and equipment maintenance are common causes. Working long hours in harsh conditions and confined spaces can cause distraction or fatigue, which can be fatal. Companies have to face inquiries, litigation, and pay a heavy price in compensation and credibility.

Some critical oilfield hazards include explosions and fires, falls, high-pressure lines and equipment, electrical and chemical, well blowouts, H_2S, environmental pollution/damage, and ergonomics. Identifying and controlling hazards is critical to preventing loss, injuries, and deaths. Oil and gas companies do take HSSE seriously. They have an HSSE policy and management system to ensure the integrity of health, safety, security, and environmental protection processes/practices. Safety inspections and audits of wells, installations, pipelines, and facilities are routinely carried out to prevent accidents and sabotages.

Digital security is an all-inclusive term that includes the safety of modern digital tools, assets, and computer systems from hacking and virus. Advanced security and protection system are designed to safeguard persons, data, and property against crime, espionage, sabotage, subversion, and terrorism. It has become increasingly challenging to protect the computer systems from theft or damage to the hardware, software, or electronic data. Security systems have moved into the next generation in automation, particularly sensing and communicating hazards and vulnerabilities.

In recent decades, environmental problems have multiplied due to unplanned development activities and human interference with ecosystems. Cases of water pollution, air pollution, global warming, and noise pollution are on the increase. The dominance of fossil fuels in the energy and transport sectors is established as the primary cause of air pollution, impacting all living beings' health and life.

5.5 Management Controls on Business Activities and Results

Dynamic organizations recognize that integration and alignment of people, processes, and information are the critical requirements for a company's success. Communication is a tool to achieve this goal. They ensure that company goals and strategies are conveyed through written discussions and presentations to their employees, leaving no room for speculation. "*What is not on paper has not been said*" is found to be the expression of true wisdom. In situations of conflict of interest, justifying one's position based on an individual's interpretation of rules is common.

Reservoir management protocols aim at doing things properly with the input of expert teams and departments in the organization. There should be no confusion about the lead and support role in case of responsibilities shared by more than one group.

Petroleum reservoirs have a fixed geographical location and a long-lifecycle. In contrast, the company staff, management, and owners are transitory. An oil and gas reservoir, particularly a giant, may see several generations of executives, managers, and employees coming from different backgrounds and mindsets during its lifetime. Many business models may be adopted to change the company's growth profile to meet domestic energy needs and international commitments. As a result, the development and management of petroleum reservoirs are ruled by different ideas, concepts, and theories. The emergence of new technologies and workstyles have their effect on the ways reservoirs are developed and managed.

Policies, protocols, processes, and best practices are management tools to maintain consistency in the business activities and results (Figure 5.7). They reduce the effect of personal judgment and interpretation on company performance. Despite an apparent overlap, they are singularly different and independent of each other.

The objective of policies is to reduce bias and discretion in decision-making. A process comprises a series of steps to accomplish a task or produce a result. A protocol is best described as the "accepted standard" for doing something in a particular order or manner. A best practice is a working method or technique that produces better results than other methods.

Policies and protocols are designed by management/governments based on their mission, vision, and values. They specify the dos and don'ts. Process and best practices, on the other hand, evolve as a result of ongoing operations.

Some examples of policies, protocols, processes, and best practices are given as follows:

Policies:

1. Foreign participation in the E&P sector
2. Replacement of produced reserves
3. Produced water management
4. Border field exploitation
5. Personnel training and development

Protocol:

1. Use of safety kit at well/operational site
2. Perforating wells in daylight
3. Stabilizing well flow before fluid sampling
4. Wax coating of core plugs for SCAL work
5. Decommissioning of surface facilities at the time of field abandonment

Decision-making	**Policy** Set of rules that guide and enable consistent decision-making. ***Examples:*** 1. Foreign participation in the E&P 2. Replacement of produced reserves 3. Produced water management	Decorum	**Protocol** Accepted standards for doing something in a particular order or manner. Has dos and don'ts. ***Examples:*** 1. Use of safety kit at well/ operational site 2. Perforating wells in daylight 3. Stabilizing well flow before fluid sampling
Delivery	**Process** Comprises a series of steps to accomplish a task or produce a result. ***Examples:*** 1. Releasing a well location 2. Booking and categorizing reserves 3. Constructing a reservoir model	Doability	**Best Practice** Working method or technique that produces better results than other methods. ***Examples:*** 1. Delivering a high quality well 2. Extracting a representative formation core 3. Controlling environment pollution

FIGURE 5.7
Management Tools to Controls Business Activities and Results.

Process:

1. Releasing a well location
2. Booking and categorizing reserves
3. Constructing a reservoir model
4. Conducting a pressure build-up test
5. Separating oil from gas and water

Best Practice:

1. Delivering a high-quality well
2. Extracting a representative formation core
3. Controlling environment pollution
4. Disposal of effluent water
5. Ergonomic health and office safety

These examples include only selected policies, protocols, processes, and best practices. In reality, they are many, and all of them are very closely followed by the concerned teams/departments in line with their functional roles and responsibilities.

5.6 Significance of Best Practices

A best practice is the process of delivering standard quality of goods and services every time, efficiently, and cost-effectively, using modern, fair, and legitimate means. Merriam Webster calls the best practice "a procedure that has been shown by research and experience to produce optimal results, and that is established or proposed as a standard suitable for widespread adoption[5]." The words "optimal results" imply a balance among the product's quality, time taken from start to finish, and the cost.

Wikipedia describes a best practice as a method or technique that has been generally accepted as superior to any alternatives because it produces superior results to those achieved by other means or has become a standard way of doing things. It is simply found to be the most prudent way to proceed. Best practices are used to maintain quality as an alternative to mandatory legislated standards and be based on self-assessment or benchmarking.[6]

Companies not interested in reinventing the wheel can save time and money on research and adopt the best practices developed elsewhere for their key activities and operations. Many NOCs, IOCs, and Regulatory Authorities can prescribe conditions, issue guidelines, or recommended practices that form the basis for doing a particular job in a specific way in a region.

A best practice is an ever-evolving process. It grows stronger by ongoing research to find better solutions emerging from the improved knowledge base, new technology, or out-of-box thinking. The effort and money spent on the research and trial and errors are worth the results.

Documentation and communication of the best practices are essential in organizations where teams responsible for delivering goods or services are subject to change due to employee transfer, promotions, or resignations. If followed meticulously, such practices help avoid reservoir management malpractices, resulting in loss of production, oil and gas recovery, and revenue.

The Reservoir management policy framework is extensive. It includes various aspects of ownership, resource development, governance models, risk management, and compliance. Policies, protocols, processes, and best practices are designed based on research and experience. These tools help an enterprise to deal with situations of recurring and sporadic nature. Specific studies and investigations are necessary to resolve issues arising out of exceptional, complex cases.

Notes

1 Wendy Hirsch website; https://wendyhirsch.com/blog/project-governance-examples
2 Silvana Tordo, David Johnston, Daniel Johnston; World Bank Working Paper No. 179 Petroleum Exploration and Production Rights, Allocation Strategies and Design Issues
3 Paul Stevens, Ph.D.; A methodology for assessing the performance of National Oil Companies;https://www.scribd.com/document/105008244/NOC-Methodology-Stevens
4 The World Bank website; Zero Routine Flaring by 2030; https://www.worldbank.org/en/programs/zero-routine-flaring-by-2030
5 Best practice; Merriam-Webster.com Dictionary, *Merriam-Webster*, https://www.merriam-webster.com/dictionary/best%20practice.
6 Wikipedia The Free Encyclopedia; https://en.wikipedia.org/wiki/Best_practice

6

Oilfield Water Management

6.1 Introduction

The primary focus of planning and development is on oil/gas production due to their revenue earning potential. Water production from petroleum reservoirs where water is an active source of the operating drive mechanism is unavoidable. Produced water is a side effect of oil and gas recovery operations and is not a significant concern in early field life. Unlike oil and gas, produced water is not a "cash cow." In contrast, proper handling and management of produced water require upfront expenditure. In later phases of the oilfield lifecycle, the aquifer water or injected water assumes a more dominating role in oil recovery. As a result, water production increases and can overtake oil production, depending on the reservoir's maturity.

High reservoir withdrawals can cause a significant increase in water production. Produced water has two implications. Firstly, the produced water must either be disposed of or be reused for injection though the timing and amount of production may not necessarily match with that of injection. Secondly, the produced water may have toxic contents that can cause damage to the environment. As a result, the management of produced water becomes quite costly. It must be disposed of safely before it can be fully utilized for re-injection into the oil reservoir, if suitable, for improving oil recovery (Figure 6.1).

Water, therefore, plays a critical role over the entire lifecycle of an oilfield. In the reservoir, it displaces oil from pores and eventually breaks through the producing well. The water produced by the reservoir must be handled on the surface. The separators first separate it at the gathering centers. The separated water is then either sent for disposal or to an injection facility for reinjection into the reservoir. Return of the water to the oil reservoir where it originated constitutes a full water cycle.

Water management is a large and integral part of the oilfield lifecycle. It consists of planning and managing the full cycle of water viz., production, disposal, and injection over the entire field life. The effectiveness of water management programs plays a crucial role in assuring higher oil recovery and economic returns. If a company can manage water issues properly, half of its job is well done.

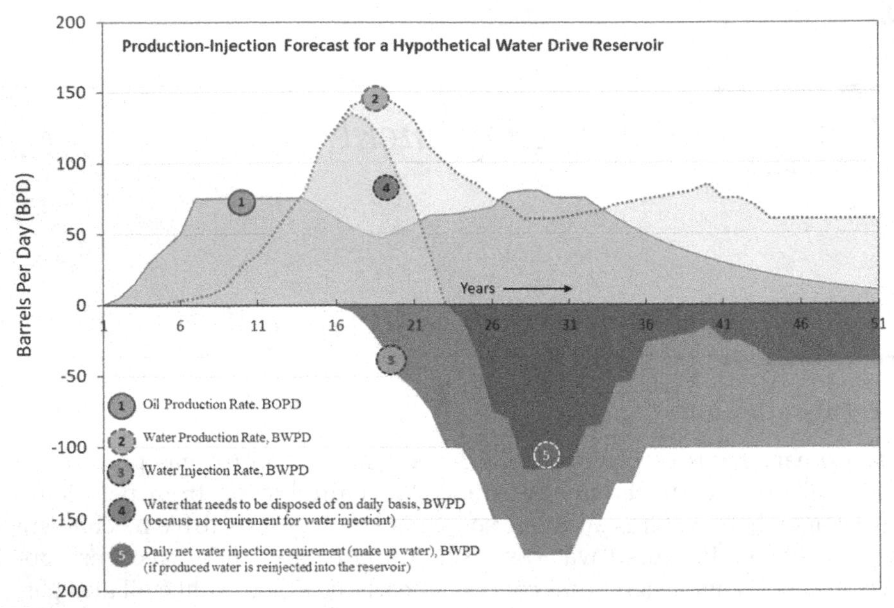

FIGURE 6.1
Typical Oilfield Production and Injection Profiles.

6.2 Produced Water Management

Hydrocarbons migrated from source rocks to reservoir rocks that were initially saturated with water. Consequently, natural water or formation water is always present in petroleum reservoirs in the pore spaces not occupied by hydrocarbons. Water is held in the pores as connate water. It may also underlie oil in the form of an aquifer due to the density segregation of oil and water.

Production of oil and gas depletes the reservoir and lowers the reservoir pressure. Water injection is considered to be an excellent option to maintain reservoir pressure above its bubble point pressure. Injection water displaces oil and gas through porous media into the production wells. Then a time comes when displacing phase water reaches the production wells. It is the beginning of water production alongside hydrocarbons. This water is known as the produced water or oilfield brine, or effluent water. It accounts for the most significant volume of by-products generated during oil and gas recovery operations. The term "effluent water" implies that this wastewater will require some treatment before reuse or discharge into the environment.

6.2.1 Sources of Water Production

Water is paradoxically the source and also the solution to many problems. No oil company prefers water production as it adds to its problems and costs. However, they realize that water is so integral and useful for oil and gas recovery that it cannot be wished away. If they need to live with water, they would better know how to deal with it.

The industry approach to water management is based on distinguishing good and bad water when producing oil and gas, as mentioned before. The water that helps displace and recover oil economically is referred to as good water. The one that travels to the producing well without any incremental oil production is called bad water. Good water cannot be shut off without losing reserves. It implies that the production of good water from the reservoir needs to be maximized. Technology is available to separate this water downhole and for reinjection into the qualified reservoir(s). These technologies help to avoid lifting and surface-separation costs but add to the complexity of operations and surveillance.

Bad water leads to excessive water production from individual wells without commensurate oil and gas production and reserves. Identification of zone(s) in a well that produces bad water and its cause is key to applying the specific water-control method. The source of most of the bad water problems can be found in casing leaks, channeling behind casing, fracture or conducting faults, water coning, thief zones, viscous fingering, and a variety of other reasons (Figure 6.2).

Water production from a reservoir generally increases as it matures. It is quite common for the water to bypass or finger through the oil with little or no oil displacement. It happens because the flood front cannot maintain its piston-like uniformity or "conformance" for the ideal displacement of oil by water due to the rock and fluid heterogeneity. Conformance is described as the flood front's ability to retain its areal and vertical uniformity while displacing oil through reservoir rock. Conformance control measures include water shut-off, gel, or polymer treatments in injection and production wells besides selective perforations for profile modification.

6.2.2 Produced Water Control Solutions

There are necessarily two primary modes to control water production from oil and gas wells: mechanical and chemical. Water production can also be controlled through customized well completion designs such as horizontal or multi-lateral wells or sidetracking of problematic wells into a trouble-free area (Figure 6.3). Mechanical or conventional solutions as they are called, cement and mechanical devices are generally used to mitigate near-wellbore problems, for example, flow behind casing, casing leaks, bottom water coning, and watered out layers without crossflow. Cement squeeze or setting inflatable packer/plug to isolate the offending interval is proven to

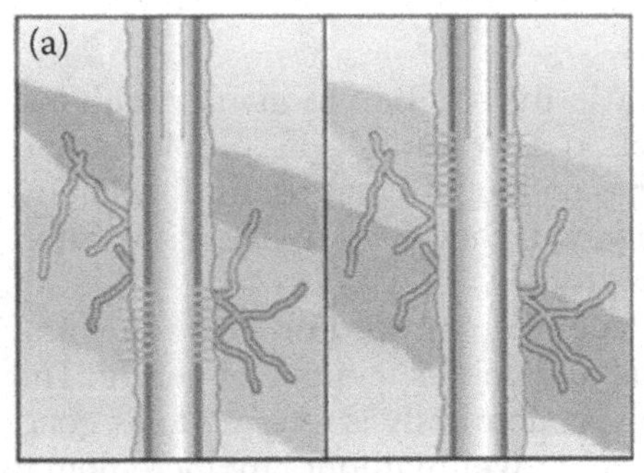

Water production due to fractures surrounding a vertical well

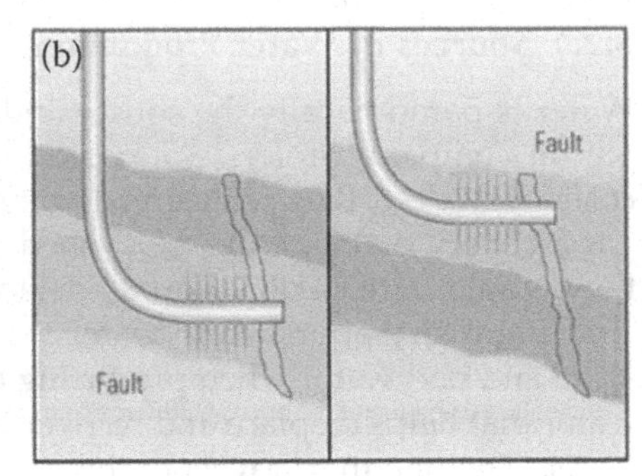

Water production due to conductive fault surrounding a horizontal well

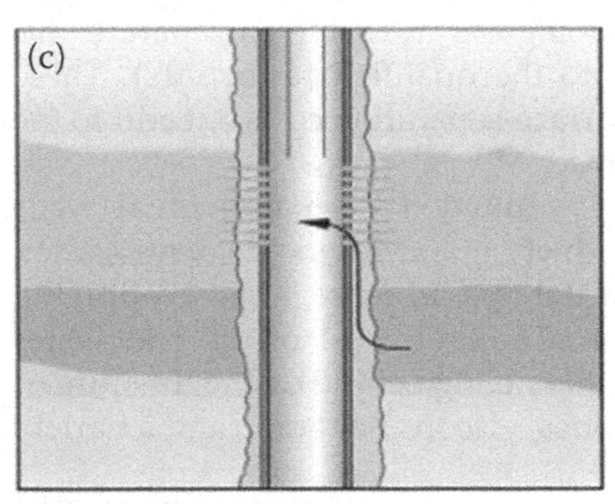

Flow behind the pipe

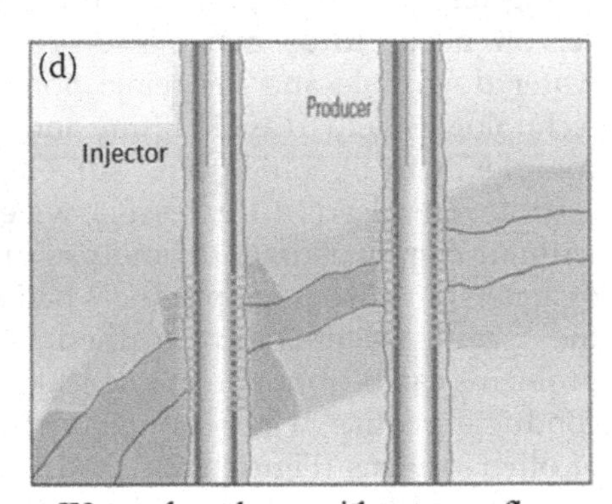

Watered-out layer without crossflow

FIGURE 6.2
Common Causes of Water Production- Some Examples[1].

shut off the water in the cased and openhole environments. After shutting off the offending perforations with cement or the mechanical packer/plug, the well is re-perforated above the sealed zone, and oil production is resumed.

Well-design and its completion depend on water problems, whether due to water coning or rising oil-water contact. The overall water production from a reservoir with robust natural aquifer support may sometimes be controlled by deliberately coproducing the aquifer in some selected oil wells. The selected wells may be completed dually with two parallel strings; the short string can produce from the zone perforated in oil and the long one from the perforations in the aquifer to weaken the aquifer's strength. It is a field-tested production strategy that has demonstrated improved oil production from wells that produced excessive water from the aquifer.

Chemical solutions are directed at shutting off the water with accurate placement of polymers or special gels in target zones. A polymer is a large molecule made up of many similar repeating units called monomers

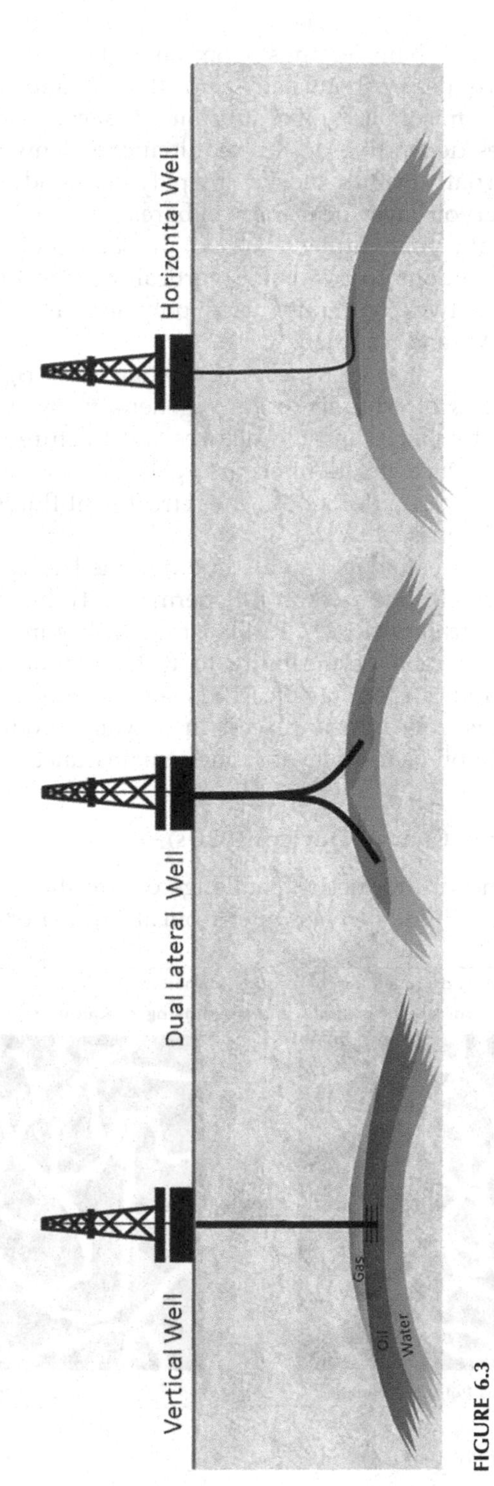

FIGURE 6.3
Well Completion Design to Control Water Production.

bonded together as a chain or rings. Commonly known, life-defining natural polymers are deoxyribonucleic acid (DNA) and ribonucleic acid (RNA). Gels are a semisolid jelly-like substance that can be designed to have variable properties depending on the requirements. Unwanted water production can be partially or fully blocked by polymer floods or gel systems in the offending reservoir layer near the wellbore.

Polymer floods and gel treatments are not the same. Polymer floods make use of polymer solutions to overcome vertical stratification and improve mobility ratio. The two commonly used polymers are hydrolyzed polyacrylamide (HPAM) and Xanthan.

Specially designed gel systems (Figure 6.4) such as "rigid gels" and organic or other types of cross-linkers are generally low volume, low-cost solutions that can be placed in the fractures and fracture-like features that cause channeling. Once gelation occurs, gels do not flow through the rock.[2] Coiled tubing can set or jet out most treatment fluids when required without risk to oil zones.

Polymer gel-type treatments can also be designed to control water production via either selective partial/full permeability blocking or Relative Permeability Modification (RPM). RPMs' strength lies in the fact that they can reduce the formation permeability to water without significantly affecting oil permeability. As a result, RPMs significantly reduce water productivity from this zone when placed in a water-producing zone. The productivity of the oil-bearing layer/zone remains unchanged.

6.2.3 Use of Inflow Control Devices (ICDs)

Coning is a significant production challenge due to the preferential movement of undesirable fluids such as water from an aquifer or gas from the gas

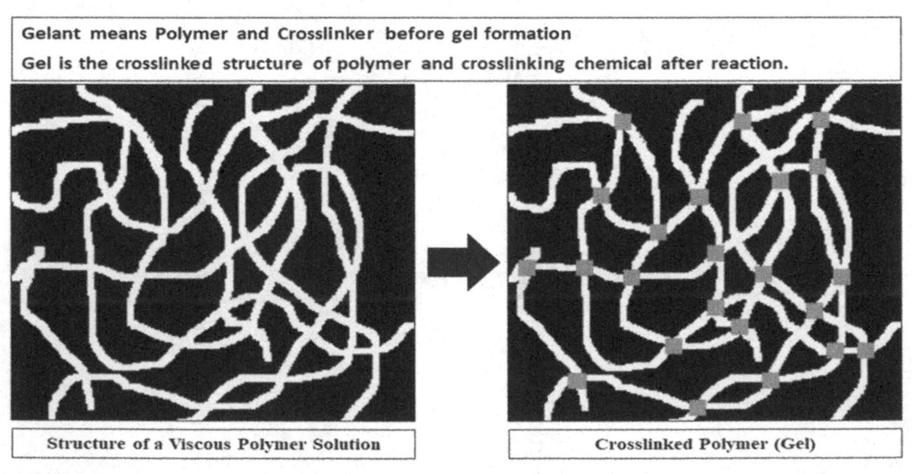

FIGURE 6.4
Chemical Structure of Polymer Gels[3].

cap. Coning tendency is a function of several reservoir parameters such as reservoir rock permeability, flow rate, pressure drawdown, and contrast between density and viscosity of fluids.

Water coning is a severe problem in reservoirs with large aquifers. One way to manage this problem is to reduce the pressure drawdown applied to the producing interval so that the water cone does not develop in the production well. It may entail producing the well at low rates, which is counterproductive from an economic standpoint. The other option is to drill a horizontal well parallel to the OWC in the oil zone. Increased reservoir contact ensures that the horizontal well can deliver more oil with less pressure drawdown. The net result is higher oil rates with minimal interference of aquifer water.

The application of horizontal wells for oil and gas recovery presents a new set of problems. The most common problem is that a few hundred to few thousand feet long horizontal sections of a well passing through a heterogeneous reservoir do not make a uniform contribution to production. Under a common pressure drawdown, high permeability sections contribute more, and low permeability sections deliver less or nothing. As a consequence of this, the water rises in the vicinity of the high permeability section. However, the water rise is much gentle compared to the cone that develops in a vertical well. With continued oil production from the horizontal well, the rising water level ultimately interferes with oil flow, limiting benefits from a horizontal well.

ICD technology has been developed to improve completion performance and efficiency. The objective of ICDs is to balance the fluid inflow from the high and low permeability zones along the horizontal section (Figure 6.5). In other words, ICDs can neutralize the effect of reservoir heterogeneity. The horizontal section is divided into several segments of constant permeability bins. These segments are completed with swell packers, which are used for zonal isolation. ICD completions are designed such that high permeability zones are subjected to lower pressure drawdowns and vice versa. This approach creates an even and uniform production profile, which can significantly delay the water breakthrough and increase oil recovery.

ICDs have been successfully used in sandstone and carbonate reservoirs both in production and in injection wells to achieve the even distribution of production and injection via horizontal wells. ICDs can also be used in vertical wells for specific purposes, although their popularity with horizontal wells has gained general acceptance.

6.2.4 Produced Water Handling

According to an estimate, three barrels of water are produced for each barrel of oil worldwide.[4] Globally, approximately 250 million barrels of water per day were produced from oil and gas fields, and more than 40% of this was reported to be discharged into the environment. The cost of handling this water can be 5

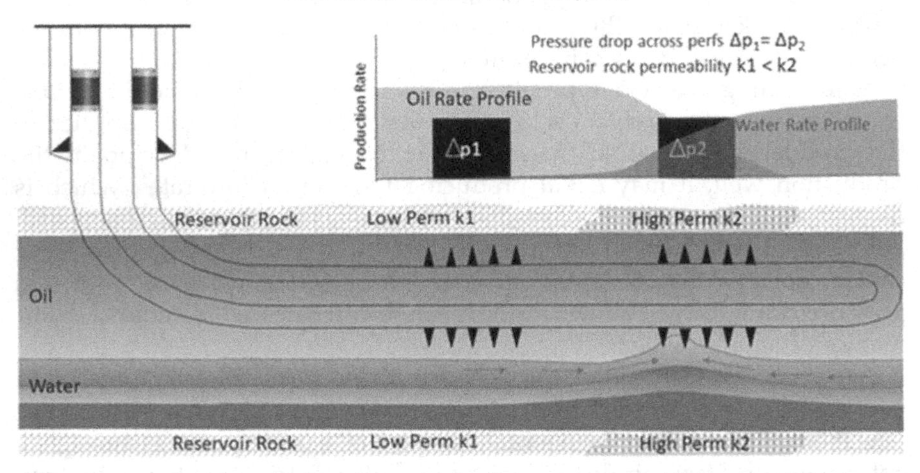

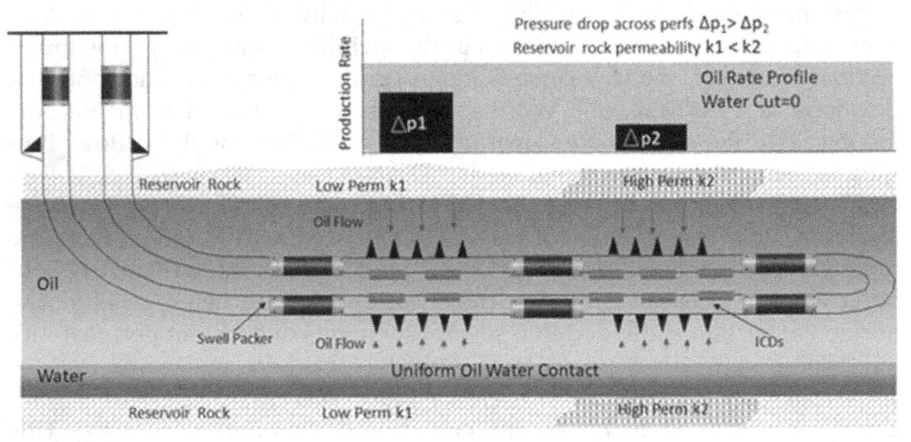

FIGURE 6.5
ICD Completion in a Horizontal Well.

to 50 cents per barrel[5] and seriously disturb oil and gas operations economics. A new estimate published in 2017 forecasts water production from the global oilfields to exceed 363 million barrels per day[6] by 2020.

In case the oilfield operator has no immediate use of large volumes of produced water, it may be decided to store it nearby or transport it to a location away from the oilfield for reuse or storage. Both of these options have contamination risks because of design failure or spills due to accidents during transportation. Disposal is usually considered an economically and environmentally sound solution for managing oilfield wastewater. However, improper design and lack of monitoring of disposal wells can defeat these wells' very purpose and contaminate the groundwater or aquatic life in the region.

Historical records of the produced water and its disposal, particularly in the early life of oil and gas fields, lack transparency. When there were no environmental regulations in yesteryears, much unaccounted untreated water from oilfields was released into the deserts, sea, and farms to irrigate crops, mainly in the arid regions. It could be ascribed to the poor appreciation of the produced water's value and its harmful effects on the land and aquatic environment. However, it is incumbent upon the operating company to follow the new stricter regulations backed up by heavy penalties in letter and spirit.

The relatively newfound concerns for the environment and deteriorating prospects of water availability for humanity in the future have mandated modern research to convert wastewater into a resource. Current treatment technologies are targeted at turning the "off-spec water" into water that is fit for reuse with a minimal negative impact on the environment.

6.3 Composition of Produced Water

Water chemistry is an important branch of chemistry that can predict the problems arising from using the produced water based on its analysis. This knowledge arms the professionals to resolve water-related issues and manage risks.

Generally, produced water is a mixture of inorganic and organic compounds, dissolved or produced solids, and dispersed oils. Additionally, it can contain a range of chemicals (acids, corrosion inhibitors, surfactants, etc.) used in oilfield operations, dissolved formation minerals, and gases (CO_2 and H_2S).

The magnitude of its organic and inorganic constituents varies and typically depends on the reservoir's age, lithology of the formation, and the type of hydrocarbon produced. The water produced from natural gas reservoirs is much smaller as compared with oil reservoirs.

6.3.1 Inorganic Constituents

Dissolved solids in produced water can have the following ionic composition:[7]

Cations

- Sodium (Na^+), Calcium (Ca^{2+}), and Magnesium (Mg^{2+}): Sodium, Calcium, and Magnesium are common constituents of produced waters.
  Sodium does not have any significant side effects unless it

precipitates as Sodium Chloride (NaCl) from oilfield brines with very high salinity.
Calcium combines with Bicarbonate, Carbonate, or Sulfate ions to form adherent scales or suspended solids.
Magnesium can form a soluble scale known as Magnesium Sulfate ($MgSO_4$). This scale is usually seen with the Calcium Carbonate scale.

- Ferrous (Fe^{2+}), Ferric (Fe^{3+}), and Manganese (Mn^{2+}): The presence of Ferrous, Ferric, or Manganese ions in water is indicative of its corrosion potential. Waters residing in formations with Iron or Manganese content for long periods can also produce corrosive compounds. Such waters, if used for reinjection, cause reservoir damage by plugging the formation.
- Barium (Ba^{2+}) and Strontium (Sr^{2+}): Barium and Strontium can combine with Sulfate to produce Barium Sulfate or Strontium Sulfate; both of these are insoluble and tough scales to remove.

Anions

- Chloride (Cl^-): Chloride is a significant and essential ingredient of oilfield brines sourced from sodium chloride. The corrosion potential of the water increases as it gets saltier.
- Sulfate (SO_4^{2-}): Sulfates are a part of naturally occurring minerals in some rock formations that contain water. Sulfate combines with Calcium, Strontium, or Barium to form stubborn scales. Sulfur-reducing bacteria use sulfur in Sulfates as their food source to produce H_2S.
- Bicarbonate (HCO_3^-) and Carbonate (CO_3^{2-}): Oilfield brines have bicarbonate ions but not necessarily carbonate ions. Water with higher concentrations of bicarbonate and carbonate ions is more alkaline than with low levels. These ions can react with calcium, magnesium, iron, barium, and strontium ions to form insoluble scales.

6.3.2 Dissolved Gases

Dissolved gases in produced waters can have a variety of detrimental effects which can hurt the bottom line of operating companies-

CO_2: Being an acidic gas, CO_2 lowers the pH of the water. Consequently, the water becomes more corrosive as well as prone to calcium carbonate scaling.

H_2S: Dissolved H_2S, like CO_2, increases the corrosion potential of water. The sudden presence of H_2S in an otherwise sweet oil reservoir is an indication of souring due to Sulfate Reducing Bacteria (SRB). Souring can cause massive corrosion of subsurface and surface tubular, material, and equipment.

O_2: Dissolved O_2 promotes greater corrosivity of water and the growth of problematic aerobic bacteria. Oxygen goes into the system and reacts with the dissolved ferrous or ferric ions in water to form ferric oxides. This substance can plug equipment, filters, and flowlines if the produced water is used for injection purposes.

6.3.3 Treatment Chemicals[8]

Oilfield waters may also contain various types of chemicals used in oilfield operations, such as spent acids, corrosion inhibitors, reverse emulsion breakers, and biocides that are very harmful at levels as low as 0.1 parts per million. Corrosion and scale inhibitors can also make oil/water separation less efficient.

6.3.4 Naturally Occurring Radioactive Material

NORM leaches into the produced water from some formations. Radium-226 and Radium-228 isotopes are highly radioactive (Radium-226 has a half-life of nearly 1600 years) and can build up as a scale/sludge that accumulates in water separation systems.

6.4 Produced Water Quality

Produced water quality depends on the geographic location, type of hydrocarbon production, and the oil-bearing formation's geochemistry. Produced water can be reused if its quality and composition meet government regulations and laws. In most cases, the produced water quality is poor. Therefore, treatment of the produced water becomes mandatory before this water can be disposed of or reused.

The quality of several water samples or their blends can be ascertained by the relative degree of plugging when a given volume of a particular class of water is passed through a membrane filter of a specific pore size. It is only a qualitative test, which is performed to compare the quality of different waters.

Oilfield operations require more detailed and in-depth investigations to check the quality and composition of the produced water. For this purpose, many water samples are periodically collected from various oilfield locations, for example, producing wells, flowlines, and gathering centers, to check the quality of produced water. A detailed water chemistry analysis is performed to determine the composition of produced water and evaluate fundamental properties (Table 6.1). This analysis is a prerequisite to deciding the strategy for further use of the produced water.

A four-pronged strategy (Figure 6.6) to manage produced waters is generally attempted. The first line of action is to reduce or avoid water production at the level of the subsurface. The surplus amount of water that comes to the surface can either be released in open areas after ensuring that it conforms to environmental regulations. Or it can be directly injected into disposal wells with proper safeguards for the land and aquatic environments. This water can also be pumped back into the reservoir where it originated or other reservoirs that may need pressure support after suitable treatment. Depending on the quality of produced water, some may be reused for oilfield or other operations as the industrial need for water is everlasting.

6.5 Produced Water Handling

Production from a well needs to be separated into the products of interest, viz. oil and gas. It is typically done at the crude oil and gas gathering center using a train of separators and free water knockout tanks. Here, water and other impurities are removed from the mixed stream of fluids delivered by production wells at the surface. The use of chemicals and heat application is quite common to break the oil-water emulsions, if any, before entering into the separators.

The change in pressure and temperature conditions between the reservoir and surface can cause deposition of scales, paraffin formation, and changes in the pH value of produced water. Further, exposure of the produced water to oxygen, whenever it happens, leads to the precipitation of iron compounds and elemental sulfur.

Large volumes of water, even with traces of oil content, render it unsuitable for surface disposal. The safest options from economics and environmental perspectives are to dispose of this water into shallow subsurface reservoirs that are devoid of hydrocarbons and not connected with any aquatic bodies such as groundwater or nearby lakes, rivers, or oceans.

Field Development teams decide whether to use idle wells with no future use or drill new wells for subsurface disposal of the produced water. The water separated at the gathering center can be diverted to the disposal wells via an injection pumping unit.

TABLE 6.1

Main Properties of Produced Waters[9]

Properties	Range
1. Specific Gravity: Water with higher specific gravity has higher Total Dissolved Solids (TDS).	1.014–1.140*
2. Conductivity: Conductivity is an electrical property. It measures the capability of water to conduct the flow of electricity.	4200–58,600* µS/cm(Micro-Siemens per centimeter)
3. Surface tension: The property of a liquid surface that allows it to resist an external force.	43–78* dynes/cm
4. Salinity: Salinity describes soluble salts' concentration in water, such as sodium, magnesium, and calcium. Sodium chloride is the main constituent of seawater.	Brine: >50,000 mg/L* Brackish water: 1,000–30,000 mg/L* Seawater: 30,000-50,000 mg/L*
5. Total Dissolved Solids (TDS): It represents the total concentration of dissolved inorganic salts/substances in water.	Ground water: 607 mg/L* River water: 207 mg/L* Seawater:35000 mg/L* Produced Water: 100-400,000 mg/L#
6. Total Suspended Solids (TSS): It accounts for particles larger than 2 microns in a suspended state in the water. Particles smaller than 2 microns are considered dissolved solid.	1.2–10,000* mg/L
7. Turbidity: It means that the water is not clear. It may result from many undissolved and suspended solid particles that are generally invisible to the naked eye.	182* NTU(Nephelometric Turbidity Units)
8. pH: It is a scale that specifies if the water-based solution is acidic or basic.	4.3–10*
9. Pure water is said to be extremely pure if it has a pH value of 7. If the pH is less than 7, it means that water is acidic and corrosive. If the pH is more than 7, the water is basic and has a higher scaling tendency.	
10. Total Oil: Refers to the measurements of free and dispersed oil in water that collectively lends an "oily" property to the water.	2–565* mg/L
11. Volatiles: Volatiles consist of BTEX (Benzene, Toluene, Ethylbenzene, and Xylene), which are highly water-soluble and volatile aromatics that enter the air, soil, sediments, and groundwater due to accidental oil spills and improper oil-related waste disposal. Their potential acute toxicity poses a health hazard to humans and aquatic life.	0.39–35* mg/L

* Figures are indicative. Can change with geographic location, geochemistry of formation, and type of hydrocarbon production

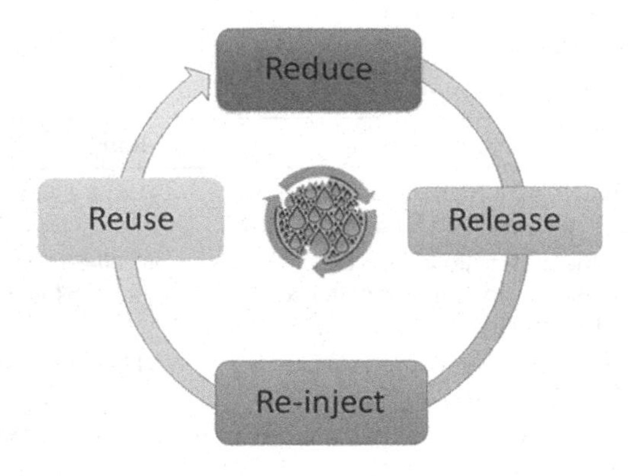

FIGURE 6.6
Produced Water Management Strategy.

6.5.1 Produced Water Treatment

Produced water may have to be discharged in an onshore or offshore environment, disposed of in the subsurface, or reused for injection purposes. It may also be used for irrigation or other industrial and non-industrial purposes (Figure 6.7). Treatments must ensure that produced water specifications conform to the prescribed criteria before discharge into the offshore or onshore environments.

The general objectives for the produced water treatment are to make it suitable for discharge, reuse, or reinjection by removing free and dispersed oil and toxic chemicals, suspended particles, sand, turbidity, dissolved organics, bacteria, microorganisms, algae, carbon dioxide, hydrogen sulfide, dissolved salts, sulfates, nitrates, contaminants, scaling agents, excess water hardness, NORM, and so forth. If the water is to be reused for irrigation, then water sodicity may have to be adjusted by adding calcium or magnesium ions to balance sodium concentration.

A detailed discussion of the treatment technologies is outside the scope of this book. However, a mention of some generic techniques (Figure 6.8) is necessary for the closure of the subject.

The produced or effluent water must be treated to achieve the quality specifications for injection.

6.6 Injection Water Management

Water injection into an oil reservoir is carried out with the following objectives:

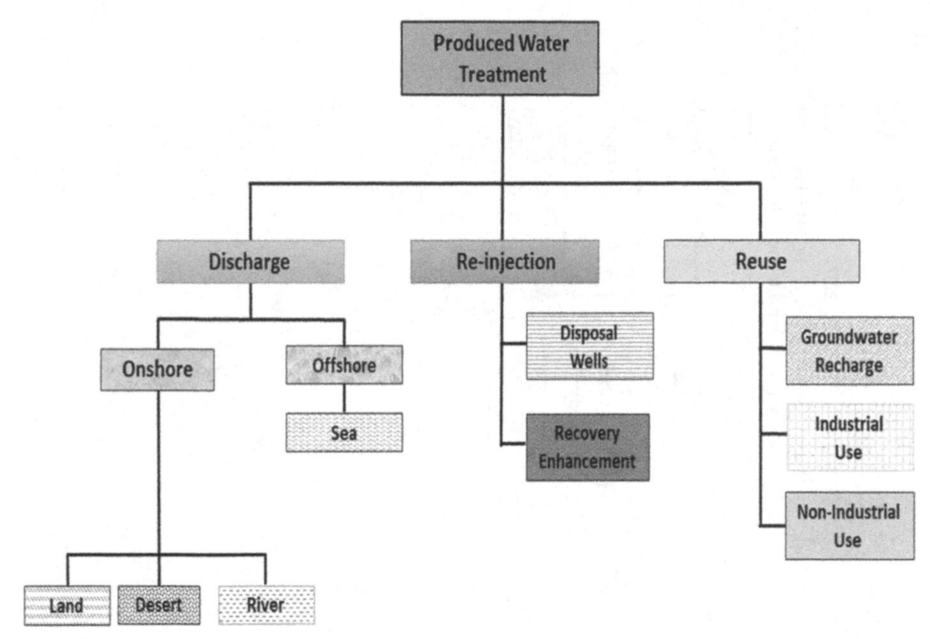

FIGURE 6.7
Produced Water Treatment Objectives.

- Maintain reservoir pressure at or above bubble point to arrest declining reservoir pressures and to improve well productivity
- Reinject the produced water back into an oil reservoir to circumvent disposal
- Execute a secondary recovery plan to improve oil production recovery
- Complement the natural water drive

Water injection objectives can be realized with proper integration and implementation of subsurface and surface strategies. Subsurface plans deal with identifying producer and injector locations and estimates of oil, gas, water production, and injection rates along with wellhead pressures. These plans constitute the design basis for surface facilities required to inject water and process additional oil and water volumes due to water injection. The critical elements of a surface plan include the capacity and location of water injection facilities and the piping network. New production facilities or upgrades of existing facilities may also be necessary to process the additional crude oil and water production due to water injection plans. The integration of subsurface and surface plans is vital for ensuring investments in drilling and surface facilities to realize the planned production capacity. Any mismatch in timing, for example, delay in the drilling of injection and

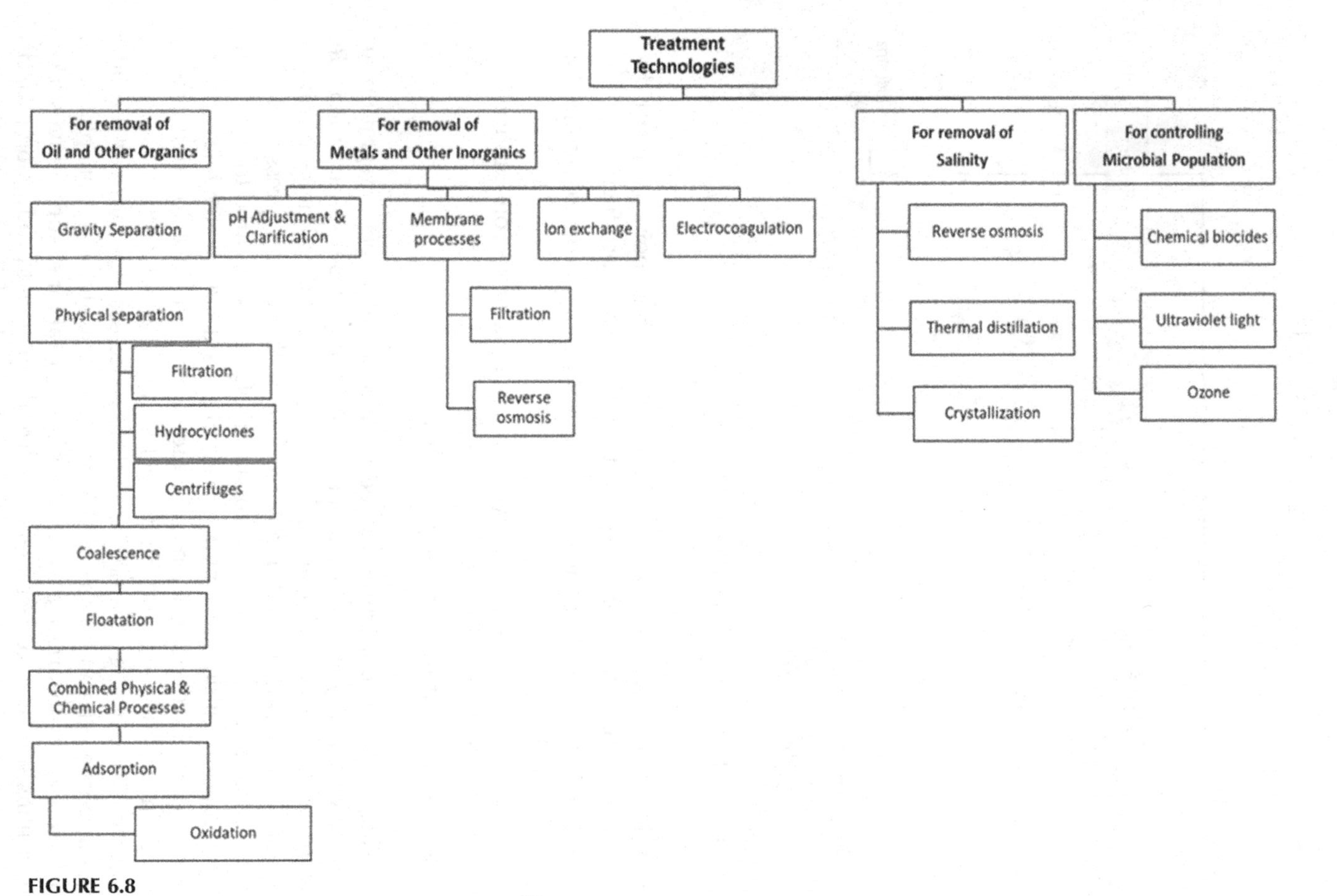

FIGURE 6.8
Produced Water Treatment Technologies at a Glance[10].

production wells or delays in construction of surface facilities, is counter-productive to the operator's goals.

6.6.1 Source of Injection Water

The source of water for injection is decided based on the availability, quantity, and quality of water. Water may be available as produced or aquifer water from the same or neighboring oilfield(s) or as a river or seawater nearby. A thorough understanding of the volume and quality of water required for injection is necessary to ascertain the suitability of source water-any deficit in quality results in loss of well injectivity and the consequent drop in productivity. Restoration of injectivity may involve a large number of injection wells with costly field operations and treatments.

Available source waters can be screened by evaluating the quality and cost of treatment required to make it compatible with the reservoir rock and fluids. Water quality specifications can be developed based on extensive coreflood studies in the laboratory where the water of different quality is injected in formation cores to assess the effect of water quality on reservoir rock. High permeability reservoir rock is more tolerant and can accept somewhat poor injection water quality. Preventing the formation damage by maintaining a good standard of water quality is good rather than remediation after injecting poor quality water.

Seawater is the most common source of water for injection in the case of offshore or nearshore oilfields. Aquifer water from a similar geological setting as that of the oilfield may be a good option to choose injection water. The water from aquifers is expected to be clean and compatible in terms of salinity with the oil reservoir that needs injection support. It may require minimum treatment before injection. However, care must be taken to assess that the aquifer water withdrawn for injection into the oil reservoir does not significantly impact oil recovery. River water is not a very common source of injection. Nevertheless, seasonal changes in quality, volume, and filtration/treatment of river water must be factored into injection planning.

One of the critical quality requirements for injection water is filtering. The water may contain suspended solid particles that can enter the pore spaces and plug it. Consequently, a higher injection pressure is needed to continue injection that may initiate the growth of fractures. With time, further extension and growth of these fractures with plugged interfaces can cause injection water to move wayward and lead to an early water breakthrough in surrounding oil producers.

Produced water is generally a source of injection water at some point in the life of the oilfield. It is commonly referred to as PWRI (Produced Water Re-Injection), and its handling and treatment are essential components of injection strategy. The objectives of the produced treatment and technologies are provided under item 6.4 above. In case the produced water volume is insufficient, makeup water from other sources is required for injection. If so, water from different available sources must be evaluated to check the suitability needed for injection.

6.6.2 Need for Pilot Testing

Pilot testing is the first step to take the findings of the bench-scale core flood experiments to the field. It offers an opportunity to test the assumptions, analogies, and concepts used in the development plans. Many hypotheses, ideas, and strategies can be verified with the use of reservoir models. Yet pilot testing provides a realistic assessment of operational difficulties, opportunity for de-risking, and assurance for the success of the injection plan.

The water injection pilot test, if conducted for a reasonable period, can help evaluate the following:

- Injectivity into the formation
- Reservoir response time to water injection
- Injectors to Producers Ratio
- Interaction of the injected water with the reservoir rock and fluids
- Operational issues
- Potential risks associated with the full-field implementation
- Incremental oil gain
- Ability and efficiency of water to reduce oil saturation
- Estimates of OPEX and CAPEX

Generally speaking, the pilot area should be "representative" of the reservoir. Sometimes the constraints of suitable space on the land surface and high reservoir heterogeneity can force a judicious mixing of good and bad reservoir portions. Hence, the final pilot area represents the overall reservoir performance. Thus, reservoir performance and oil recovery when upscaled to field level have no bias. Field Development teams can, however, choose other strategies for selecting the waterflood pilot areas and their expansion. Highly heterogeneous reservoirs may sometimes require more than one pilot test to be conducted. The pilot area is usually small because of the operating company's interest in limiting the pilot test's cost and time. It may also be necessary to drill additional observation wells between injector and producer wells to monitor the reservoir response to water injection (Figure 6.9). The number and location of observation wells are determined by the type of waterflood pattern, presence of geological faults in the vicinity, and the heterogeneity between injector(s) and producers in a pattern. The observation wells facilitate important data gathering about the movement of the flood front.

The design and architecture of a waterflood pilot are usually kept simple. The complexity arises when the pilot test results are used to book

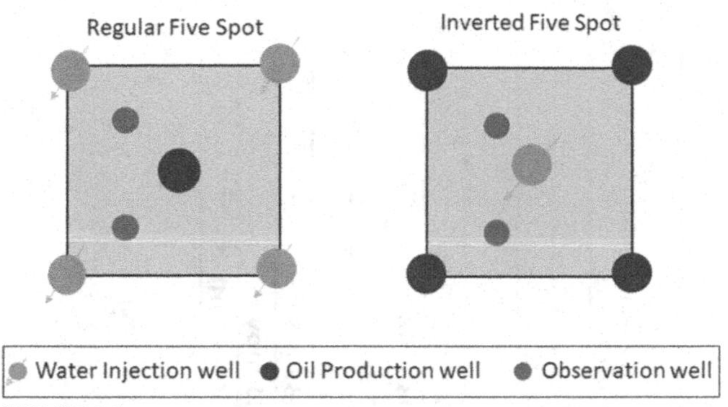

FIGURE 6.9
Regular and Inverted 5- Spot Water Injection Pilot Patterns.

waterflood reserves *en bloc*. Therefore, many companies disallow this practice and have guidelines that link stepwise reserve bookings as the waterflood project progresses from pilot to field-scale implementation.

6.6.3 Injection Water Quality

Injected water should meet its objectives of maintaining reservoir pressure or sweep the oil out of the pore spaces without causing any adverse side effects. It is possible only if the injected water quality standards/specifications can be maintained consistently. Decent injection water quality can preempt:

- Formation plugging or permeability reduction caused by suspended solids or dispersed oil in the water
- Scale formation that can result from the precipitation of salts or minerals
- Bacterial growth causing a loss in injectivity due to the buildup of biofilm
- Clay swelling and migration of fines in sandstone reservoirs can reduce formation permeability

Seawater contains about 30,000 to 35,000 mg/l of dissolved solids (Table 6.2). Two of the most prevalent ions in seawater are sodium and chloride, which make up about 80% of the total dissolved solids. Sodium, potassium, calcium, and magnesium ions combine with chlorides to form salts. These ions are primarily responsible for the salinity of the seawater. Calcium Bicarbonate, Carbonate, or Sulfate combine to form scales. Magnesium Sulfate is a soluble scale that usually coexists with the Calcium Carbonate scale. The presence of Iron in water is indicative of its corrosion potential. Such waters, if used for reinjection, cause reservoir damage by

TABLE 6.2
Typical Seawater Composition

Dissolved Solids	
Sodium	9690 mg/l
Potassium	343 mg/l
Calcium	479 mg/l
Magnesium	1250 mg/l
Barium	<0.1 mg/l
Iron	<0.5 mg/l
Sulphur	1050 mg/l
Aluminum	<0.2 mg/l
Silicon	1.3 mg/l
Ammonical Nitrogen	<0.3 mg/l
Chloride	17410 mg/l
Sulphate	3276 mg/l
Nitrate	<30 mg/l
Bicarbonate	165 mg/l
Total Dissolved Solids	**33694 mg/l**

Dissolved Gases	
Oxygen	4.5 ppm
Carbon di Oxide	Nil
Hydrogen Sulphide	Nil

Other Seawater Parameters	
Ph	8.03-8.58
Temperature	20°-30° C
Specific gravity	1.023
Conductivity	45.32 mS
Corrosion (Carbon-steel)	22.1 mpy
Bacteria	>106 per ml

plugging the formation. Barium and Strontium can combine with Sulfate ions to produce insoluble and tough scales difficult to remove.

Seawater can also contain dissolved gases such as O_2, CO_2, and N_2. These gases can get into seawater from the atmosphere to balance the ocean and the atmosphere. O_2 reacts with Iron to form an iron oxide, which can plug the filters and equipment. It also proliferates aerobic bacteria. Dissolved nitrogen in seawater is a growth-limiting nutrient for living organisms. N_2 is used up by some microbes that convert it into a much more useable form known as ammonium (NH_4^+). Microorganisms most easily consume ammonium through "assimilation." It is further converted into nitrite and nitrates as a part of the nitrogen cycle to support the marine ecosystem. Dissolved gases can be removed from seawater by chemical scavengers, vacuum deaeration, or by stripping methods.

Three standard classes of microorganisms found in the oil field are algae, fungi, and bacteria. Seawater, if lifted away from the shore, can help reduce the concentration of algae. Bacteria (0.2–10 microns in size) present the most severe problem and are controlled using biocide treatments. Injection water, if not treated properly, bacteria can accompany the injected water into the formation. Aerobic bacteria can survive in the presence of oxygen and multiply. Their growth results in the build-up of biofilm, causing permeability damage. Anaerobic bacteria such as SRBs survive and proliferate by breathing in sulfate available in the reservoir rock. They breathe out H_2S gas, which can induce Microbially-Influenced Corrosion (MIC) in the wells and surface facilities. H_2S is also a human health hazard, as discussed in Chapter 4.

Impurities in seawater should be removed/treated to make it suitable for injection. Pre-filtration systems using cartridge filters and ultrafiltration technologies can remove the bulk of dissolved and suspended solid particles. Sulfate removal from seawater requires nanofiltration membranes with a pore size of the order of nanometers ($1nm=1x10^{-9}$ m) and can remove up to 95% of salts in water. The nanofiltration system works mainly for the liquid phase and separates a range of inorganic and organic substances from water. The membranes are produced from various materials, including ceramics, cellulose derivatives, and synthetic polymers. Operators know that the "salinity shock" caused by low salinity seawater injection into a reservoir with higher salinity formation water can yield extra oil recovery. Treatment of injection water for removing salts, solids, and other impurities is limited by the reservoir's ability to tolerate impurities and treatment costs, not necessarily by technology availability.

Treatment technologies used for produced water are also suitable for seawater. It is filtered and treated to achieve the specifications for injection water provided in Table 6.3.

TABLE 6.3

A Sample of Treated Seawater Parameters

Parameter	Description	Quality of Injection Seawater
1. Total suspended solids (TSS), mg/l	TSS is a water quality parameter that specifies the amount of suspended solids trappable in a filter. Media filters should bring the TSS to the specified level.	<30
2. Particle Size (<5 microns)	Media filters should be capable to remove at least 98% of the particles of larger than 5 microns (1 micron = 1 μm = 10-6 m)	<98%
3. Total Oil-In-Water (OIW), mg/l	OIW is a parameter that defines the character of oil-water emulsions consisting of oil droplets dispersed in water.	<40
4. Dissolved Oxygen (DO), ppb	DO is not the oxygen component of the water molecule H_2O. It is essential for aquatic life and its reproduction system. A higher DO level can make the drinking water tastier. However, DO in injection water promotes corrosion.	<10
5. General Aerobic Bacteria (GAB) Planktonic, per ml	Aerobic bacteria are seen on the surface of the liquid, need oxygen to survive, and produce more energy. Planktonic bacteria cannot move on their own. They survive in the sunlit portion of the upper ocean, can freely float in the fluids, and spread through the system. They are important for life on Earth and its atmosphere.	<10,000
6. General Aerobic Bacteria (GAB) Sessile, cc	For aerobic refer to point 5 above. Sessile bacteria are attached to a surface.	<100
7. General Anaerobic Bacteria (GAnB) Planktonic, per ml	Anaerobic bacteria grow in the absence of oxygen. They live at the bottom of the liquid and produce little energy. For planktonic refer to point 5 above.	<10,000 per ml
8. General Anaerobic Bacteria (GAnB) Sessile, cc	For anaerobic bacteria refer to point 7 above. For sessile refer to point 6 above.	<100
9. Sulphate Reducing Bacteria (SRB) Planktonic, per ml	SRB occurs naturally in sulfate-bearing surface waters and seawater. These microorganisms breathe sulfate and reduce it to H2S causing reservoir souring increasing the possibility of hydrogen blistering or sulfide stress cracking. For planktonic refer to point 5 above.	<1
10. Sulphate Reducing Bacteria (SRB) Sessile, cc	For SRB refer to point 9 above. For sessile refer to point 6 above.	<100
11. Corrosion Rate, MPY (Mils Per Year where 1 mil=1/1000 inch)	MPY is the unit for the corrosion rate in a pipe, pipe system or other metallic surfaces. It expresses the weight loss of a metal surface in a year.	<2 for general corrosion <5 for pitting corrosion

6.7 Produced Water Disposal

The oil industry is witnessing exponential growth in volumes of produced/effluent water. It is primarily due to the ongoing efforts of operating companies to improve the oil recovery factor. The age-old practice of the disposal of produced water in large evaporation ponds is no longer acceptable. Produced water not used for injection to recover additional oil needs to be disposed of under social, environmental, and legal obligations. In the areas with limited disposal capacity or injection/pressure maintenance projects not being concurrent with water production, reuse of produced water for other sectors such as agriculture and construction is an option.

The following criteria can help decide a reservoir's suitability for subsurface disposal of the produced water. The reservoir should:

- Have no hydrocarbons
- Occur at shallow depth (for economic reasons)
- Be large enough to accept disposal of the estimated volumes of produced water
- Have geological continuity and connectivity (areal and vertical)
- Have adequate storage capacity and injectivity (porosity and permeability)
- Not have significant and conductive faults
- Not be connected with subsurface or surface water bodies (groundwater reservoirs, nearby rivers, lakes, and ocean) to avoid pollution risks
- Be able to absorb the estimated increase in reservoir pressure due to the disposal of produced water

Idle wells in the field without any present or future use may be good candidates for disposal wells. If not, new wells need to be drilled in places where they do not interfere with the main reservoir's development plan. Drilling new disposal wells requires as much care and caution as any oil/gas wells. Disposal well locations need to be picked prudently to achieve their objectives. Particular attention should be paid to the well construction. In case the reservoir selected for disposal of produced or effluent water is shallower than the oil-producing reservoir, the drilling company already knows what the best well design is for a disposal well. Disposal wells are typically permitted the pressure and rate limits that do not fracture the formation when injecting the produced water.

Disposal well design requires the following salient points to be kept in mind:

- The expected average volume of produced water to be disposed of in each well
- Assurance about the well integrity (casing, cement, tubing, and packer)
- Limiting rate and pressure so as not to fracture the target formation

Produced water disposal has both cost and environmental implications. However, continued pumping of large volumes of water into a shallow reservoir can overpressure it. Therefore, a constant watch is kept on the disposal reservoir's pressure to avoid loss of reservoir integrity due to sustained injection of large water volumes over a long period. Any breach in the reservoir seal is counterproductive because the injected water does not stay in the intended place. It can mix with groundwater or other marine water sources in the vicinity endangering the life of land or aquatic species.

6.8 Environmental Concerns

Impurities in the produced water can contaminate and pollute the soil, air, vegetation, and water bodies, whether marine or inland.

Sodium chloride (NaCl) is the predominant salt in the produced water. A higher concentration of NaCl harms the quality and fertility of the soil. Sodium being a natural dispersant promotes swelling and dispersion of the soil, disturbing its structure. As a result, water's ability to reach and move through the soil is considerably limited, creating soil erosion conditions. Salts also restrict the ability of seeds and plant life to draw nutrients and water from the soil. High concentrations of chloride ions are toxic and not conducive to the growth of some biological species.

Free or dispersed oil and highly water-soluble, toxic, volatile aromatics (BTEX-Benzene, Toluene, Ethylbenzene, and Xylene) can easily damage the ecosystem. Inhaling these toxic substances or consuming the contaminated water can pose serious health hazards to human and aquatic life. These substances enter the water and significantly restrict the supply of oxygen to sea plants/animals and contaminate sea mammals' food stock. Oil can also damage the water repelling properties of birds' feathers and destroy their insulation capability exposing their skin to harsh weather conditions.

The severity of environmental damage caused by the pollution depends on the nature, type, and amount of oil or other toxic substances that pollute the environment, apart from the weather conditions at the time of the incident.

6.9 Water Management Process

The scope of the water management program extends from subsurface to surface. Life cycle water management can be divided into three separate yet interdependent modules-injection water management, production water management, and disposal water management. Under each of these modules, operational activities are directed at overall value-addition, not forgetting the Company's Social Responsibility (CSR) towards the environment. The reservoir development strategy must address the critical water management questions presented in Figure 6.10.

Production, injection, and disposal strategies are closely linked with each other. Many water drive reservoirs are rate sensitive. Oil recovery from such reservoirs is negatively affected if they are subjected to high withdrawals. Apart from the rate of withdrawal, well type, completion, and perforation policy are the other crucial factors that determine the rate of water movement through the reservoir.

High fluid withdrawals are associated with a decline in reservoir pressure or oil rates after some time, if not immediately. This situation calls for water injection either in a particular pattern or on the periphery. A detailed modeling study followed by a successful waterflood pilot paves the way to implement a water injection program. However, where, how, and how much water will be injected determines the sweep efficiency and the resultant oil recovery factor. Usually, high oil gains in the early phase of the waterflood are offset by high water cuts later. Aggressive conformance management with and without rig workovers is essential to maintain the oil rate.

Re-injection of produced water into oil reservoirs offers dual benefits. Firstly, it offers a safe solution to manage the produced water. Secondly, it helps to increase oil recovery. An operator can use this option to turn a cost-intensive effluent water disposal program into a value-added recovery plan, subject to its techno-economic feasibility.

Prima facie, the produced water is usually compatible with reservoir(s) in its neighborhood, simplifying the implementation. The water can be collected from gathering centers and injected into needy reservoirs after minimum treatment. In case of excessive water production, chemical or

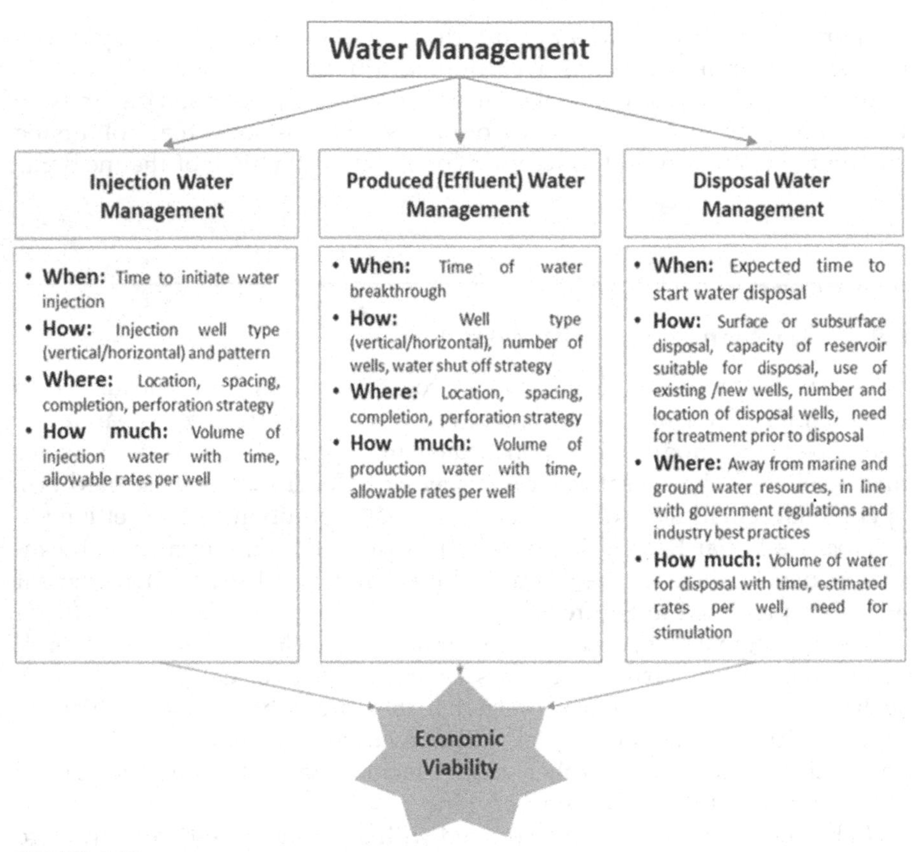

FIGURE 6.10
Typical Water Management Questions.

mechanical methods may be used to control it in the subsurface, while the surplus water can be injected back into the oil reservoir after necessary treatment to achieve the quality that is required for injection. Should this water not be sufficient for injection, makeup water can be lifted from the sea if close by or from other water sources. Any surplus produced water volume will require disposal into wells.

This water management process (Figure 6.11) must be reviewed periodically by subsurface and surface teams together, particularly for reassigning injection and production well allowables under constantly changing reservoir conditions. Any loopholes or pitfalls in water injection and disposal plans must be plugged well in time. For any significant changes in subsurface or surface plan modifications, company management should be on board.

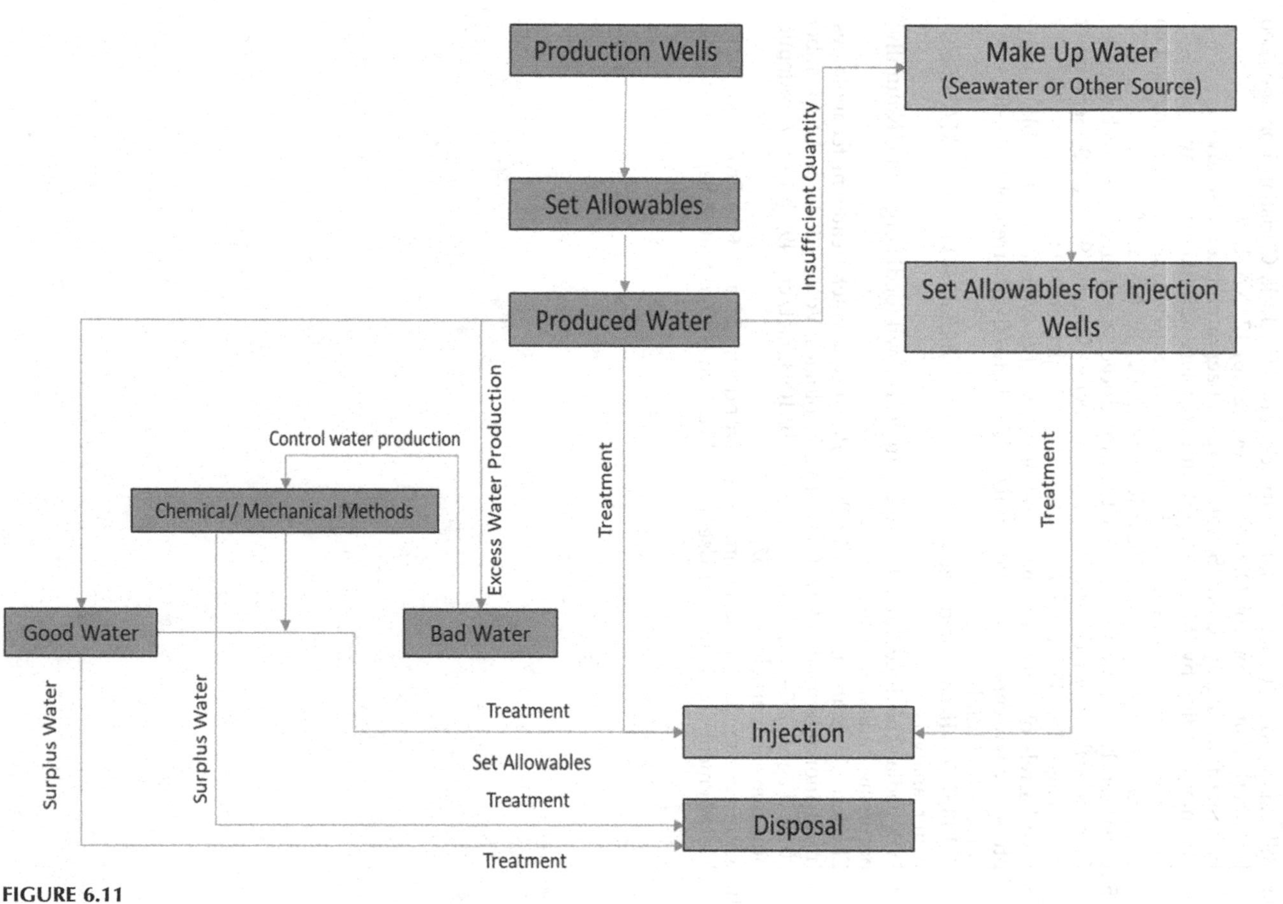

FIGURE 6.11
Oilfield Water Management Process.

Notes

1 Bill Bailey, Mike Crabtree, John Elphick, Fikri Kuchuk, Christian Romano, Leo Roodhart, Water Control, Oilfield Review (2000) 12, No. 1, 30
2 Polymer flooding by Randy Seright; http://baervan.nmt.edu/randy/
3 Polymer flooding by Randy Seright; http://baervan.nmt.edu/randy/
4 Bill Bailey, Mike Crabtree, John Elphick, Fikri Kuchuk, Christian Romano, Leo Roodhart, Water Control, Oilfield Review (2000) 12, No. 1, 30
5 Ebenezer T. Igunnu, George Z. Chen, Produced water treatment technologies, *International Journal of Low-Carbon Technologies*, Volume 9, Issue 3, September 2014, Pages 157–177, https://doi.org/10.1093/ijlct/cts049
6 Five barrels of water for one barrel of oil; CNBC website, Sustainable energy; https://www.cnbc.com/advertorial/2017/11/13/five-barrels-of-water-for-one-barrel-of-oil.html
7 OILFIELD WIKI, Oilfield water analysis; http://oilfieldwiki.com/wiki/Oilfield_water_analysis
8 Wikipedia The Free Encyclopedia; https://en.wikipedia.org/wiki/Naturally_occurring_radioactive_material
9 Ebenezer T. Igunnu, George Z. Chen, Produced water treatment technologies, International Journal of Low-Carbon Technologies, Volume 9, Issue 3, September 2014, Pages 157–177, https://doi.org/10.1093/ijlct/cts049; https://academic.oup.com/ijlct/article/9/3/157/807670
10 Katie Guerra, Katharine Dahm, Steve Dundorf; Oil and Gas Produced Water Management and Beneficial Use in the Western United States; September 2011.

7

General Reservoir Management Practices, Aberrations, and Consequences

7.1 Introduction

Each activity that can influence oil, water, and gas production from the reservoir, timing of production, injection, and cost becomes a part of reservoir management practice. It includes asset planning, data acquisition, field development activities, contracting, operations and maintenance, plan reviews, capital projects, and overall decision-making about various activities.

Reservoir management mandate originates from a company's business objectives and operating philosophy in a given regulatory, fiscal, and market environment (Figure 7.1). While the operating company works to maximize its benefits, its objectives of cost savings, efficiency, and value-addition must be backed by sound reservoir management principles and best practices. Reduction in cost and time using inferior quality of material or increased risk to operations is not acceptable. Similarly, efficiency at the expense of HSSE and profitmaking by indulging in unethical practices are not appropriate. Reputed business enterprises that care for their brand value and image organization use fair and legitimate means to achieve results.

The use of best business practices in operational areas and other activities is crucial in value creation. Costs and benefits cannot be generalized. Both are closely related and specific to the location and geological complexity of reservoirs. Appropriate business practices, their designs, and subsequent modifications evolve with time and ongoing efforts based on the experience. Good reservoir management practices balance oil and gas operations with cost and value components.

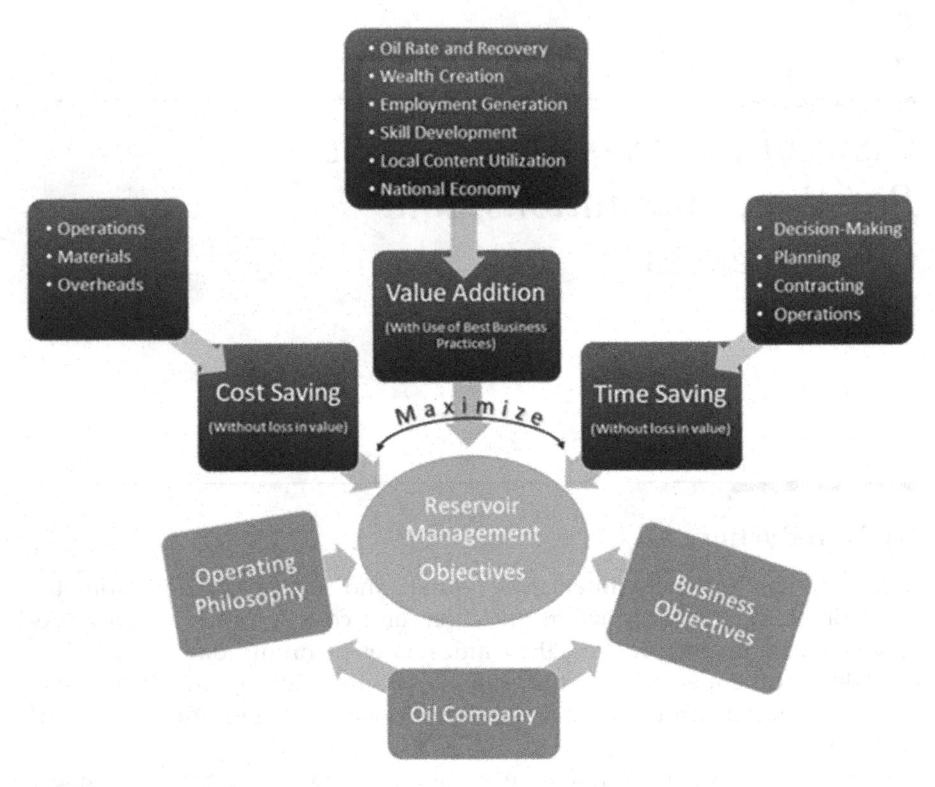

FIGURE 7.1
Value-System of the Petroleum Reservoir Management Practices.

7.2 Planning

Planning is an essential part of reservoir development and is based on organizational goals, business guidelines, and a roadmap. The management of petroleum reservoirs is complicated because the underground reservoirs are not visible, and their size and details are a matter of prognosis and interpretation than substantial evidence. Proper technology and human intelligence are necessary to locate, characterize, and produce these reservoirs. Interpretation of the data collected in this process is subject to many uncertainties causing risks to financial investments. These uncertainties are the primary source of delays in project planning and execution.

Planning and execution are two critical functions of an organization that cannot afford to work in isolation. They require a different mind and skillsets. The task of planning is predominantly imaginative. They have to make wide-ranging assumptions depending on the time scale of plans.

Therefore, a planner must be knowledgeable and unbiased with a focus on objectives and ownership. He/she must possess an excellent talent for quantitative analysis and strategic thinking coupled with precise communication, prioritization, and an abundance of soft skills.

Planning may adopt a top-down or bottom-up approach. The top-down approach consists of setting production goals by the management without an adequate understanding of the resource size, its production capacity, and the range of associated uncertainties. Of course, it has management's full backing and assurance to allocate necessary resources. This approach can sometimes lead to a frequent revision of targets and compromised project economics.

The bottom-up approach is better connected with ground reality. It has an accurate assessment of reservoir potential and hands-on experience of operational difficulties. It has employees' full buy-in and participation in the delivery of company objectives. In the absence of a vision and exposure to the business environment, this approach is introverted and might miss big-time expansion and growth opportunities. In reality, excessive emphasis on either of these approaches can be tricky. The planners have to devise a critical path picking up the strong points of the two strategies.

The planning team must have a good knowledge of the strengths and weaknesses of the organization. It should also have a clear understanding of planning premises, both internal and external, and controllable and uncontrollable. A correct assessment of these factors helps in the formulation of realistic plans based on the reservoir data. The planning team is usually an extended arm of the company management, and its assessment may sometimes be biased.

However, planning under uncertainties is about imagining possible scenarios. The future is always uncertain. Therefore, the plans are based on forecasts with many assumptions supported by available information, knowledge, and experience. The accuracy of these assumptions is fundamental to the correctness of the estimates. Any further constraints or changes in premises require that plans be periodically adjusted to align the forecast with actual performance.

7.3 Implementation

If planning is hard due to assumptions, execution is complicated by constraints. Building a plan requires imagination. Executing a plan to perfection requires experience and determination. That is where the "constraints" come in. All projects have constraints, whether internal to an organization or external. Budget constraints, schedule severity, quality specifications, resource availability, and risk perceptions can affect a project's quality.

Implementation has to face many unseen and unforeseen challenges not considered by planning, such as unexpected operational failures, work strikes, the resignation of key personnel, fire, blow out or other kinds of accidents, and so forth. Good managers always keep an eye on the assumptions and constraints throughout the project's lifecycle and work proactively at their mitigation. A critical analysis of the success or failure stories is documented as the lessons learned. The idea is to learn from experience and not repeat the same mistakes.

A common problem with the projects that require participation by multiple teams in the asset is the communication gap. This problem can be quite severe in large organizations where coordination among stakeholders is relatively tricky.

7.4 Experimentation

Improvements may require experimentation. Trial and testing of new ideas can bring a step-change in oil and gas recovery, cost-saving, and operational efficiency. Yet, there has to be an organized and systematic approach to experimentation.

Many plans may not go as anticipated. For example, a waterflood development plan with vertical wells may experience early water breakthrough. The company may switch to horizontal wells with ICDs to improve the oil rate and reduce water production, where the company has no previous experience. Such a step-change upsets the conventional paradigms and can be quite testing for the teams. Nevertheless, the results of the experiment may be worth all the pain.

The application of new technology is another example of experimentation. Technology can make a massive impact on business operations and their outcomes. Extraction of oil and gas both onshore and offshore is not possible without the use of technology. New E&P technologies enable reserves depletion at a higher rate and provide access to new oil and gas reserves. The companies that go into field-wide application of new concepts/technologies without a full assessment of consequences may also be caught in a predicament later.

Most IOCs, NOCs, and service providers have their in-house Research & Technology departments that either develop or scout for new technologies to meet their exploration, production, or reserve growth objectives. They find and pilot test a relevant and appropriate technology in the field to evaluate its potential for commercial application.

One of the weaknesses of the R&T departments of the NOCs is that they cannot fast-track the proceedings after the successful pilot test. They work at a relaxed pace, and their schedule is usually tied with their subcontractors plan. This practice causes excessive delays in the field-wide application of new technology. Field activities are often synchronized with the delivery of targets. Failure to take necessary actions on schedule can lead to missed company targets and sidelining of R&T departments from short-term plans of the organization.

7.5 Reservoir Characterization

Reservoir characterization is the first and fundamental step in the understanding of a petroleum reservoir. This understanding is gained from geologists' data based on their insights into basin analysis, rock types, and analogs. Seismic surveys may be carried out in prospective areas to determine the location and size of the structure. Following this, a few successful exploratory wells may be drilled to establish the presence of oil or gas. After discovery, the appraisal wells are judiciously placed to delineate the size and volume of hydrocarbon discovery. Reservoir characterization is the process that starts with the exploration of hydrocarbons and continues till the late development phase. The process involves reconstruction of the external (reservoir extent and geometry) and internal features (reservoir quality and heterogeneity) of a petroleum reservoir after synthesizing the available data relating to geology and properties. Seismic data, formation cores-conventional and sidewall, rock cuttings, well-logs, borehole images, well tests, and sustained production performance provide a wealth of data that provide deeper insights into the reservoir architecture and its production behavior. All this information is an essential input for the characterization of the reservoir. A reconstructed model is considered "representative" if it can reasonably reproduce the actual reservoir's static and dynamic behavior.

Seismic is an initial major scientific investigation into reservoir depth and architecture. Seismic data is used to create subsurface structural and stratigraphic features such as faults, folds, unconformities, and pinch-outs. On a small scale, seismic data can also decipher relatively minor faults and fluid contacts. It allows the geophysicists to mark the prominent seismic reflectors and create a base map with locations of lease/concession/field boundaries in UTM coordinates.

Well-logging is one of the trusted methods of formation evaluation. The logs record and present the variation of a specific reservoir property in a continuous record with depth. Their main aim is to ascertain the presence of hydrocarbons in an oil or gas well. Open and cased hole well logs play a vital role in picking the new well locations for drilling, setting various types of casings, deciding the interval for perforation, assessing the well potential in terms of in-place volume, reserves, and production potential of the wells.

The log interpretation offers a qualitative and quantitative estimation of many important petrophysical properties of reservoir rock and fluids. These properties comprise lithology, porosity, pay thickness, permeability, fluid saturations, water salinity, and density.

Routine Core Analysis (RCA) and Special Core Analysis (SCAL) are the two standard techniques used for reservoir characterization based on the laboratory tests on thin sections, rock cuttings, and small core plugs extracted from full-size bottom-hole or sidewall cores. The RCA involves a chemical and mineralogical description of cores and porosity and permeability measurements in horizontal and vertical directions. SCAL is performed on the representative rock sample plugs extracted from the full-size cores, whole cores, or the composite cores made up of several core plugs.

The most crucial SCAL measurements include the wettability index, capillary pressures, and resistivity index. Two-phase relative permeability measurements by steady/unsteady-state methods are essential for the SCAL data used in reservoir engineering calculations to calculate moveable oil and reserves.

Reservoir characterization is not complete without a full understanding of the behavior of fluids present in the reservoir. Oil, gas, and water samples collected at the bottom hole and surface conditions can provide valuable insights into their composition and behavior at varying reservoir temperatures and pressure conditions.

A PVT (pressure, volume, temperature) study consists of laboratory experiments to measure oil, and gas properties, such as P_b, B_o, B_g, R_S, μ_o, and μ_g in flash and differential separation processes. Flash and differential are two standard methods used in the petroleum industry to remove gas from liquids. In flash liberation, gas evolved due to a decrease in pressure is allowed to remain in equilibrium with oil and not removed from the cell. Therefore, the overall hydrocarbon composition in the PV cell remains unchanged. The bubble point pressure of an oil and gas mixture is determined by Constant Composition Expansion (CCE), known as flash liberation. This experiment is performed in a PV cell at a constant temperature.

On the contrary, in a differential separation experiment, the pressure on the hydrocarbon contents in the PV cell is reduced in stages, and liberated gas is removed from the cell at each stage. Consequently, the composition of hydrocarbons in the PV cell is subject to continual change, making them progressively richer in heavier components. Therefore, it should be clear that the differential process of separation starts only after the liberation of gas at bubble point pressure.

Well tests, also known as pressure transient tests, are a powerful technique that records the variations in bottom-hole pressures due to production or injection rates. The resultant pressure signal can provide a proper assessment of a well's production capacity, damage to formation permeability, presence of a sealing fault, or fractures near the wellbore. Pressure build-up, falloff tests, interference, and pulse tests are standard well tests that provide an excellent assessment of essential reservoir characteristics.

Data acquisition, processing and reprocessing, and most importantly, data integration drive the reservoir characterization process (Figure 7.2). The need for data is ever-lasting; the data acquired from drilling wells and reservoir performance are of greater importance in the appraisal and early development phases. The entire production infrastructure comes into existence based on this data. In the late development and abandonment phases, most of the information is used to plug the gaps in characterization or narrow down the uncertainties. A multi-disciplinary team consisting of geophysicists, geologists, reservoir, and petroleum engineers with investigative talent, skills, and experience works together to describe the reservoir. It may take a few years and drilling many wells to reasonably characterize an oil and gas reservoir.

While thick, coarse-grained, consolidated sandstone reservoirs such as those in the middle-east are an operator's dream, the ones with low-resistivity, thin beds, unconsolidated sands, and low permeability sands can pose serious development and production problems. Low-resistivity sands have a resistivity in the range of 0.5 to 5 ohm-m. Such sands were not developed in the past, owing to the belief that they were water-bearing. Small grain size and presence of clays and electrically conductive substances like pyrite can contribute to low resistivity. The occurrence and development of low-resistivity sands are now reported worldwide. Detection of thin sandstone reservoirs with a typical thickness of 0.2 to 2 ft is challenging because logging tools have a vertical resolution of 2 to 8 ft.[1] Improved tool resolution and data processing techniques are successfully more hydrocarbons from such reservoirs.

Some hydrocarbon-bearing sandstone formations may be unconsolidated with loosely packed sand grains. Such reservoirs may experience sand production in the course of the extraction of hydrocarbons. Sand production can cause erosion and corrosion of downhole and surface equipment, forcing the operator to shut in wells. Stand-alone screens and gravel packs are the most suitable methods for oil and gas production from unconsolidated sands. However, the cost and simplicity of installation make the screens the first operational choice.

Tight oil and gas sands are characterized by low porosity (3%–10%) and low permeability of a millidarcy and less.[2] Recovery of oil or gas from such sandstone reservoirs requires special techniques such as hydraulic fracturing. Therefore, the oil and gas production cost from tight sands is higher than the oil and gas produced by conventional methods.

Carbonate reservoirs are usually more heterogeneous and complex than sandstone reservoirs. The characterization of carbonates with fractures and fissures can be genuinely problematic. Experts' advice to modeling such reservoirs may vary drastically. Some experts recommend building a dual porosity-dual permeability (DP-DP) model to represent this kind of

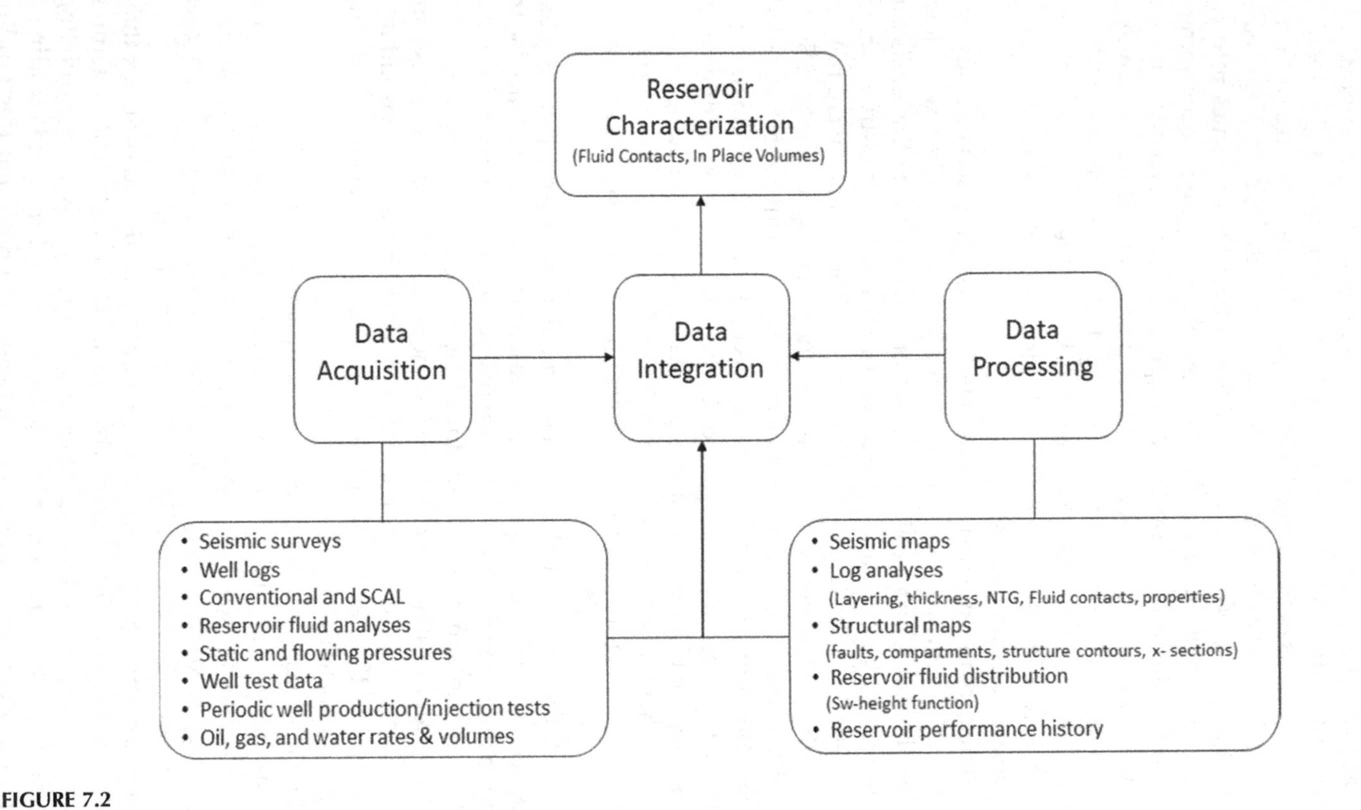

FIGURE 7.2
Multi-disciplinary Data Integration for Reservoir Characterization.

reservoirs, while others don't agree. These disagreements can lead to serious questions about the production profiles and recovery estimates. Many peer reviews on this subject have ended in a stalemate with a reconciliatory approach rather than an explicit agreement adopted in the concluding session to move forward. However, a 3D, 3-Phase, DP-DP high-resolution full-field model of a large-sized reservoir can be a real "monster" that is challenging to feed and run. The teams have to struggle hard to provide separate input data for the model's matrix and fracture systems; they have to make assumptions that are difficult to validate. These models are also a massive drain on geomodelers' time and computing resources.

RCA and SCAL data are fundamental to oil-in-place and recovery factor estimates. Most companies use the service labs to generate these data by paying a substantial amount of money. However, a study based on the review of 50,000 SCAL experiments claims to have found that about 70% of the SCAL data were untenable[3] either due to damage caused to cores or due to inconsistencies in testing procedures or experiments data quality. Cores may be damaged due to weak rock's failure during core recovery, core preservation, and core handling at the well site, during transportation, or in the lab. The use of oil-based muds and the invasion of drilling muds with chemicals into the formation can alter cores' wettability.

There are instances when the cores are incorrectly plugged or when plugs are not representative of the cored section. Very high or erratic values of irreducible water saturation (S_{wirr}) also render the SCAL results unusable. SCAL results are also dependent on the methods used by the lab. For example, capillary pressure curves generated by a particular lab using different methods (mercury injection, porous plate, and centrifuge) may be significantly different from each other. Similarly, the S_{or}, S_{wirr}, and capillary pressure curves generated by different service labs using the same method can differ significantly. It is essential to assess the service lab's capabilities and limitations before awarding work to them. There must be clarity about the test procedures and proper audits of their work and expertise. More or less, the same discussion holds good for PVT analysis of reservoir fluids that are captured from wells, transported to the lab, and then subjected to analysis. Cost without considering quality should not be the sole criterion for awarding the laboratory work to a service lab. SCAL and PVT parameters have a huge potential to impact oil recovery. The combined errors in their results due to inadequate equipment, improper test design and procedures, lack of expertise, and poor planning can influence the estimates of oil volume and recovery factors and risk the project economics seriously. Plug selection based on CT scans assures the selection of representative and non-damaged plugs. Similarly, plug photographs before and after tests indicate any significant changes in cores' conditions. A real-time quality check of core condition is a good practice that can help avoid expenditure on the poor quality of tests.

Pressure transient well tests are an excellent direct method to assess the well productivity, wellbore damage, stimulation, boundary and fault locations, reservoir size, and so forth. A proper systematic and in-depth analysis of pressure transient tests can serve as a useful aid to characterize reservoirs. These tests are mostly a record of the bottom-hole pressure of an oil or gas well with time. The pressure is a "signal" generated by a reservoir due to the flow rate changes. The engineers look at this problem as an "inverse modeling" problem where reservoir (model) parameters are computed from the model response (pressure). A "direct modeling" problem consists of a model (reservoir) of known properties where model response (pressure) is generated by causing a change in flow rate. The direct modeling problem is straightforward and has no room for uncertainties. On the other hand, inverse modeling is quite challenging due to non-unique, undistinguishable pressure responses generated by two or more utterly different reservoir models. There are circumstances when an infinite-acting homogenous reservoir can be misinterpreted as a naturally fractured reservoir or a finite-acting reservoir with artificial boundaries[4,5] due to the similarities in their pressure behavior. Similarly, drawing a straight line on the semi-logarithmic and log-log pressure plots or choosing an appropriate type curve is not intuitive. Simple mistakes in the well-test data interpretation can lead the team to select the wrong reservoir model and design a different reservoir development strategy.

This problem has persisted over the decades since the first generation of Horner's method to analyze pressure buildup test data to the modern-day derivative approach. It is interesting to see the growth of analysis techniques during this time; however, the solutions are still not perfect. The published literature about the idealized cases for various reservoir models and their non-unique pressure response can confuse the less experienced engineers into their unintentional misuse. In such cases, the best practice for well test analysis involves building a small simulation model that can generate the pressure response for comparison with well test data. Modern well test analysis software facilitates this process to reduce uncertainty in test data interpretation.

7.6 Reservoir Simulation

The use of models is common in consumer and fashion goods industries to promote their products' sales. Life-sized mannequins are quite popular to display or fit clothing, and so are replicas of human figures to teach the methods of first aid, CPR, and advanced airway management skills in a computer simulation. In the oil industry, laboratory or physical models such as core floods are as common as mathematical models. Because oil and gas extraction is a costly activity, the companies want to do it the right way the very first time. A simulation model has a set of mathematical equations representing the fluid flow through porous media. Under certain assumptions and boundary conditions, it acts as a dummy reservoir to test early development and production concepts at a low cost. The most advantageous option tested on the model can then be selected for execution in the field.

The idea of reservoir simulation dates back to the use of tank models in extensive use to forecast oil and gas reservoirs' performance. A tank model is the simple representation of a homogeneous reservoir containing oil and gas, and its production/injection and pressure behavior are governed by the Material Balance Equation (MBE). Fluids flowing into the tank (water influx or water injection) tend to increase the reservoir pressure, and those flowing out (production of oil, gas, and water) cause a drop in reservoir pressure. The process of matching the model performance and volumes of oil, gas, and water with those of the reservoir is called history matching. It is an essential step in model calibration.

History matched models can provide reasonable estimates of original oil-in-place and the strength of natural aquifer if present. The thumb rule is that a good history-matched model can be used to forecast the reservoir's performance for a period equal to that of the production history. The tank models are known as zero-dimension models and cannot deal with reservoir anisotropy.

Research indicates that the human mind has difficulty in analyzing more than four variables at a time.[6] The problems of rocket science, weather forecasting, and oil exploration/exploitation have multiple variables that a human mind cannot handle precisely. Physical models have been traditionally used to understand the mechanics and behavior of the invisible underground reservoirs. However, these tools are not suitable to address complicated reservoir management questions about infill drilling, pattern versus peripheral water injection, the strength of aquifer or gas cap, development options and strategies, and the impact of various reservoir parameters on its performance, rate, recovery, and so forth, due to their inadequate size, poor representation of heterogeneity, and a large number of variables. These are essential reservoir engineering issues that need to be resolved before committing enormous capital expenditure and organizational effort. Therefore, the decision makers look for a robust technical rationale that they can use to move forward decisively. Reservoir simulation, reservoir modeling, mathematical modeling, and numerical simulation are different names assigned to a stream of reservoir engineering specializing in building computer models capable of representing an oil and gas reservoir's behavior under static and dynamic conditions. The viability of any proposed drilling, production, and injection scheme can be tested *a priori* in the model without actually implementing it in the field. The term static model refers to a geological or geo-cellular model of the reservoir in its virgin state, that is, without any production or withdrawal from the reservoir or any change in reservoir conditions from the original. The geological model contains the lithology, facies, significant faults, and pinch-outs. It is populated and calibrated with a seismic structural interpretation based on seismic surveys, petrophysical data (porosity, permeability, and water saturation) acquired from the well-logs, and fluid properties. The dynamic model represents the behavior of a reservoir as a result of production with time.

Depending on the number of directions and dimensions (x, y, z) chosen to describe the reservoir, the models can be classified as 0-D, 1-D, 2-D, or 3-D. Water and oil are treated as two separate immiscible phases. Gas is the third phase, which is assumed to be soluble in oil but not in water. The model dimensions (1D, 2-D, 3-D) and fluid phases (oil, water, gas) present in the production stream add to a reservoir simulation's complexity and run time. For example, a 1-D model takes much less time to run than a 2-D model. Similarly, a 2-D model is more efficient than a 3-D model. Likewise, a single-phase model runs faster than the two-phase model, and a two-phase model is quicker than the three-phase model.

Reservoir simulation models aim at solving fluid flow problems both in time and space. The models can be steady-state (time-invariant) or transient (time-dependent). They can also be radial, cross-sectional, 2-D, and 3-D areal. A summary of the type of reservoir simulation models and their application is presented in Table 7.1.

Modelers' desire to make a perfect model is curbed by what they can do with the software and hardware. These limitations force them to build a simpler assumption-based model than the exact representation of the reservoir. However, in the last three decades, both software and hardware have made remarkable progress. Modelers use these advancements for constructing high-resolution models that can be run on superfast computers or a network of computers in parallel.

A deterministic or stochastic procedure may be adopted to manage the uncertainty in reservoir modeling. The deterministic model assumes with certainty a fixed value of a parameter, such as reservoir properties at a given location in space. The very concept that future behavior can be predicted based on a data set representing past behavior is deterministic by nature. For a given input, such models produce the same output. Probabilistic or stochastic models assume a random distribution of parameters. Randomness is described as the absence of any pattern in the past or the predictability of future events. In these models, the variables have ranges of values in probability distributions instead of a single value. Monte Carlo simulations use this concept for probabilistic estimation of oil-in-place where porosity, thickness, areal extent, and thickness can be set as probability distributions rather than a single value.

Deterministic simulation models, particularly with a good history match, may tend to "blind" the study teams towards other geological possibilities

TABLE 7.1

Type of Reservoir Simulation Models and Their Application[7]

Model Type	Application
1. Tank Model (Zero-D)	• These models have no spatial dependency, are only time-dependent. • Good for quick answers • Used when pressure gradients in the reservoir small • Needs minimum reservoir data
2. 1-D Model	• This model is a function of time and one dimension in space (X/Y/Z) • It is good for evaluating the performance of reservoirs that have variations in reservoir parameters • It is good for studying the effect of heterogeneity in the direction of flow • Not preferred for full-field studies as they cannot model areal and vertical sweep
3. 2-D Areal Model	• This model is a function of time and two dimensions in space (XY, YZ, XZ) • Most commonly used models • Generally used for entire reservoir studies • Used when areal flow patterns dominate reservoir performance • Determine the influence of areal heterogeneity on reservoir behavior • For selecting the optimal waterflood pattern • Used for production forecast for oil, water, and gas, the requirement of wells and surface facilities, production and injection allocation, the timing for installation of artificial lift, setting production targets, evaluation of development options, and so forth
4. 2-D Cross-Sectional and Radial Models	• Used to simulate peripheral water injection, crestal gas injection, gas and water coning, and related sensitivities • To develop well functions or pseudo functions for use in 2D areal or 3-D models • To study the effect of gravity and heterogeneity on displacement and sweep efficiency in a miscible process • Multi-layer reservoirs without any crossflow
5. 3-D Models	• Used when the reservoir geometry is too complex to be defined by 2-D models, for example, the reservoirs with extensive barriers causing the cross-flow through intermittent windows/breaches • Used when reservoirs experience cusping and coning, and so both areal and vertical details may be needed to represent the fluid distribution • To carry out sensitivity studies for reservoirs that have significant variations in thickness • Used for multi-layered reservoirs with inter-communication or crossflow between the layers either in the aquifer or in wells • Used for production forecast for oil, water, and gas, the requirement of wells and surface facilities, production and injection allocation, the timing for installation of artificial lift, setting production targets, and evaluation of development options, etc.

and realizations. Reconstructing a new static model to test a new geological realization and history matching it all over again is a considerable effort and may not necessarily give the expected results. Therefore, there is a tendency to reject alternative hypotheses even if some excellent leads are available from the data. The modeling teams update/customize these "fit for purpose" models to meet their specific requirements, such as sensitivity analysis, establishing incremental benefits, or estimating long-term recovery factors.

The practice of implementing the "most likely" realization in geo-models and its calibration with production data is the most time-consuming and laborious task. The task is further complicated by the uncertainty associated with many variables that are input to the model. Matching a long production history of a giant brownfield with hundreds of production and injection wells can be a daunting assignment. Hit and trial methods to achieve a good history match by making manual adjustments to various variables are inefficient. Assisted history matching (AHM) is a relatively new technique that considers many probabilistic realizations within the bounds of uncertainty and geological realism. The program is trained to automate simultaneous checks on production behavior such as oil rate, bottom-hole pressure, and water cut for each case. The realization with the most representative geology and history match can be taken forward for production forecasting.

A good reservoir simulation study must be based on clear objectives. It is always better to design simple models that can provide insights into the recovery process. However, reservoir heterogeneities and the mix of depletion processes in matured reservoirs force companies to build detailed 2-D or 3-D reservoir models. 2-D or 3-D models are quite common to evaluate reservoir performance, decide about infill drilling, or make production forecasts. Correctly history-matched good-quality 3-D full-field models can accommodate rock and fluid complexities and provide excellent guidance for reservoir development and management. Simulation results from these models can be viewed on workstations and straightaway exported to the spreadsheets.

However, these models can sometimes be unwieldy for the computing system used to run these models. There is no use in building an exact and complex reservoir model if the model cannot be run successfully in a reasonable time. Model run time is an essential factor when considering the simulation model design. Run time is the time taken by the computer/workstation to complete a simulation run. The following factors control it:

- Size and degree of complexity of the model
- The period for which the simulation run made
- Modeling capabilities of the simulator
- Size of the simulation time step

- Settings of the run controls
- Calculation algorithm and
- Capacity and configuration of the simulation hardware used to run the simulation

Run times longer than 4 to 5 hours are not ideal. Assuming that the companies have eight work hours in a day in most parts of the world, simulation engineers must make about two model runs a day to support the critical operational and investment decisions. Models taking longer runtime can be frustrating, and their results are likely to be ignored by the teams. Fortunately, some companies allow submission of model runs remotely, meaning that the engineers can submit the simulation jobs from their laptops from any location and not necessarily their office computer/workstations. This facility allows the model runs to be monitored even after office hours. In case the run stops, crashes, or is aborted due to errors or some other reason, it can be restarted or resubmitted remotely using the internet facility. The companies that do not allow this facility yet are working on overcoming their data security worries.

Reservoir simulation is a great tool to assess the impact of uncertainties of various parameters on reservoir performance. It is typically done by carrying out the sensitivity analysis where the value of a single input parameter is changed between its lowest and the highest bounds. For example, the impact on OOIP due to uncertainty in S_w can be evaluated by altering the value of S_w between its maximum and minimum recorded values in the reservoir. If this alteration causes a significant impact on the OOIP and history match, then clearly model is sensitive to S_w. It also implies a need to collect more data on S_w to narrow its band of uncertainty. In case there is no or insignificant effect, the uncertainty in the value of S_w is not a significant issue, and there is no need to spend too much money on collecting further data on S_w.

It is a good practice to use a Tornado Chart to understand the effect of constituent parameters on the calculation of OOIP (Figure 7.3). Some key points that will be obvious from this example are:

- Oil FVF (B_o) has the highest and average water saturation (S_w) the lowest impact on the OOIP
- The OOIP calculations are also sensitive to the area (A), porosity (Ø), and effective pay thickness (h)
- The OOIP, in this case, is more sensitive to an increase in areal extent (A) and porosity (Ø) than the decrease.
- Any possible reduction in average water saturation (S_w) has a higher impact on the value of OOIP than any increase in its value.
- Because of the above, it is essential to reduce uncertainty in the measurements of B_o, A, Ø, h, and S_w in that order.

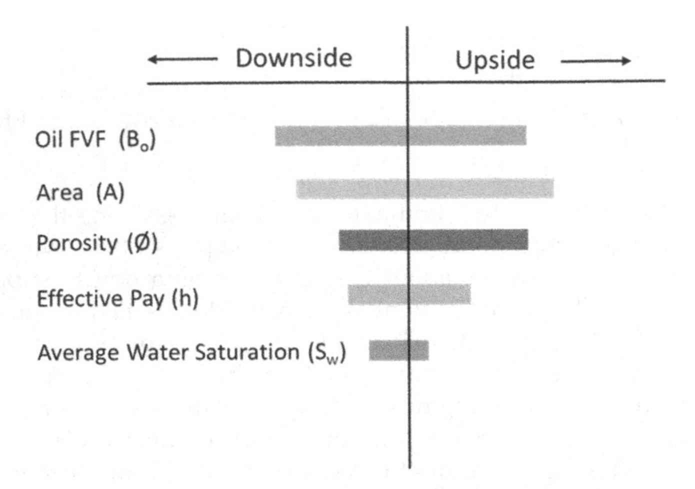

FIGURE 7.3
An Example of Tornado Chart Showing Effect of Uncertainty in Parameters on Oil in Place.

Many oil companies do not have strong in-house simulation capabilities to build their models. But their teams have enough expertise to manage and run those models. Some companies who can construct their simulation models in-house also engage their geo-modelers deeply involved in the company's operational activities to support drilling and completion operations. For companies that drill many wells every year, shouldering the dual responsibility of monitoring drilling operations and managing models can be very demanding and affect the simulation quality. A better option is to isolate reservoir simulation teams from daily operational activities.

Some consulting companies tend to rubbish the findings of simulation studies. Many individuals do not trust reservoir simulation, and their only objective is to attack the model's weaknesses rather than capitalize on its strength. However, one must remember that all classical reservoir engineering methods have their limitations too. And they may not provide as much insight into reservoir behavior as a good simulation model can.

Modern pre-processing tools help in assembling and processing various data (core, PVT, and well data, etc.) to quickly construct the simulation models. Similarly, post-processing applications coupled with simulators are equally powerful. With these tools, one can dissect the model to gain insight into the reservoir anatomy and analyze its performance at any time step as required. These functionalities make the model construction and investigation processes efficient and transparent. That is why reservoir simulation is so popular not only for long-term forecasts but also as an investigative tool.

Placing too much emphasis on reservoir simulation and reliance on its results would not be correct either. Reservoir simulation is undoubtedly a powerful tool to find solutions to complex reservoir development and management problems. But it works best when the amount and quality of data fed into it are adequate. Like any study based on inadequate and poor data quality, reservoir simulation will also give wrong answers. In the final analysis, the "man behind the machine" controls the quality of input and output of the studies. It is particularly true in reservoir simulation because the pertinent calculations take place in the background.

One must be careful in using full-field models, particularly those of large-sized fields, to make decisions on individual wells in the field. The problem is that the static models in their original size are usually huge, with millions of grid cells. If used directly as dynamic models, they can choke most of the computer systems available today. Upscaling is an essential procedural step that allows averaging the static model's fine-grid properties for a relatively coarser grid dynamic model. This procedure is so implemented that the two models, viz, fine-grid static, and coarse grid dynamic models, are practically identical. Even in the static model, inter-well properties are generated based on geostatistical and other modeling algorithms that honor the specific well data yet may not be the exact representation of the geology. These approximations and inaccuracies, however small, can have a significant impact on individual well-performance but not on the overall reservoir performance.

Reservoir modeling is undoubtedly a powerful tool for integrating data with knowledge. But the experience indicates that large-size models have many constraints and limitations. Because of this, many consultants now favor small and agile "fit for purpose" models over large lifesize models. One or more small cross-section or sector models can be used to have quick insights into a particular phenomenon or the role of a geologic feature due to the simplicity of concept and hassle-free quick workflows, efficient model runs, and precise results. However, combining the production profiles obtained from such models is not very convenient and straightforward.

Historically the subsurface and surface segments have always been modeled separately. However, new machines, combined with powerful hardware and software, attract the modelers to build the more demanding integrated asset models (IAM) that challenge both human and machine intelligence and their capacity. IAM is a generic name for a single computer model combining the subsurface and the surface models. The subsurface segment includes fluid flow from the reservoir to the wellhead, and the surface segment replicates fluid flow through the flowline network and facilities. As usual, the final choice of selecting the final modeling option continues to rest with the user, and he/she alone will be responsible for managing the quality and results.

Modeling requires a software platform, a simulator, for constructing a simulation model. A variety of simulators are available today that deal with specific fluid types and recovery processes. Black oil simulators assume that oil and gas compositions do not undergo material change with the reservoir's depletion and are typically used to simulate oil and gas reservoirs. Many black oil simulators are available today that are patented and marketed by different vendors such as Eclipse E100, VIP, IMEX, Nexus, INTERSECT, tNavigator, Exodus, and many more. Oil companies usually buy or rent the licenses of the simulators they want to use. Other simulators can simulate much more complex oil recovery processes such as thermal, compositional, and miscible.

Experienced geoscience and engineering professionals recognize the strengths and weaknesses of both classical and simulation methods. Most of the time, they use a simulator available to them readily or the one they can use efficiently. The selection of one or more simulator(s) is part of the oil companies' modeling strategy. Large oil companies may have licensing options under which any simulator can be requisitioned for deployment depending on the user team's choice.

7.7 Realistic Production Forecasting

Realistic production forecasting is the crux of reservoir modeling or classical reservoir engineering methods. The company invests capital in oil and gas fields in the hope of earning a profit. However, it takes time, particularly in the early development phase, when the oilfields lack the production infrastructure. Production forecasting is a technique that, based on the field operating and management strategies, allows the transfer of underground oil and gas reserves into the estimates of daily oil, gas, and water rates. These estimates can then be translated into an expenditure and revenue table, which can be used for detailed project economics. This methodology permits the company to evaluate and rank viable operating strategies.

Several assumptions have to be made for generating long-term production forecasts based on a clear understanding of the following factors:

- Resource availability and its timing
- Knowledge of the reservoir and its performance behavior
- Organizational capability to successfully manage field operations

The forecasts also assume that there is no impact of any unforeseen incidents on oil and gas production/development activities.

Companies use these forecasts to mobilize resources (funds, material, equipment, and workforce) and develop sales agreements. These forecasts are also used to set the company's annual and 5-year production targets. In a nutshell, production forecasts are an integral part of business planning and form the basis for its short and long-term business activities. Therefore, it is a general practice to update these forecasts when necessitated by circumstances or any assumptions.

Production forecasting for greenfields (new) and brownfields (mature) varies in principle and quality. The biggest challenge in green fields is that there is very little performance-based data lacking the proper calibration of models/methods. Due to this, production forecasts can have numerous uncertainties and may even be erroneous. On the contrary, an excellent long production history match for mature fields boosts confidence about the production forecast's assumptions and correctness.

Today, engineers can choose from a wide variety of available forecasting tools, both on the analytical and modeling sides. The use of internally developed spreadsheets that include Buckley Leverett, Stiles, Dykstra Parson, Craig, DCA, and any other methods besides third-party applications such as MBAL and GAP can fast-track the process. It is not necessary to build a large and complex model. If one can represent the most dominant mechanism of fluid displacement in the reservoir with the help of a single well or a small sector containing a few wells, it may be good enough. This model can be scaled up appropriately for production forecasting for the part or entire reservoir. It is an excellent practice to cross-check the results using more than one tool to generate and compare production forecasts.

Most of the production forecasts are based on deterministic methods/ models, generally but not necessarily, close to the P50 production forecast. The term P50 forecast refers to the most likely forecast corresponding to a P50 figure of the EUR. Accordingly, the P10 production forecast corresponds to the P10 (high) figure of the EUR, and the P90 production forecast corresponds to the P90 (low) figure of the EUR. Sometimes production forecasts may be overly optimistic or pessimistic based on recent operational successes or failures. This tendency can make the production forecasts lopsided.

The production forecasts can be generated at the well level, gathering center level, reservoir level, field level, asset level, and multiple asset or company level. Aggregation of production forecasts at well and field levels provides a low estimate. Well-level production forecasts suffer from inherent conservatism. Also, it is difficult to calibrate well performance against expected RFs due to a variety of reasons, some of which are-

- Wells may be flowing to the GC through shared flowlines
- Production from stacked reservoirs in a well may be commingled and vary in fluid quality
- Actual well spacing may vary than planned, and fluid flux from well to well and pattern to pattern may differ from that envisaged

Aggregation of production forecasts for multiple stacked reservoirs at the field level can suffer from inaccuracies resulting from inhomogeneity in fluid types and production commingling. Wells producing through common flowlines may interfere, and the sum of production from individual wells may not add up to the field production.

Aggregation of production forecasts at GC and reservoir level is a more plausible option. Applying DCA at the GC level makes sense because that is where production from the wells/reservoirs/fields is gathered and most reliably measured. Aggregation of production forecasts at the reservoir level may sometimes include the performance of "mature" and "immature" areas or "prolific" and "less-productive" areas. Predictions may be skewed in favor of the area, which has a more considerable influence on the overall performance.

Production forecasts at multiple-field or asset level and company level are combined by arithmetic additions assuming no dependency between fields. This assumption may not necessarily be true.

7.8 Production below the Bubble Point

The crude oil may be saturated or undersaturated with associated gas when discovered. Fully gas-saturated oil holds all the gas it can at a given temperature. On the contrary, undersaturated oil has a lesser volume of dissolved gas than required to make it fully saturated. The reservoir pressure of an undersaturated oil reservoir is higher than the bubble point pressure. For producing the oil and gas, the system must traverse from original/initial reservoir pressure P_i to bubble point pressure P_b and below (Figure 7.4). Bubble point pressure is a function of temperature, gas solubility in oil (R_s), oil gravity, and gas gravity. Most recovery processes are isothermal or nearly isothermal. However, a drop in reservoir pressure caused by depletion can alter the fluid properties and composition, affecting the bubble point pressure value.

Bubble point pressure is an important landmark in the journey of production of an undersaturated oil reservoir. Above the bubble point pressure, the associated gas is wholly dissolved in oil, and the mixture of the two fluids is in a single phase. During depletion, oil is produced by the wells naturally if the reservoir pressure is higher than the hydrostatic head of the fluids column in the well. With continued production from the wells, reservoir pressure decreases.

If the reservoir does not have the pressure support of natural aquifer or water injection, production wells may require pumping to maintain the oil rate at an economic level. As long as the reservoir pressure is above the bubble point pressure, the producing and solution GORs will be equal and constant. Any further drop in reservoir pressure below the bubble point liberates associated solution gas out of the oil. With ongoing production,

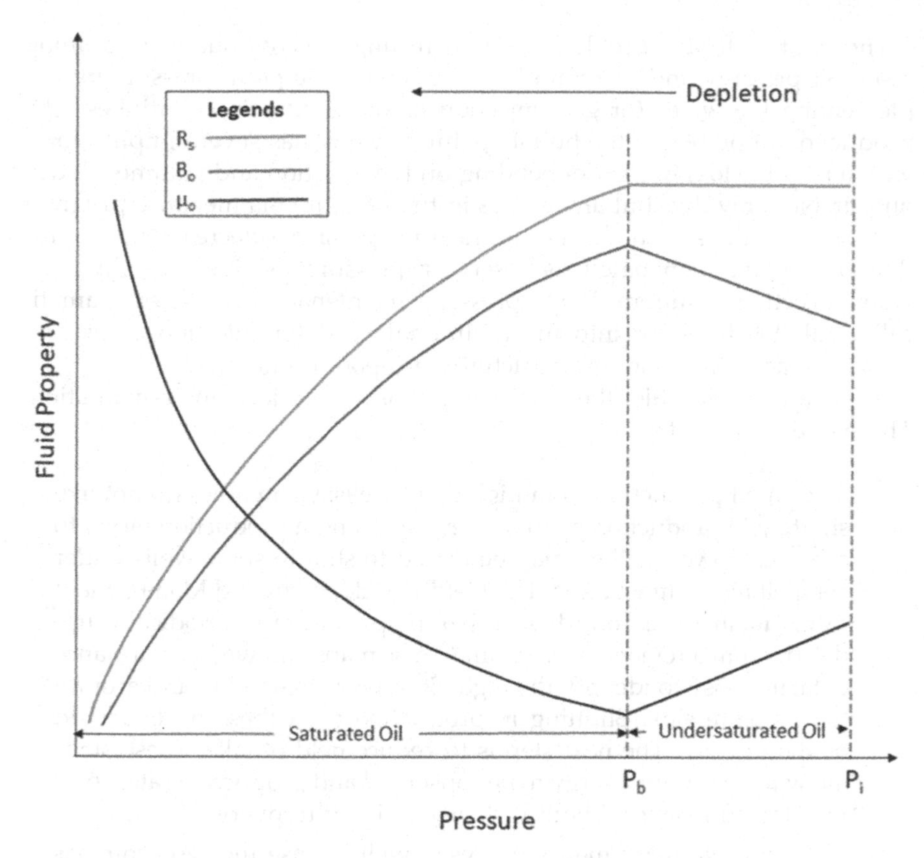

FIGURE 7.4
Depletion Characteristics of an Oil Reservoir.

reservoir pressure continues to drop, and oil releases solution gas that grows as a separate phase in the reservoir pore space. With the gas separating from oil in the reservoir, oil viscosity (μ_o) increases. It causes the effective oil permeability to decrease and the remaining oil volume to shrink, as indicated by B_o (Figure 7.4).

However, the evolved gas phase remains immobile until the reservoir's gas saturation exceeds the value of critical gas saturation. During this period, gas provides most of the drive energy for oil production. Once the gas saturation crosses the critical gas saturation (S_{gc}) mark, the gas having a much lower viscosity than oil flows to production wells in preference to oil. As a result, the oil rate starts to decline, and producing GOR increases rapidly. The result is that the excessive amount of gas that provides drive energy and is rich in valuable lighter hydrocarbons is produced at the cost of oil rich in heavier hydrocarbons is trapped in the porous network. The oil production rate is further reduced because there is a reduction in the oil's mobility after losing gas and increasing oil viscosity.

The solution to this problem lies in arresting the continuously declining reservoir pressure and maintain it above the bubble point pressure by implementing the water (or gas) injection as the production continues. The production of oil below the bubble point pressure has severe implications, with a possible loss in EUR depending on the duration and volume of such output. Not only this, but any delays in the pressure maintenance program also worsens the situation. The increase in gas and a decrease in oil production continue unabated as reservoir pressure declines sharply. The water injection requirement for pressure maintenance increases dramatically. All this translate into the additional need for injection wells and higher capacity injection infrastructure with poor economics.

In situations like this, the companies look for options for remediation. These options include:

- Shut-in oil production from high GOR wells: Companies do not like shutting in production because they have a tight production target to achieve. However, they may be forced to shut in some wells under compelling circumstances. The Field Development and Management teams maintain a record of isobar maps that are periodically updated. From a review of these time-lapse maps and well performance data, it is easy to identify the high GOR wells that fall in and around the pressure sink. Shutting in production from these wells should be the priority. The next step is to reduce field offtake. These steps allow the reservoir energy to be conserved and production rates to be maintained longer without further loss in oil recovery.
- The above strategy may work very well in case the reservoir has poor connectivity or is compartmentalized. But if the reservoir has good connectivity, shutting in production from the high GOR area is not a long-lasting solution. Continued oil production from the normal GOR area will soon result in the expansion of pressure sinks manifested by an increase in GOR of wells, which had normal GOR before.
- The above two options are only time-gaining tactics and not the real solution to the problem. Besides, these options result in the loss of production due to wells' temporary shut-in or reduced field-offtake. Accelerating the implementation of a water injection (pressure maintenance) program is the best option to arrest the decline in reservoir pressure. If the commissioning of a water injection project will take time and cannot be accelerated, the possibility of implementing a dump flood or other innovative ideas must be examined.

The above remediation efforts take time during which the company may have to revise its production targets downwards.

There is at least one published evidence where the operator was permitted "on a case by case basis, to produce below the bubble point locally to the producer well in order to accelerate production or, where pressure support is not economically viable to produce below the bubble point more extensively."[8] But there are many unpublished cases where many oil wells in the fields are produced below the bubble point because the operator cannot shut in wells in the face of stiff production targets.

The majority of hydrocarbon-bearing countries have an undisputed right to the ownership of their natural resources unless negotiated otherwise under some commercial arrangements. Production below bubble point pressure is a subject where the "in principle" position of the government or NOC is quite clear. Any oil production below the bubble point is not acceptable as the NOC is committed to maximizing the oil recovery by implementing the best industry practices that do not cause reservoir damage to national wealth. However, much indiscretion is exercised by oil companies about oil production below the bubble point in practice. Such situations usually arise from delays in pressure maintenance projects due to subsurface uncertainties, lack of clarity about project definition, and midterm re-scoping projects. In such cases, the companies' usual defense is that of "deferred production" or the production that can be realized later, and not a "permanent loss" in oil recovery. Such arguments only dilute the gravity of the matter. The fact is that the production target was missed, oil recovery was compromised, and the precious reservoir energy was wasted in liberated gas.

However, a gas phase in water drive reservoirs can increase oil recovery, depending on the oil quality, by 7% to 12%.[9] Reservoir pressure in the water drive reservoirs can be allowed to drop below the bubble point. The formation of the separate gas phase in the reservoir provides extra drive energy, reduces residual oil saturation, and improves displacement efficiency. Laboratory core flood tests and sensitivity studies with a simulation model can help determine the optimum operating reservoir pressure below the bubble point to gain the recovery advantage. While the practice of microscopic level studies on core samples is technically sound and indicative of benefits at the reservoir level, their direct application for quantitative estimation of performance and recovery has limitations.

7.9 Venting or Flaring of the Produced Associated Gas

Increased production of unplanned associated gas below bubble point can cause various operator problems because natural gas is a viable energy source. Natural gas is a conventional heat source for manufacturing cement, bricks, ceramics, tile, glass, steel, paper, and many food products. It is also used at many industrial facilities for incineration, supplied to households for cooking and heating the homes in cold countries. In the absence of a ready market or pipeline infrastructure, the operator is forced to dispose of this gas by venting or flaring to avoid fire or explosion.

Venting is more harmful than flaring due to the reported global warming potential of methane being 28-36 times higher than that of carbon dioxide.[10] Venting involves the release of associated produced gas, which is rich in methane, into the atmosphere. Gas flaring systems, on the other hand, allow complete combustion of hydrocarbon gases in the presence of air to produce water vapor (H_2O) and carbon dioxide (CO_2). Incomplete combustion can generate carbon mono oxide, total hydrocarbons, particulate matter, black carbon, and volatile organic compounds. Emission of these harmful substances and greenhouse gases by venting or flaring poses serious air pollution risks that negatively impact our environment and ecosystem. They must be addressed by a comprehensive development strategy that includes the plans to utilize the produced gas. Most oil companies recognize the importance of the World Bank initiative of zero gas flaring, and some have it included in their KPIs as one of the performance measures.

Natural gas can also be stored in underground depleted oil reservoirs, aquifers, and salt caverns. The storage plays an important role in assuring the gas supply when needed, mainly because its demand may fluctuate in cold countries depending on weather conditions. However, this option requires suitable reservoirs for gas storage, upfront investments, time for drilling wells and building gas handling facilities, and finally, the market of interested gas consumers in the future.

7.10 Disposal of Produced Water

As discussed in Chapters 5 and 6, the produced water can be reused, reinjected, or released after treatment because it has many impurities that can harm the environment. In the olden days, many countries discharged large volumes of produced water into the deserts and open areas without treatment due to a lack of awareness on the part of operators and governments.

Disposal of significant volumes of produced water without treatment into shallow reservoirs is a common practice. Many disposal wells are openhole and have primitive completions with minimum safeguards to save cost. These wells are repeatedly stimulated to maintain the injectivity and not monitored to check for an increase in reservoir pressure. Such practices are not healthy and need to be rectified as soon as possible.

7.11 Slippage in Pressure Maintenance (Water Injection) Program

Pressure maintenance programs are critical to improving oil recovery. If implemented in time and correctly, these programs have the potential to double the oil recovery. Water injection is the most common pressure maintenance method for reservoirs produced under depletion drive or weak aquifer support.

Pressure support by an aquifer to the reservoir is evaluated with the fluid withdrawals (voidage rate) from the reservoir as follows:

Rate of Water Influx = Total Reservoir Withdrawal or Voidage Rate

Rate of Water Influx = Rate of Oil Production + Rate of Gas Production + Rate of Water Production

$$W_e = Q_o B_o + Q_g B_g + Q_w B_w \tag{7.1}$$

The left-hand side in the above equation represents the water influx rate, whereas the right-hand side is made up of the oil, gas, and water withdrawal or voidage rate.

If the rate of water influx ≥ the voidage rate, the reservoir is operating under an active water drive

The rate of water influx < the voidage rate, the reservoir is operating under partial water drive

As long as the reservoir performs under an active water drive, there is no need for any pressure maintenance program. Because the natural aquifer is strong enough to provide pressure support, the reservoir pressure experiences no or low-pressure decline. Oil recovery from such reservoirs is high. The most common problem these reservoirs face during their lifecycle is that of water production.

For reservoirs operating under partial water drive, the voidage rate is greater than the water influx rate. As a result, the reservoir pressure declines faster. Such reservoirs may need pressure support depending on the extent to which the aquifer can compensate for the loss in reservoir energy caused by reservoir voidage. Pressure maintenance by water injection can play an important role to improve reservoir performance and oil recovery.

In case there is no water influx into the reservoir (due to the absence of aquifer or aquifer permeability being too low), and there is no free gas cap, the reservoir will operate under depletion drive. Such reservoirs will have poor oil recovery without pressure maintenance programs.

Excessive withdrawals from a reservoir can turn an active infinite aquifer into a partial aquifer. Conversely, little withdrawal from an oil reservoir can make a partial aquifer respond as an active aquifer. Therefore, extensive studies are essential to optimize the reservoir withdrawals commensurate with aquifer strength and capacity.

7.11.1 Four Fundamentals to Watch

All oil companies recognize the importance of pressure maintenance by water or gas injection operations. Our discussion here will be about water injection, which is more common and prevalent globally. The most important questions that relate to the performance of water injection are as follows:

- Timing of water injection (when?)
- Water quantity (How much?)
- Water quality (What?)
- Location of injection (where?)

All oil companies in the E&P business know the importance of each of the above questions. Yet, there are slipups when it comes to the execution in some or all of these aspects. Waterfloods often fail to deliver expected production rates and sweep efficiency due to inadequate understanding of the reservoir and its behavior and the slippages in four key factors listed above.

Delays in water injection projects are not uncommon and can push the water injection requirement and project costs upward. Lead time for a full-fledged water injection project from pilot testing to commissioning is about 7 years. Any mid-stream changes in the project scope can affect the project completion date and cost significantly. Inordinate delays can result in excessive loss of solution gas, oil recovery, and project economics. Given these problems, the companies that can afford to create injection facilities upfront can consider initiating water injection since the inception. However, injecting water early in an oilfield's life without proper assessment of the aquifer has its fallouts, if there is one. Injection facility design in terms of capacity and injection pressure may not be adequate. Nevertheless, in an offshore environment, the decision to start a waterflood must be made in the early stage of the field development planning.

The quality of injection water is an equally important issue. Monitoring and maintaining a consistent quality of water according to the specifications is critical but not easy. Injection of off-spec water for long durations, particularly during injection plant upsets, can clog the reservoir pores and damage reservoir permeability. It is essential to keep a constant watch on TDS and suspended solids that plug the reservoir pores. Formation of inorganic scales, tricky emulsions, and deposition of asphaltenes in the tubing and reservoir can adversely affect production and injection significantly. The presence of iron sulfide and iron oxide in injection water can induce the formation of H_2S and make matters worse. A proactive approach for maintaining a good injection water quality is essential to avoid the negative impacts of poor water quality, which can damage the reservoir. The reversal of this damage to the well/reservoir is sometimes not possible by acidizing, fracturing, re-completing, sidetracking, and new well-drilling programs. Similarly, repeated pigging and surface piping replacement may not fully solve the surface piping network's problems.

The location and pattern of injection wells have a bearing on the sweep efficiency. The idea of a peripheral flood is to inject water at the periphery of the oilfield. This method is used to displace oil towards the crest of the structure. It works well in reservoirs, which have good areal and vertical connectivity. Large reservoirs with tight flanks where pressure sinks are far away from the injectors are not suitable for peripheral flooding. Years of water injection in flanks may not address the pressure sinks at the crest due to poor voidage replacement and poor areal and vertical conformance.

The pattern waterflood consists of injecting water in patterns all over the reservoirs. Areas with pressure sinks can be attended on priority to bring high productivity wells back online. Pressure sinks indicate good productivity wells in the area, and their early restoration to original productivity helps the company's production objectives. However, pattern waterflood may require frequent workovers for water shutoffs as the distance between injectors and producers is considerably reduced, and one producer may be connected with several injectors.

The presence of thief zones adds further complexity to waterflood operations and management. Thief zones are high permeability streaks within the reservoir that act contrary to the objective of waterfloods. They steal water from injection wells and convey directly to production wells without displacing much oil. Interwell tracers play a vital role in the characterization of thief zones. Waterflood projects require extensive monitoring with PLTs and production testing. PLTs help to map the distribution of injected water into different reservoir layers connected with producers. It is direct evidence and measure of vertical conformance.

Production tests are essential tools to make the first pass diagnosis of waterflood performance. Waterfloods in both sandstone and carbonate reservoirs require each production well to be tested at least once a week, particularly after water breakthrough. It is quite a challenging task if the fields are large with hundreds of wells and if several oilfields are operating under waterflood simultaneously. Despite the big business, it is quite difficult to organize so many portable test units to test such a large number of wells every week. Therefore, though not ideal, the most common approach to manage this workload is to test the wells every month or sometimes, even a quarter. This frequency of production testing is not enough to diagnose well, problems. The volume of data generated by these tests is enormous. It must be processed, validated, and stored in the corporate database in the right format.

Other routine measurements of surface and bottom-hole pressures, saturation logs, PVT sampling, routine and special core analysis, and open-hole logs in new wells can provide useful insights into waterflood performance.

It is a human tendency to go after the direct and assured benefits rather than the indirect and uncertain ones. Accordingly, drilling, workover, and hook up of oil wells usually receive a higher priority over water injection wells. The significance and contribution of water injection in the form of increased production from the reservoir are not obvious to all compared to the visible oil barrels delivered by a newly drilled/remediated oil well. Operations teams are driven by the same mindset when they seem over-concerned about the oil wells and production facilities than water injection wells and injection facilities. They do not fully appreciate that they stand a better chance to achieve higher oil production by paying more attention to the water injection system.

The teams have to deal with some difficult questions regarding choosing between water injection quality and quantity. It is always preferred to have long-term matrix water injection in sandstone reservoirs. Sometimes, it may not be possible to achieve matching injectivity despite high well-productivity. This problem could often be solved with high-quality water. In other sandstone reservoirs, sustained injection of cold off-spec water can cause near-wellbore fractures due to thermal stresses and pore networks' plugging. The development of fractures can cause conformance issues as the injected water may not necessarily go into the targeted zones. However, with proper surveillance tools, the companies can manage a reasonable control over water movement.

7.12 Data Management

Data is a single or set of values collected from observations made from an experiment, investigation, test, analysis, event, or process. Data consist of facts in raw form. When processed intelligently, it transforms into information. Data by itself is a scalar quantity because it has only the magnitude. Speed, mass, and volume are some common examples of scalar data.

On the other hand, information is a collection of data or facts that is more valuable than its constituents. Information is a vector quantity because it has both- a direction and magnitude. Velocity, weight, and force are typical examples of vector quantities.

Data is an essential input to start a business and drive its growth. It helps to improve or diversify the business. It is a key factor in decision-making, and nowhere, its impact is felt more deeply than in the petroleum industry.

Everything in the petroleum industry, particularly the E&P sector, is data driven. The problems and scope of data management in the petroleum industry can be gauged by the cliché "data is always finite and contain measurement errors." Crucial decisions about the development and management of oil and gas operations are made based on the daily oilfield data.

These data come at a high cost and demands technology, time, and human effort before it can be used for decision making.

Data can be unstructured or structured. Humans can read and interpret unstructured data such as an image or the meaning of a block of text. On the other hand, computers are capable of reading massive amounts of structured data in a short time. Large volumes and cost characterize data in the E&P sector. These data can be raw as well as interpreted. It can be organized in separate databases categorized by disciplines such as drilling, operations, reserves, analogs, corporate, finance, and so forth. These databases contain a variety of raw data, for example,

a. Seismic data
b. Geological data
c. Well logs
d. RCA and SCAL data
e. Pressure measurements
f. Well drilling and completion data
g. Production tests, fluid sampling
h. PVT analysis data
i. Daily production and injection data
j. Flowline and facilities data
k. Material, equipment, and cost data
l. Sales/export records
m. Expenditure and revenue data
n. Employee details and a host of other types of data

Interpreted seismic data can include seismic conversion of time-to-depth maps and seismic attribute maps. The other types of interpreted data would comprise log-analysis reports, formation tops, structure and isopach maps, and cross-sections. A comparison of k_h and k_v from conventional cores and well-tests, well completion details, bottom-hole pressures corrected to the datum, estimates of oil and gas in place, reserves, and so forth are all examples of interpreted data. Data may also incorporate G&G models, reservoir simulation models, spreadsheets, documents, standard operating procedures and protocols, valuable minutes of meetings, and even PowerPoint slides.

Information technology has revolutionized the way data is now acquired, processed, and managed. This cultural change is not complete yet because the people managing the change cannot fully cope with computers' overwhelming capabilities and users' expectations.

The truth is that the data quality and systems used to enter and maintain the data are as perfect or imperfect as the people who manage these systems.

7.12.1 Data Quality

Data quality is defined by its accuracy, validity, consistency, verifiability, and completeness. Large oil companies have data management teams and processes to manage data in collaboration with functional units. Technical teams have the responsibility to capture and perform a quality check on the data. The importance of high-quality data that can be used for decision-making straightaway cannot be overemphasized. Extracting data from databases and checking/interpreting it every time an investigation or decision needs to be made could be quite frustrating. Simulation studies are a good example whereby 50% of the time is spent collecting, correcting, and organizing the raw data before actual modeling activity can take off.

Data quality is a crucial success factor in the data-driven business of oil and gas. High-quality data can reduce costs, increase efficiency and revenue, and deliver results that can create a brand image. The most common data quality issues are as follows.

7.12.2 Excessive Data

The volume of data generated by oil companies is enormous. The companies that clamored for more data in the past find it difficult to manage it today. New tools and technologies such as real-time measurements of pressures or measurements while drilling (MWD) can generate enormous data on pressures, drilling rates, and other related parameters compared to old practices. Also, the redundant data continue to stay in the databases forever because segregating these data from the usable data is seldom a priority. It can sometimes make the data search painfully slow.

7.12.3 Duplicate Data

Multiple entries of the same records, either due to human error or incorrect logarithm, negatively affect the computation and storage time or produce the lopsided view. Inadvertent use of spaces, hyphens, improper use of upper or lower case can result in duplicate entries, inaccurate reporting, and reduced user acceptance.

7.12.4 Poor Data Organization

The value of data lies in its quick recovery. If the users cannot quickly extract the correct data from the company database, they will make little use of it and create their databases, which is not desirable. On the contrary, good data organization will promote more extensive use of the company database and shared data-driven decision-making.

7.12.5 Inconsistent Data

Data may be inconsistent in values because it comes from different sources. Pressure measurements from RFT and SBHP measurements in two neighboring wells may be off by several hundred or a thousand psi without any plausible explanation. Data may also be entered or reported in an inconsistent format, for example, dates in dd-mm-yyyy and mm-dd-yyyy, depths in meters and feet, and production rates in barrels/day and m^3/day. These seemingly minor errors can be the root cause of poor decision-making, costing millions of dollars.

7.12.6 Incorrect or Incomplete Data and Misspellings

Data may also be poorly defined or sectioned into the wrong category. For example, there may not be any depth or date references to bottom hole pressure measurements. Or else, the month of February might carry 30 days! It is common to see inconsistent, misspelled, or incomplete reservoir or well names, which can suppress many data during search operations. Errors of this type may sometimes render the most impactful data useless.

7.12.7 Data Accessibility

In the E&P business, the data are as necessary as the resource itself, if not more. It may be bought and sold for cash or favors. Data wield power. The people on top of a readily available, updated, and comprehensive data set of their fields or reservoirs are always in demand. Some smart geologists or reservoir engineers maintained their clean dataset in their notebooks in the old days without sharing it with their teams. They used this information uniqueness to build their own "data kingdoms" and promote self-importance over high-quality team-based decision-making. It severely affected team performance because the team was deprived of the benefit of other members' collective cross-functional wisdom and effectively relied on the limited experience and knowledge of a single individual.

Computers and shared databases demolished this culture and promoted team-based decision-making. Data is a crucial ingredient for a company's core business activities such as drilling and production of oil and gas; or commercial activities such as tenders, contracts, mergers, acquisitions, and so forth. Companies spend a fortune to collect data and spend millions on protecting it. Therefore, they have to ensure that data is accessible to the right people with the correct privileges. Only designated people who can extract value from data or add value should be authorized to view data relevant to their functional role. Edits are usually allowed by the SMEs (Subject Matter Experts) in the quality control or validation phases. After the data goes into the final database, only the system administrator is authorized to change if required after the due process. Rights to view sensitive data are granted to a limited number of professionals according to their roles and responsibilities.

Some of the geoscience and reservoir engineering probabilistic versions may be valid for a long time over the field life, while others may be obsolete early. All versions of the data need to be kept in the databases for future use, record, and reference.

7.12.8 Data Security

Oil companies collect vast amounts of data, but not everyone in the organization needs access to everything. Identity governance ensures that both management and employees have access to the relevant data to perform their role effectively and nothing more.

The oil companies have to be very careful against cyber-attacks from professional hackers who can steal or corrupt the company databases with critical information. A company that is not able to protect its data loses the faith and trust of the people.

7.12.9 Poor Data Recovery

Users typically spend 30% to 40% of their time looking for the data they need for their work. To make matters worse, the less proficient users cannot easily find the exact information they want at the time they need. It can push the users away from using the databases because they are not confident if their time is well spent when they need to extract data from the database.

7.12.10 Data Purging and Archival

Oil and gas company databases are usually large, depending on the size and history of their assets. These databases have unstoppable growth as the enterprise's business activities pour in additional gigabytes of information every passing day. Large databases suffer from performance issues. The software database size increases with every transaction. They can be slow, show incorrect records, and are corrupted or crash frequently. It is an excellent practice to keep it trim and compact.

Data purging removes obsolete or superfluous data from the system to avoid unnecessary crowding of the active database. The purged data is not permanently deleted from the system; it is backed up and archived in a separate storage device to be recalled later if required. At this point, let us differentiate between the backup data and archiving. Database backup is usually a periodic short-term measure mandated often by both the organization and the government. It ensures that the operational database is always functional, and critical/essential business data is protected from accidental system failures, outages, or crashes. Database backup is performed with a database management software that creates duplicate or multiple copies of the same data stored locally or on a backup server.

The archiving aims to move inactive or unused data in the primary database to a storage device for recovery when required. It is like placing your unused furniture in the basement. Data archiving reduces the disk space in the primary storage systems and improves data management and system performance efficiency.

7.13 System Downtime, Outages, and Failures

Information technology is the main driving force of modern business organizations. Enterprises have enormously benefitted from the speed, accuracy, and consistency that the technology has brought to their business. They can now afford the luxury to operate from anywhere, anytime.

With all these benefits realized by the enterprises and their customers, IT has established itself as a critical business enabler and raised somewhat unrealistic expectations regarding speed and perfection. Organizations are spending huge sums on upgrading their IT infrastructure to minimize system downtime, failures, and crashes. Despite these investments and a trial of various advanced solutions, downtimes and outages are still a valid and real threat to the industry. IT failures can sometimes last for a few hours to a few days and have the potential to bring the business activities of an enterprise to a grinding halt.

Despite the critical role of IT teams in oil companies, they are seldom in the organization's mainstream. The charge of IT functions is administered by a Director whose primary role is to govern its core business activity. It effectively means that there is always a lag between the mainstream and IT functions of the company. Users expect their company to be at par with the IT industry elsewhere, but clearly, the company's priority is the oilfield operations.

Software-hardware compatibility is an essential aspect of the overall system performance. It is critical to ascertain if a software application is hardware-compatible and proficient enough to run in different browsers, databases, operating systems, and networks. The compatibility of different versions, internet speed, and configurations with various browsers and operating systems are essential for achieving high system efficiency. Installing today's software on yesterday's hardware may not be very efficient and productive.

7.14 Database Management

E&P operations and activities use and generate a lot of multi-disciplinary data that should be available to a variety of end-users simultaneously for the transaction of business. These companies require the support of a large pool of staff, which must QC and validate various datasets.

Extensive inventories of data cannot be efficiently managed without the help of computers and databases. The database is a structured set of data stored safely and securely in a centralized computer system. End-users can access this database to retrieve the selected data when required to perform their work. The quality of data and databases can have a significant impact on the organizational culture and efficiency today. E&P companies use and generate much data about planning, operations, customers, wells, facilities, contractors, sales, cost, staff, and so forth. All these data go into a central database with typically three corporate, project, and personal layers.

The corporate database is the authorized database of the company. It stores only the relevant and authenticated data of high quality, which can be viewed, downloaded, or exported to other authorized staff members' applications. The database is managed by one or more administrators who can populate it with new data after following the due process.

The project database is project-specific. It contains data relevant to a particular project, which usually has its budget and schedule. The regular inflow of new data characterizes this database as the project progresses through time and multiple versions of the data due to several interpretations or uncertainties. All the team members who participate in the project have "read" or "view" access to the database. "Write" access is limited to a few key members who are authorized to make changes to the data. The project database can sometimes be vast due to the project size involving hundreds of wells, many reservoir sublayers, and extended project duration. Project data is moved to the corporate database after completing the project or at stages when the data is final and no longer subject to change.

Team or operational databases are widespread in the upstream sector and contain data relevant to a single application/software. G&G, petroleum, and reservoir engineering professionals may generally use their discipline-specific applications such as IRAP RMS suite, Petrel E&P Platform, and so forth, for geological modeling, OFM and Surfer, and so forth, for production data management; and Eclipse, VIP, Nexus, and so forth, for reservoir simulation. These applications provide the flexibility to store, QC, analyze, and view multiple versions of individual or shared interpretation of the data. Such databases can be shared by a limited number of inline professionals working in the same team or department. These databases are highly "volatile" because they can be added, removed, or edited as required. The information is not easily shared with other applications due to associated uncertainties and incompatibilities between vendors' applications. The team/operational database data may be migrated to the project database after the working team takes a final view of the dataset.

Personal databases can be in the form of text files, reports, or excel spreadsheets and are maintained on personal computers to fulfill individual workers' requirements. The information contained in these databases is user and subjectspecific. The private databases may not be sophisticated enough to communicate with the corporate or operational databases correctly, yet they are custom-made to deliver the results/output.

7.15 Integrated Planning

Planning is a process to prepare the road map for a safe and successful journey to achieve the desired goal. Plans may fail because something happened which was not thought of. It only indicates that either the idea was not right or the preparation was not good enough to succeed.

An excellent integrated plan has clear objectives and uses the input of all stakeholders and affected teams. It is simple, time-bound, participative, flexible, balanced, and comprehensive. It is based on proper analysis of data and examines all economic, social, and environmental costs and benefits. The integrated plan recommends suitable options to deal with vital organizational problems of rising costs of development and production, optimal utilization of available resources, improved system efficiency, and operational excellence. Planning teams have a crucial role to play in this critical mission.

Oil companies in the upstream sector recognize the importance of planning discipline. It is just that it is not as effective or as integrated into some companies as it should be. The perceptions of the company executives and the asset teams are usually at variance about the effectiveness of planning teams in the organization. Company executives may think highly of them and their effectiveness, but the asset teams may not necessarily share this view. To be effective, the planning team must not only be clear about what it is supposed to do and deliver, but it must also know the capability and constraints of the groups that participate in the plan. It helps them to make plans that are based on the ground situation. Organizational structure can play an important role in enhancing the effectiveness of the planning team in the asset.

In some E&P companies, planning is seen as a dormant group that wakes up at the time of target setting or performance reviews. In other companies, it works as the corporate messenger who must complete an impossible plan before the end of the day. Some companies have such an elaborate and bureaucratic planning process that can keep the entire group and asset teams busy filling data sheets leaving little room for creative or productive work.

In some other companies, the planning team is not aware of the current situation of projects and activities, and therefore they pass on their workload to the asset teams.

Planning units of large E&P companies may also suffer from the lack of integration and open communication between the management and staff members, resulting in a loss of trust and transparency. This loss can result in frequent plan changes, which may upset the schedules and the material requirements across the asset. The organizations have a better chance to succeed with a stable plan than to rush with a quick half-cooked plan. Changing plans mid-stream is also risky as it can dilute the personal accountability and commitment of officials responsible for delivery.

The planning team is the nodal agency for annual, 5-year, or longer term plans that receive varied inputs from different groups across the organization. These groups may sometimes be using different software, vendor databases, or spreadsheets that make the planning team go through several data organization and verification steps before finally getting to an integrated plan.

Probabilistic information can improve the quality of decision-making if adequately integrated with the plans. Unfortunately, there is an acute shortage of good planners who have adequate experience preparing short- and long-term plans under uncertainties. Expert planners know their reservoir potential, organization capabilities, operating environment, costs, and schedules very well. They will not let any sub-standard inputs go into the plan.

Integrated asset planning has the potential to eliminate most of the ills of planning discussed earlier. It results from seamless coordination across functional teams committed to delivering their services/products as per the agreed quality, cost, and schedule provided in the FDP. If executed correctly, it can save many downtimes, inefficiencies, and redundancies in material and equipment.

Integrated planning is based on a precise FDP which aims to maximize the economic oil recovery. Good organizations do not compromise with HSE activities. Integrated planning ensures that these activities have sufficient funds, and the extra time and budget are also earmarked to take care of any contingencies. It is a good practice to announce the plan's freeze date, which may be 30 to 45 days before the execution starts.

7.16 Field Development Planning

Field Development Planning is a process that combines inputs of multidisciplinary teams to prepare and optimize a practical high-value development plan that, on implementation, generates value, including profits.

The process is characterized by Front End Loading (FEL), a commonly used methodology in project management. FEL permits the optimum allocation of capital and human resource, reduces critical uncertainties, and presents a holistic view of the development plan, including project economics and risk assessment. The term FEL derives its name from intensive planning in the early phase of the project or the front end. At this stage, it is possible to make changes in project design at a relatively low cost. In general, all E&P companies follow a rigorous process of planning before approving a capital subsurface or surface project. This process, though, may have a different name, such as Pre-project planning (PPP), Front-End Engineering (FEE), or Front-End Engineering Design (FEED). The entire FEL methodology can be split into three stages-

FEL-1: Opportunity identification or pre-feasibility assessment
FEL-2: Concept selection or feasibility assessment
FEL-3: Project definition or preliminary engineering and design

The main task in stage FEL-1 is the opportunity identification and assessment phase, where a company's business objectives are evaluated against prevailing market opportunities. A "first pass" estimate of production potential, investment, and commerciality based on a few discovery and appraisal well data are good enough for an experienced team to decide whether and how to go about the project. An essential component of this pre-feasibility assessment is the unbiased statement of uncertainties and risks and their mitigation plan.

FEL-2 is the feasibility assessment of the "doability." This phase incorporates the field development strategy and concept selection for its development. Practical concepts/options are shortlisted and compared based on estimated recovery, conceptual engineering, and costing forecasts. Preliminary techno-economic evaluation and comparison of shortlisted options form the basis to decide the most likely alternative for development. This phase highlights and ranks the key uncertainties and quantifies the associated technical and financial risks. HSE is a significant aspect that the FDP must consider and address.

In the FEL-3 phase, the focus is on delivering a development plan presented to the management for approval and future implementation. The development plan usually consists of the following major topics:

1. Executive summary
2. Data availability, quality, limitations, and acquisition plans
3. Summary of geophysical, geological, and reservoir engineering work
    - Details of 2-D and 3-D seismic surveys, quality of seismic data processing, structure, fluids, inter-well heterogeneity

- Geological Setting, depositional environment, stratigraphy, faults, bed thickness, well correlation, petrophysical analysis, porosity, permeability and water saturation, fluid saturation-height model, fluid contacts, uncertainties in data or interpretation
- Structure, type, and nature of faults, lateral and vertical reservoir connectivity, compartmentalization, initial reservoir pressure, reservoir rock, and fluid quality, PVT properties of reservoir fluids, SCAL data, $k_{ro}/k_{rw}/k_{rg}$ curves
- Drilling history and complications, drilling & completion practices
- Field status with type and number of wells, well status, cumulative oil, gas, and water production, and the latest well rates
- Volumetric estimates of the original oil and gas in place, P10-P50-P90 range of the estimates
- Reservoir Performance history, operating drive mechanism, well productivity
- The method used for performance forecast (classical RE method or reservoir simulation), performance history match, uncertainties
- Discussion of development Strategy and options, oil and gas recovery schemes under natural depletion and assumed pressure maintenance schemes by water or gas injection, and so forth, the scope for infill development

4. Oil, gas, and water processing facilities

- Discussion of production handling facilities, location, capacity, performance
- Need for facility upgrade or addition, timing, and cost
- Discussion on design criteria and operating conditions
- Quality of construction of the oil and gas facilities, quality or material defects, compromised safety requirements, weather conditions

5. Technology needs in case any new technology, equipment, material, or expertise required

6. Oil, water, and gas production forecast for various development options, CAPEX, OPEX, techno-economic evaluation of forecast cases- Profit, NPV, IRR, Payout Time

7. FDP scope of work & schedule, main uncertainties in the FDP
8. Uncertainties and risks, Mitigation plan
9. HSE issues, if any
10. Human resources requirements
11. Recommendations and Conclusions

Execution is the next stage after FEL-3, which includes the detailed engineering and design of wells drilled and their completion. Approximate location and capacities of the facilities to handle produced oil, water, and gas streams are provided to the construction company who provides the final cost estimates. A trunk pipeline or alternate mode of transports is also operational by this time. Buyers for oil and gas are identified, and contracts for the sale of hydrocarbons are in place. Many times this phase can run parallel to FEL-3 activities to save time and accelerate revenue generation.

Theoretically speaking, an onshore field's commercial production starts in the "Operate" phase of the Field Development Process. But practically, several appraisal wells may already be in production using a temporary or Early Production Facility (EPF) to process the crude.

Reservoir monitoring and performance evaluation are the key activities to validate the development plan's assumptions and design. Well-surveillance is a value-driven diagnostic program that is at the center of reservoir monitoring and performance evaluation. Flowing and shut-in wells are used to run various surveys to measure pressure, temperatures, rates, cuts, and so forth, along the borehole. These surveys reveal the source of problems in oil production and recovery from the reservoir. The success in operations depends on the experience and learnings from the reservoir. Teams with open minds and a culture of shared understanding stand a better chance to maximize the value. Knowledge management is, therefore, a vital component of this phase.

Each of the above stages has a Stage-Gate at the end where an assessment is made if the project is qualified to move forward to the next step. If not, should it be recycled or terminated? Gate A and Gate B are review gates (Figure 7.5), where appraisal activities are carried out to assess the reservoir potential and its commercial viability. Some new wells may be drilled and tested to achieve this objective. Based on the review results, the team may proceed further, or it may have to go back to do more work and meet again. Gate C is a decision gate that must be crossed successfully to ensure further funding of the project. This clearance is contingent upon fulfilling the three essential preconditions:

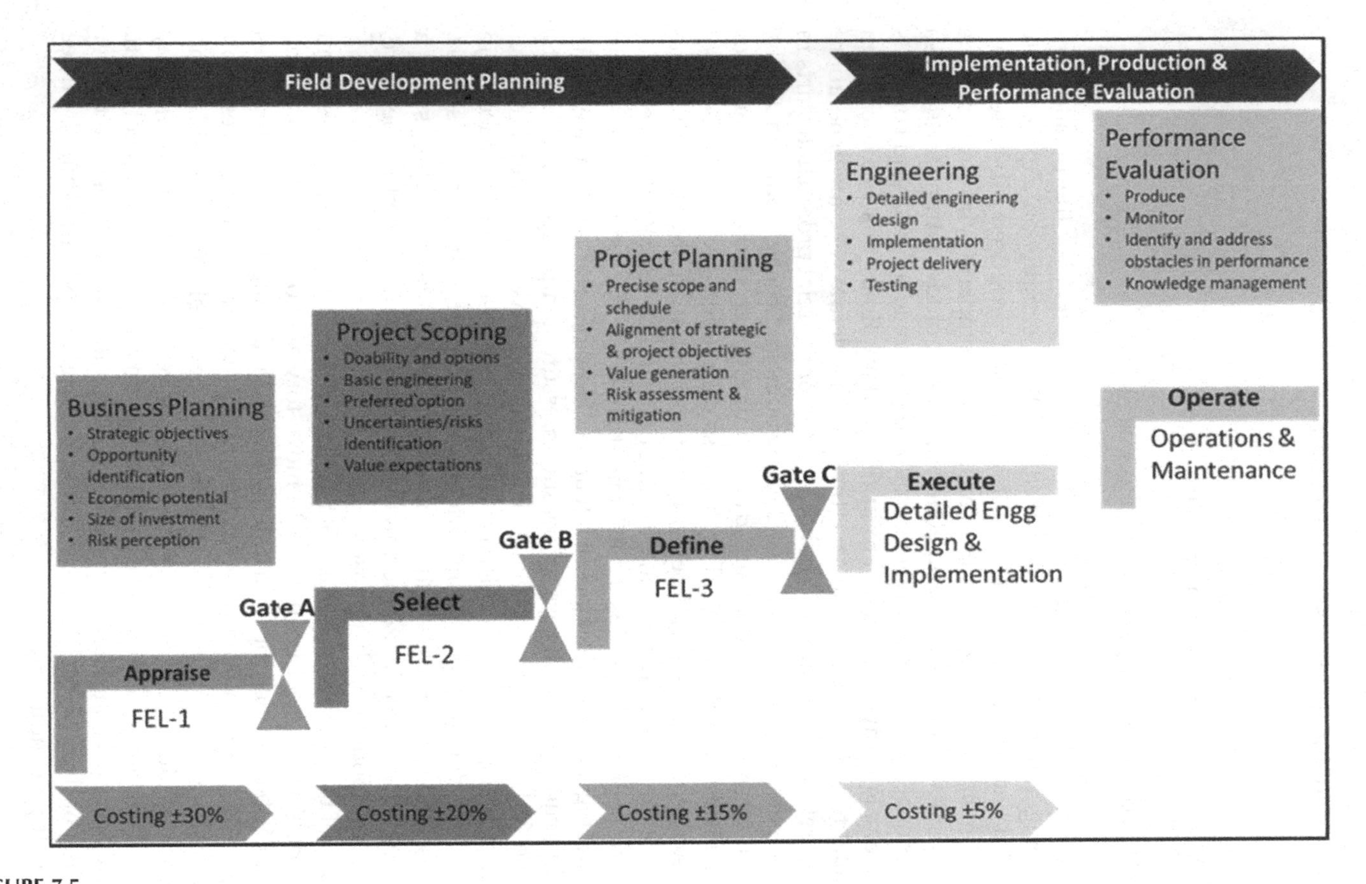

FIGURE 7.5
Field Development Planning Process.

- Is the project strategically crucial to the business?
- Are the project plan and schedules in agreement with the project goals?
- Can the project commercially sustain itself over its life-cycle?

The FEL methodology is repeated whenever a conceptual development plan, an initial development plan, and the final development plan for a reservoir/field are prepared. The development plan's quality improves significantly as technical information from newly drilled wells and reservoir performance becomes available. These data further help to refine oil-in-place, reserve, production, and cost estimates. The reiteration of this process reduces the overall uncertainties and squeezes the cost estimates within tolerance. Cost estimates can vary widely between ±30% in the "appraise" and ±20% in the "define" phases approximately in different companies. But detailed engineering and design help bring this estimate down to ±5% of the total cost by the time the "execute" phase starts.

7.17 Design and Execution of the FDP

A good field development plan results from continuous collaboration and interaction between subsurface, drilling, planning, production operations, and facilities teams. It consists of specialists in geophysics, geology, reservoir engineering, petroleum/ production engineering, drilling engineering, facilities engineering, logistics, and finance. It also includes information about significant projects, CAPEX, OPEX, the workforce's status, and future material, equipment, and technology requirements. FDP is the single-most consolidated document that provides a uniquely integrated technical, strategic, and commercial approach to field development planning. It incorporates the field's original oil and gas in-place, possible production profiles under different scenarios, associated number and type of wells required for each option, recoverable reserves, well design and construction, timing and capacity of surface facilities, technology requirement, economics, environmental impact, associated uncertainties, and risk assessment.

The preparation of the FDP involves forecasting oil, gas, and water production under various development options to maximize oil production or recovery in line with the company's strategy and business objectives. Development options include the subsurface development plan's design and cost and the expenditure plan for surface facilities and oil and gas transportation infrastructure.

A subsurface development plan can be conceptual or detailed, depending on the reservoir's maturity and data availability. It optimizes the number, type, and location of wells needed to deliver the production and economic targets. Well completion strategy is a critical element for the success of this plan.

The surface development plan incorporates the flowline network, surface facilities, and sale/export pipeline infrastructure required to handle and transport the production from subsurface to the sales terminal. Each set of the subsurface plan with a corresponding surface plan is subjected to a detailed economic evaluation. The option that yields the maximum financial recovery with the highest NPV represents the optimized version of the FDP (Figure 7.6). In real-life, there may be many more scenarios than presented in this figure.

The organization's original goal is to maximize the NPV over a project life of 25 to 30 years. Experience shows that the subsurface hardware and surface facilities/infrastructure can usually survive long with the standard maintenance regimen. 30 years is a long time. No one expects that the business practices and plans will stay unchanged over 30 years. The FDP will also require periodic reviews/updates with revised production profiles and economic evaluation based on the following justifications-

- Improvement in the knowledge and understanding of the reservoir behavior
- Access to a massive volume of data obtained from drilling and production
- Potential for loss or growth of reserves
- Deviations in reservoir performance from that projected in the FDP
    - Owing to an inaccurate forecast
    - Due to under or over withdrawals imposed by organizational/ national need
- Changes in management, specialists, technology, and mindsets
- Rise in cost of services, material, equipment, and salaries
- Variations in oil prices
- Changes in national and international policies, market conditions

The FDP qualifies for a revision or update after 5 to 7 years for an active oilfield, which drills many wells annually. The need for introducing new programs such as artificial lift may also necessitate a review of the FDP.

Designing and executing FDP is a joint effort where the experts from functional teams carry much weight. Despite a rigorous process for the design and optimization of the FDP, several pitfalls can affect its outcome. An FDP may not achieve its objectives due to significant data gaps and poor

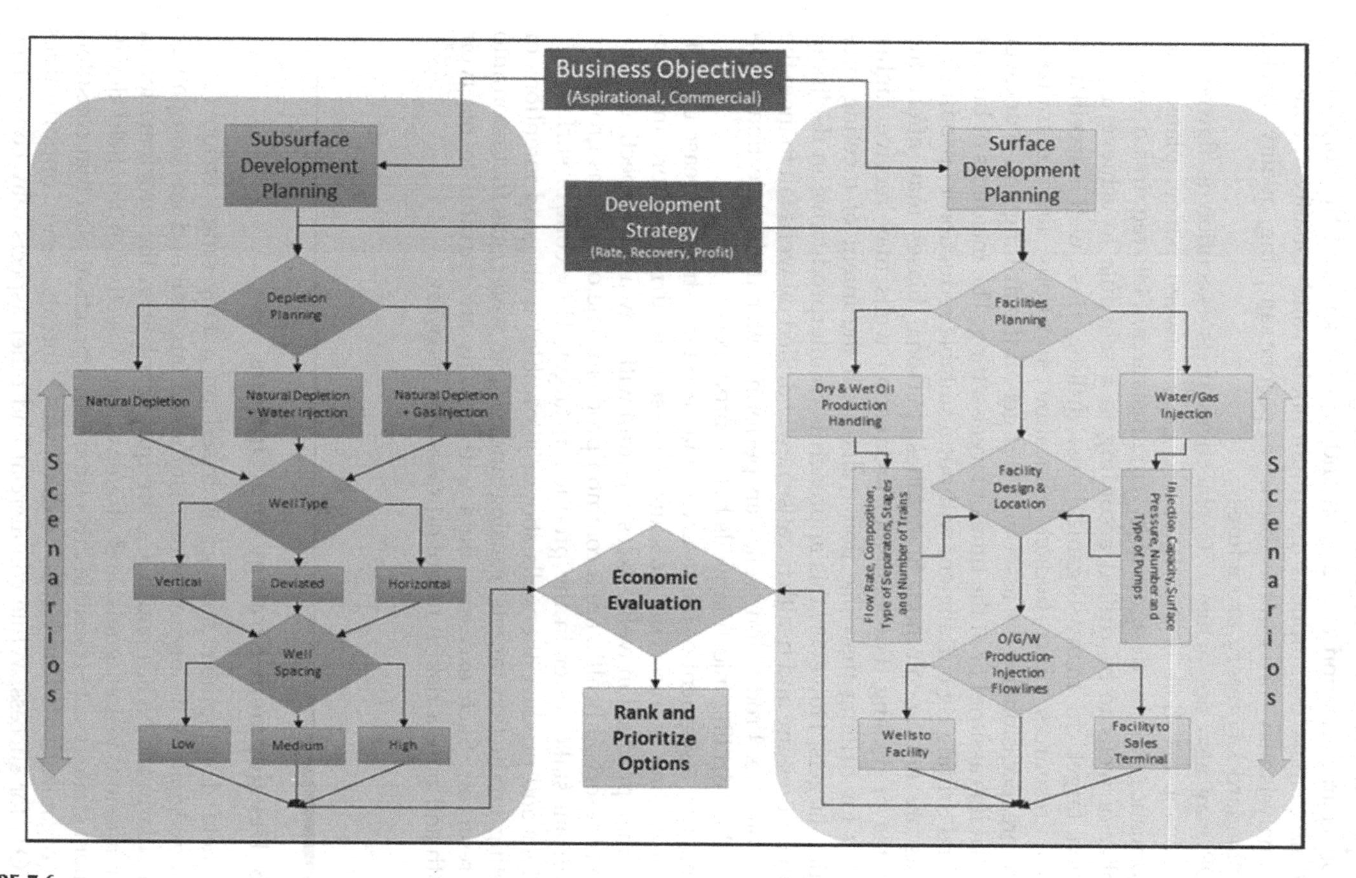

FIGURE 7.6
Field Development Plan (FDP) Optimization.

data quality. Poor data acquisition strategy can cause differences in reservoir characterization and understanding of reservoir behavior, forcing the team to make assumptions that might not be valid. Quality of core and PVT analysis may suffer in the absence of proper sampling, quality checks, and standard laboratory procedures.

FDP may also not achieve its objectives if its execution is flawed. A common observation is that there are gaps between policy, plans, and performance, not by design but by circumstances. FDP requires a certain number of wells to be drilled according to a schedule to achieve a production target. In many organizations, drilling rigs, well surveillance, human resources, and budget allocation functions are centralized. Therefore, occasionally the asset may not get all the drilling and workover rigs that it is assured of at the time of target-setting. The incidents of budget cuts and drilling rig curtailment without a suitable reduction in production targets are not uncommon. Sometimes, personal biases may affect the schedule of drilling. For example, production wells may receive higher priority for drilling than injection wells, with the intent of meeting production targets. However, this approach is counterproductive in declining reservoir pressure and health because newly drilled oil wells fail to meet the target oil rate. Production wells can perform better if the injection is expedited by drilling injection wells before producers.

Similarly, a segment of the reservoir may exhibit a sharp increase in GOR and a decline in reservoir pressure necessitating drilling more injection wells than production wells. Excessive fluid withdrawals, irrespective of the state of reservoir health, are a common problem. The operations group is a significant stakeholder in the production target. They control surface operations and produce the wells above "allowable" with good intentions of meeting production targets, causing undesirable water/gas incursion into the reservoir. Such excessive fluid withdrawals are harmful to reservoir health and pose a risk to future oil rates and recovery.

7.18 Development Strategy and Options

The field development strategy is based on a long-range vision of the business goals of an organization. Almost all large oil and gas companies, particularly the NOCs and IOCs, are quite explicit in the proclamation of their mission, vision, goals, and strategy. The objective of the field development strategy is to translate this vision into a practical plan that considers the organization's strengths, weaknesses, threats, and opportunities (SWOT) for success. While commercial and other aspects may come from specific company objectives, the field development strategy's technical aspects are managed by identifying standard options, associated issues, and their impact on the plan (Table 7.2).

Three factors, namely, production rate, long-term oil and gas recovery, and profit, are at the center of the development strategy. All elements of the development plan presented in Table 7.2 are directed at enhancing one or all of these factors. Each of these elements has additional layers of technical and economic options and issues that control the quality of operations. The involvement of specialists and SMEs ensures that a proper plan is designed and implemented to achieve the development objectives.

The field development plan's scope includes integrating multi-disciplinary information to generate long-term oil, gas, and water production forecasts. This forecast is tied to a set, type, and the number of locations considering all possibilities, starting from depletion planning with or without pressure maintenance, application of artificial lift methods, well stimulation, infill drilling, and EOR, which may be required at some stage in the field's life. Some of these methods may not find a place in the conceptual or initial development plan if they are not of significance at that stage. However, they may take center stage in the subsequent development plan to increase oil/gas production and recovery.

The FDP considers the following steps to generate a production forecast for a reservoir on production:

1. Based on available field data, list the possible depletion strategies that are in conformance with the company's production objectives-
    - Produce under natural depletion
    - Produce under pressure maintenance by water or gas injection
2. In the case of a decline in reservoir pressure, choose the most viable arrangement for pressure maintenance:
    - Pattern injection (Line drive, 5-spot, inverted 5-spot, etc.)
    - Peripheral injection
3. Evaluate the available field data to decide the type, number, and location of wells that are best suited for the reservoir under the chosen depletion strategy:
    - Conventional Wells (vertical)
    - Non-Conventional Wells or NCWs (deviated, horizontal, multilateral, or a combination)
    - In the case of Conventional and NCWs, decide about the optimum well design

The type of wells to be drilled is determined by special conditions prevailing in a reservoir such as the k_v-k_h ratio, water or gas coning, and the occurrence of more than one oil-bearing layer in a reservoir. However, the

TABLE 7.2

Various Elements of the Field Development

Element	Options	Issues	Impact
Depletion	Natural depletion	Above or below P_b	Production and Reserves Growth
	Pressure Maintenance above Pb	By water or gas injection	
	Natural depletion + Waterflooding	Above or below P_b	
Drilling	Vertical	Lower PI	Drilling and Completion Operations and Practices
	Deviated	Subsurface congestion and collision	
	Horizontal	k_v-k_h ratio, Length of horizontal section, subsurface congestion, and collision	
Production	Natural flow	Uses reservoir energy	Production Rate Management
	Artificial Lift	ESP, Gas Lift, Timing ESP or Gas Lift infrastructure	
Well Completion	Single	Comingled if more than one zone	Ease of Operations and Project Economics
	Dual	Parallel strings if more than one zone	
Perforation	Full zone perforation	Higher production rates	Production & Injection Control and Efficiency
	Selective perforation	Production from or injection only into the desired interval	
Production Plateau	Low oil production rate	Long plateau period	Rate and Recovery
	High oil production rate	Short plateau period	
Infill Drilling	Well Density	High well density	Bypassed Oil Recovery
	Location	In between existing oil producers	
Water Injection	Pattern	Quick pressure maintenance, oil recovery, and water breakthrough	Reservoir Pressure, Production, and Reserves Growth
	Peripheral	Relatively slower pressure maintenance, oil recovery, and water breakthrough	
Production Maintenance	Interventions	Rigless and Rig Workovers Requirement and Capability	Production Rate Management
	Stimulation	Type, nature, and frequency necessary to restore production	

(*Continued*)

TABLE 7.2 (Continued)

Various Elements of the Field Development

Element	Options	Issues	Impact
Well Surveillance	Rate, Pressure & Cut Measurements	Reliability and Accuracy of data	Corrective Measures
	Performance not as expected	Problem Identification, Diagnosis, and Solution	
Facilities	Dry and Wet Production Capacities	Definition of dry and wet crude (water or salt content), processing route	Sale/ Export commitments
	Operational Efficiency	Downtime, Maintenance schedules	
	Pure or mix of seawater and produced water injection	The required volume of injection water, Injection above or below the fracture pressure	
Water Disposal	Disposal for reservoir pressure maintenance	The preferred option, treatment of effluent water necessary, injection through injection wells	Produced Water Management
	Disposal into EWD wells	Waste of resource (water), presence of suitable reservoir, capital and rigs required to drill EWD wells for disposal, environmental risks	

decision to go for this option must be made after ensuring that the company can drill, complete, and monitor these wells.

The productivity index of wells will determine the number of wells to achieve the plateau rate of production. Consider high, medium, and low well-spacing scenarios to ascertain their impact on oil rate and recovery.

Locations of wells will be decided by the "sweet spots" characterized by the high value of $Ø*h*S_o$, that is, the product of porosity, thickness, and oil saturation.

4. Check for the number of drilling rigs available for drilling and their capability to drill-specific well types annually. Based on this decision, a suitable schedule for drilling before, during, and post-plateau period needs to be developed.
5. Decide about the most suitable type of well completions based on their advantages and disadvantages:
   Open hole completion
   Cased hole completion
   Single completion

Dual completion
Perforation policy

6. Decide what the most effective way to overcome the decline in production is:
   - Drill additional/infill wells
   - Stimulate the wells
   - Open other layers in the wells if available
   - Additional perforation

7. Evaluate the need and timing of artificial lift methods and establish the gain in production rates:
   - Electrical Submersible Pumps (ESPs)
   - Gas Lift (continuous or intermittent)

8. Generate long-term subsurface production forecasts for 25 to 30 years using the most likely scenarios/options/variants. It is a deterministic production forecast based on the "Most likely" oil-in-place. Use the oil, gas, and water streams of this forecast to proceed with facility design in consultation with production operations and project teams.

9. Oil production from the subsurface will be directed to the GCs for processing. Decide based on well-hydraulics:
   - Adequate number of GCs required to process the oil
   - Ideal throughput to each GC
   - Location of GCs
   - Number and locations of manifolds and the wells to be connected to each of them
   - Flowline network (length and diameters)

10. Evaluate what type of separators and number of trains of separators will be required to process the oil efficiently? Account for the cost of demulsification facilities at the GC.
11. Consider the cost of a booster station for pumping the produced associated gas to consumers if sales agreements already in place or for the gas lift if applicable.

12. Consolidate subsurface and surface scenarios to shortlist the top five that should be subjected to economic evaluation.

13. List the assumptions that are made to generate the production forecast.

14. Identify the new technologies, equipment, or expertise necessary to achieve, sustain, or improve the oil and gas production rate and recovery.

15. Identify key risks and uncertainties that can have an impact on the production profiles. Suggest a time-bound mitigation plan for resolution.

16. Highlight if there are any HSE risks and how they can be resolved.

It is a standard procedure that most oil companies use for techno-economic analysis of the selected options.

Offshore fields use different rig selection concepts, drilling, production, storage, and transport than onshore fields. Although the broad framework of the offshore and onshore fields' reservoir management principles remains the same, selecting appropriate drilling, production, transportation, and storage systems has a significant bearing on the cost and management of offshore field development activities. Water depth, load capacity of the seabed, ocean currents, possible dangers posed by the weather, wind, waves, and oceanic currents are significant factors that affect the drilling capability, well completion, hydrocarbon processing, and transportation arrangements.

7.19 Offshore Drilling Systems

Underwater E&P operations are more expensive as well as risky. The selection of an offshore drilling system depends on the water depth and prevailing sea and sub-sea conditions (Figure 7.7). Offshore drilling rigs are divided into two categories: those that are immobile or fixed, and those that are mobile can move from location to location for drilling.

Fixed Platforms: These platforms are extremely stable large structures erected on steel/concrete legs secured directly onto the seabed. They are used to drill multiple directional wells by moving a drilling rig to the assigned drilling slot.

- **Drilling Barges:** A drilling barge has a drilling rig and ancillary equipment. Drilling barges can be utilized for drilling in shallow and calm water conditions such as lakes, swamps, and rivers. The barges with floating platforms and drilling equipment are towed by a tug boat to the drill site and secured adequately with anchors at the drilling time.

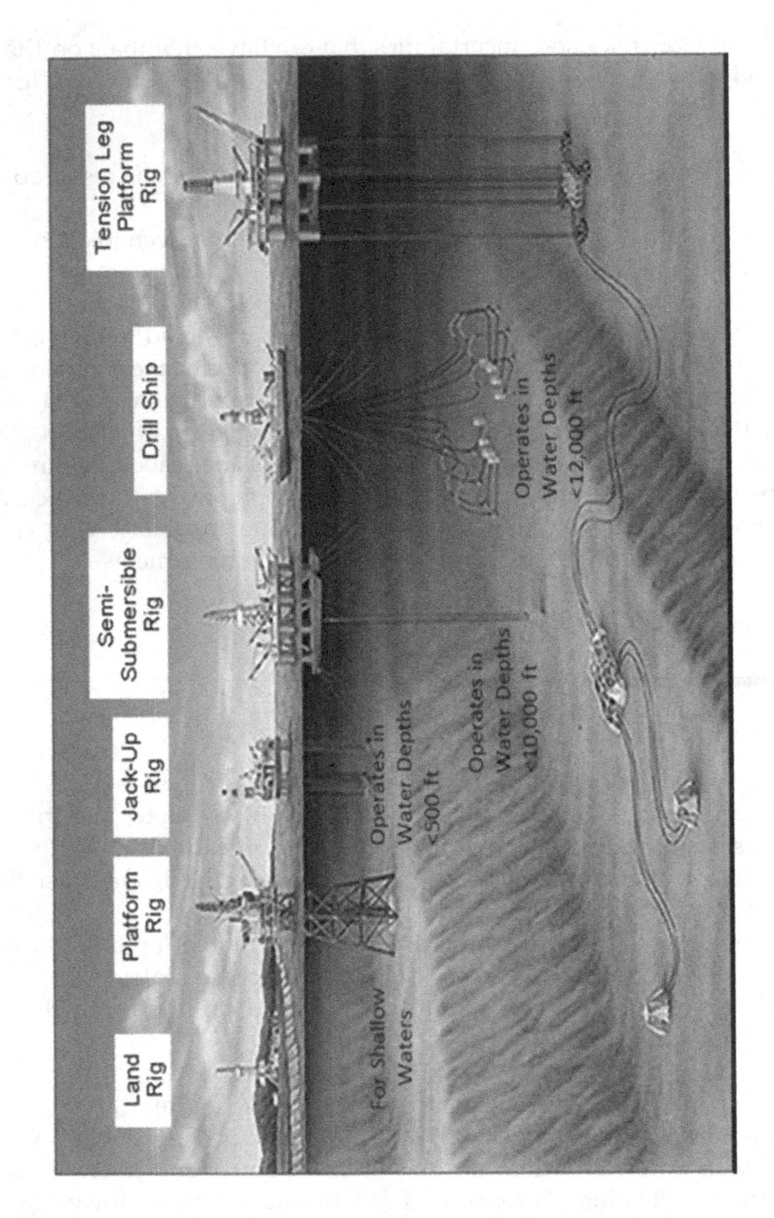

FIGURE 7.7
Offshore Drilling Systems and Drilling capability[11].

- **Jack-Up Rigs:** Jack-up rigs are also a shallow water drilling system that is more stable in water depths up to 100 meters than floating rigs. These rigs can be moved and positioned at the drill site by setting three or four legs on the seafloor. The barges are then lifted, leaving the working platform to sit above the water for drilling.
- **Submersible Rigs:** The operations of submersible rigs are also limited to shallow waters (water depth <500 ft). These rigs primarily consist of two hulls stacked over each other. The upper hull contains the living quarters, working areas, and utilities. The pontoon-based lower hull can be inflated with air or deflated as required. The rig can move from location to location, with the structure ballasted with air. The structure can be positioned on the seafloor over the drill site by removing the air from the hull. Like drilling barges and jack-up rigs, all necessary drilling equipment is provided on the platform's deck.
- **Semi-Submersible Rig:** A semi-submersible rig is the most common type of offshore drilling rig. It is designed for offshore drilling in ultra-deep waters with water depths <10,000 ft. It uses the same concept whereby the lower hull can be inflated to facilitate movement and deflated to stabilize the structure over the drill site. Either air or water can be used to fully or partially raise the structure. The lower hull is filled with water to keep the structure partially submerged above the drill site. It is also adequately moored to the seabed by huge anchors to provide stability for drilling operations.
- **Drill Ships:** Drill ships are the offshore drilling systems that can operate in ultra-deep waters ranging up to 12,000 ft. The drill ships' key distinguishing features include a moon pool, a dynamic positioning system, and self-contained drilling equipment. The moon pool allows the drill string to reach the well location through the ship's hull for drilling or well operations. The dynamic positioning system enables to steady the drill ship directly above the drill site with a satellite-supported computer network at the time of drilling. Offshore workboats facilitate the delivery of drilling material, supplies, and consumables to the drill ships.

7.20 Offshore Production Systems

Offshore fields close to the land can afford to transport oil by pipelines and process it in an onshore facility. However, oil and gas fields away from the coast need offshore platforms to process and store hydrocarbons. The selection and design of a production system are essentially controlled by the safety, cost, and water depth considerations.

Shallow water fixed production system built on steel/concrete legs is a complex of many independent platforms consisting of equipment to process, store, and transport oil. Gangway bridges interconnect these platforms to facilitate the movement of personnel and material.

Gravity-Based Structures (GBS) are fixed concrete structures, relatively more costly than steel structures but better suited for the harsh offshore environment. They are constructed in a dry dock and towed to the offshore location for placement on the sea bottom, providing a robust substructure for the topsides. A large deck on top has oil and gas processing facilities.

Concrete Gravity Based Structures are constructed in a dry dock prepared for the massive structure to be erected.

Compliant towers (CT) derive their name from their ability to comply with the wind, waves, and current forces. They are sleek and flexible, with a piled foundation supporting conventional topsides for drilling and production operations, improving its operability in much deeper water.

FPSO: Floating Production systems are classified as Floating Production, Storage & Offloading (FPSO), Tension Leg Platform (TLP), and SPAR systems. The oil and gas produced from the offshore fields that are far away from the shoreline are separated, stored, and offloaded onto another tanker or pipeline using a facility like the FPSO units. The FPSO is a floating vessel, a barge, or a supertanker (Figure 7.8), equipped with oil, gas, and water separation facilities on the deck and hydrocarbon storage facilities situated below the hull. It can hold and handle as much as 2.5 million barrels of oil. Large offshore fields may use multiple FPSO units to cater to the demand of the project.

The semisubmersible design of the FPSO has gained in popularity as the industry operations have moved into deeper waters. It consists of topsides, hull, mooring system, and risers. Topsides include modules for crude processing, power, utility, accommodation, and so forth. Hull is supported by columns that also provide strength to the structure. The mooring system provides a temporary or permanent station-keeping for floating structures. It is accomplished by mooring the vessel to the seafloor employing multiple grouped legs, piles, chains, polyester ropes, vertically loaded anchors, or dynamic positioning (DP). Risers can be rigid or flexible and are used to transport produced oil and gas, injection fluids, control fluids, gas lift, and so forth.

TLP: TLP structures are a vertically floating production system with buoyant production facilities. A simple design of a TLP has four vertical air-filled columns connected by pontoons. The structure is kept buoyant by a hull on top and tensioned tendons secured firmly into the seabed with pilings' help. This arrangement facilitates that the production wellheads are on the deck of the platform.

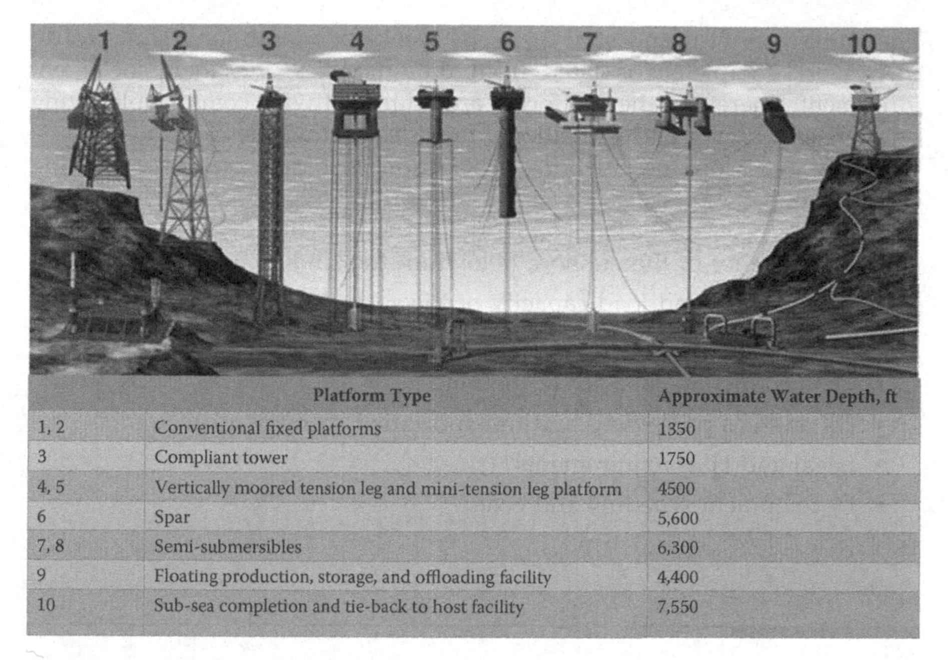

	Platform Type	Approximate Water Depth, ft
1, 2	Conventional fixed platforms	1350
3	Compliant tower	1750
4, 5	Vertically moored tension leg and mini-tension leg platform	4500
6	Spar	5,600
7, 8	Semi-submersibles	6,300
9	Floating production, storage, and offloading facility	4,400
10	Sub-sea completion and tie-back to host facility	7,550

FIGURE 7.8
Offshore Production Systems[12].

The topside of a TLP consists of a drilling rig, a production platform, and living quarters and facilities for the staff. The SeaStar version of the TLP has only one central column for a hull. A TLP is suitable to drill in water depths from 300 m to 1500 m.

Production wells deliver the seafloor's oil through rigid risers to the dry wellheads located on the deck. This oil is then processed and stored in the topside production facility, wherefrom it can be transported to the shore via tankers or alternate means.

Spar Platforms: A Spar Platform has a large-diameter single vertical cylinder supporting a deck. Spar platforms are designed to operate in water depths up to 200 m and drill wells beyond 3,000 m. They combine a topside drilling and production platform supported by a hollow cylindrical hull at the bottom. The structure is stabilized by the hull's weight, a robust network of cables, and tendons moored to the seabed.

Subsea Completions: Subsea completions may be used in deep as well as shallow water. A subsea completion refers to a system of pipes, connections, and wellheads located on the seafloor (Figure 7.8) It is a technologically advanced system where wellheads can be operated remotely to control oil and gas flow into the production facility by underwater risers and pipeline networks.

Offshore drilling and production platforms/facilities are like mini-cities. These platforms are permanently stationed on drilling or production sites.

The employees visit and stay there for weeks or month-long shifts to do their work completely isolated from their families and their social environment. Therefore, the platforms are equipped with comfortable living quarters and dining and recreational facilities around the clock.

Facility selection for offshore production is based on the following considerations:

- The distance of the offshore field from the land
- Water depth and load capacity of the seabed
- Number and type of wells
- The production capacity of the field
- Oil and gas processing and transportation strategy
- Legal and HSE requirements
- Development schedule and cost

7.21 Techno-Economic Optimization of the FDP

Techno-economic optimization is a standard method to integrate technical and economic evaluations of a field development plan. The technical assessment considers all the technological options for subsurface and surface development within an operating company's organizational capability. This exercise aims to maximize oil and gas production and recovery using resources at the company's disposal. The availability of capital in the form of the annual budget, workforce, rigs, material, and equipment decides the time-dependent activities/operations extract oil and gas that generate revenue. In this process, "time" is a common factor used to correlate oil and gas production with expenditure and income. However, an increase in oil production does not necessarily mean an increase in oil recovery.

Similarly, an increase in oil recovery may not contribute to a rise in profit. Increasing oil recovery beyond a specific limit may require additional costs that may be prohibitive from the project economics point of view. The cost of inputs viz., technology, and operations may be too high for the undertaking. Additional oil recovery from EOR and development of oil shale, offshore deepwater hydrocarbon resources, and oil sands are examples.

7.21.1 Technical Evaluation

Preparing a development plan that aims to maximize oil and gas recovery is fundamentally a technical assignment that starts with subsurface development. Reservoir Engineers have to look into different possible options varying in the modes of development and costs. Each development option consists of a blueprint of all the wells, and facility locations plotted on a map connected by the GC's flowline network. In addition to drilling schedules and well productivity, this information forms the basis for generating long-term oil, gas, and water production profiles. Well-productivity is estimated from reservoir properties and validated by field tests. In case the subsurface development option considers pressure maintenance by water or gas injection, similar information needs to be provided to the reservoir model or any other forecasting tools. A reservoir model is an excellent tool that can proficiently manage a production forecast taking into account the reservoir description, properties, events, and schedules.

Production forecast's end product for each subsurface development option is oil, gas, and water production profiles. These profiles have the type and number of wells with details of completion. In addition to production profiles, pressure maintenance options have details of water or gas injection wells and their completions.

The next step in the technical evaluation is integrating the subsurface development plan with a viable surface development plan. Oil, water, and gas production and injection profiles for each development option need to be handled and processed by surface facilities. There are many important decisions to be made for facility planning, most notable being the capacity, number, and location of the production gathering centers, size and location of the processed crude collection tanks, and size and route of the transport pipeline to the sales terminal. Each of these facilities has layers of decision-making relating to the type and capacity of separation, dehydration, and desalination equipment. The same statement applies to the flowlines that connect the wells to manifolds, separators, pumps, and tanks.

7.21.2 Economic Evaluation

Economic evaluation is the last and the most crucial step in finalizing the FDP. It provides economic justification to select or reject an execution plan because it gives an insight to the investor if the capital invested in a project will earn a profit.

An E&P project's economic value is determined by the net cash flow over its life cycle. Net cash flow is a common finance term used to explain how the income is higher than the expenditure.

$$\textit{Net Cash Flow} = \textit{Cash Inflow} - \textit{Cash Outflow} \tag{7.2}$$

Positive net cash flow indicates that income from a project is more than the expenditure. Negative cash flow, on the other hand, means higher investment than income.

7.21.3 Standard Economics/Finance Terms

Net cash flow is an excellent economic indicator of various phases of an E&P project lifecycle. Discovery and appraisal phases are essentially the negative cash flow phases because capital is invested in drilling exploratory and appraisal wells. There is no revenue generation and income in these phases in the absence of any production or sale of oil and gas. The early production phase is characterized by high capital expenditure (CAPEX) on constructing facilities and production infrastructure and drilling many development wells. It is followed by sustained long-term production and sale of oil and gas, which generates sufficient revenue to turn the negative net cash flow into positive. Most middle and late production phase expenditure is on operating expenditure (OPEX), a small fraction of the sales revenue. The abandonment phase signals the onset of project closure due to uneconomic production rates of oil and gas. Therefore, "cash outflow" or expenditure is a general economic characteristic of an oil and gas project's life cycle. However, "cash inflow" is a distinguishing feature of only the production phase, which incidentally is the most extended phase.

The process of economic evaluation starts with the collection of input data. These data include every single item that involves expenditure and income. Investment can typically be divided into CAPEX and OPEX, whereas revenue is generated from oil and gas sales. CAPEX often makes up for a large portion of its annual budget for driving the business and its growth. Drilling and completion of wells, construction of facilities, and production infrastructure are categorized as CAPEX. Buying, constructing, or upgrading any physical assets such as office buildings, housing complexes, or other major civil works are classified under CAPEX.

OPEX includes the recurring costs of maintaining office buildings, wells, artificial lift, equipment, transport, fuel, and personnel. This expenditure is critical for the maintenance of production after the wells and facilities become operational. Housing and hardware rentals, software license fees, another regular spending to continue operations and services are covered under OPEX.

An economic evaluation of an oil and gas project requires the following data:

- The selling price of oil and gas
- Details of royalty and liabilities of federal and state taxes
- Commercial terms and conditions of the production sharing agreement
- Inflation rate and discount factor
- Schedules of depreciation, depletion, and amortization

The impact of oil and gas prices on project economics is ascertained by carrying out a "price sensitivity" analysis. While most of these data are likely to remain "stable" during the project life, it is incredibly complicated to predict future oil and gas prices in the current geopolitical order. Sensitivity analysis is a standard industry tool used to examine the potential impact of hypothetical changes in input data values. Relatively significant changes above and below the base price are assumed and applied to understand the sensitive nature of the selected input data. The purpose of such an analysis is to assess how fragile or robust project economics is to the price of oil.

Royalty is the owner's share of the gross oil and gas production or its equivalent amount of money.

Production sharing is a contractual agreement between the oilfield operator(s) and owner to share the oil and gas production on mutually agreed terms.

In general, tax is a compulsory contribution to state revenue levied by the government on an individual's income and business profits. Governments may also apply tax on the cost of some goods, services, and transactions, which further increases the customer's asking price.

The discount rate is the cost of borrowing by large banks from the central bank. The central bank in a country sets it. The inflation rate is the rate of decrease in the purchasing power of money.

Depreciation, depletion, and amortization are standard accounting practices for rationalizing the economic value of resources over time by balancing the costs with revenues. Before explaining the meaning of depreciation, depletion, and amortization, it is essential to know the difference between tangible and intangible assets. Tangible assets are those that have a physical form or a specific monetary value. Examples are an office building, drilling rig, or a production facility. On the contrary, intangible assets have value but no physical form. Examples of intangible assets are ideas, software, goodwill, patents, and so forth.

Depreciation is a measure of erosion in the value of a tangible asset over its estimated useful life. It is seen as a reduction in the cost of a fixed asset due to normal usage, wear and tear, the arrival of new technology, and so forth. The simple depreciation can be calculated by dividing the cost of the asset by its estimated useful life. A motor car depreciating at 10% implies that the motor car's useful economic life is 10 years.

Depletion is a familiar term for both reservoir engineers and accountants, which implies systematic exhaustion in the value/quantity of recoverable reserves of oil and gas. It represents a non-cash expense that brings down the value of the asset. For an oilfield with 100 million barrels of initially recoverable reserves being exploited at the rate of 5 million barrels per annum, the depletion rate works out to be 5% per annum.

Depreciation and amortization use the same concept of reduction in the value of an asset with time. Unlike depreciation, amortization applies to intangible assets such as software, research and development, and licenses.

Economic evaluation is necessary to determine if a project under the proposed development scenario would generate profit or incur a loss. The company can use its own financial criteria to accept or reject the proposal.

7.21.4 Economic Criteria for Project Selection

As discussed earlier, project economics is based on calculating cash outflow for a vast number of oilfield activities and cash inflow that results from the production and sale of oil and gas. Cash inflow depends on the estimated oil and gas production profiles, which have different bands of uncertainties. Standalone deterministic reservoir models usually relate with a P50 estimate, which can constitute the base case. However, decision makers are also interested in looking at the upside and downside of this forecast. Upside potential is represented by the optimistic or P10 outlook that helps to understand the growth potential and the associated risks to investments. P90 forecast represents the pessimistic or downside view, which is vital to preserving the available capital, particularly in a sluggish economic environment.

It is a standard practice to identify the key parameters that can significantly impact project economics. This objective is achieved by carrying out sensitivity studies whereby the value of the input parameters CAPEX, OPEX, discount factor, oil price, and so forth are changed one at a time to assess the magnitude of their impact on the overall project economics. This process is repeated for all parameters that have considerable uncertainty or are prone to significant variations.

Several economic criteria can be used to accept or reject an oilfield development project proposal. The most common determinants used to rank and prioritize various field development options or compare several projects are:

- Net Present Value (NPV)
- Internal Rate of Return (IRR)
- Profitability Index (PI) or Profit-to-Investment Ratio (PIR)
- Payout or Payback period

NPV is the most straightforward criterion to invest in a project. If the NPV, being the sum of the present value of future net cash flows, is greater than zero, it would be better to invest in this project than do nothing. NPV is a project screening tool and not a prioritization tool. It can be used to filter the projects that make money but not to rank them. NPV should not be used for inter se comparison of projects because a large project can have a bigger NPV than a smaller project but requires a more significant investment.

The IRR provides a measure of profitability and is often used for ranking the projects in a project selection scenario. The plans can be listed from the highest IRR on top to the lowest IRR on the bottom. This ranking forms a sound basis for the selection of the final option. The project with the higher IRR is selected over the plan with the lower IRR. However, IRR alone should not be used to decide on a project's worth. A lower IRR project and a higher NPV qualify for selection in the "no capital constraint" case, increasing the shareholders' profits. In many countries, IRR is also an essential screening tool to clear the project proposals requiring government approval or funding.

The Profitability Index (PI) or the Profit-to-Investment Ratio (PIR) is the ratio of the present value of future cash flows and the initial investment.[13] It measures the monetary value created by a unit investment and is another useful screening tool for ranking investment projects. Based on the definition of PI:

The projects with a PI < 1 destroy the monetary value
The projects with PI = 1 are neutral and break-even monetarily
The projects with a PI > 1 enhance the monetary value

Quite clearly, the higher the profitability index, the greater the investment's potential to create wealth.

The payout or payback period is the time a project takes to recover the investment cost. This criterion facilitates the acceptance of an investment project based on a shorter payback period. For example, out of the two oil gas projects with a payback period of 3 and 5 years, respectively, the one with a 3-year payback period is preferred for execution because it recovers the invested money in a shorter period.

7.21.5 Economic Optimization

The economic evaluation mainly deals with expenditure and income elements of the development plan. The expenditure is split under the heads of CAPEX and OPEX. Wells, facilities, and surface infrastructure, which have an "economic use" life longer than 3 years, are accounted for under CAPEX. Expenditure on activities to maintain the production and facilities, office buildings, and employees' salaries are covered under OPEX.

Oilfield income is generated from the sale of oil and gas to consumers. Water production requires disposal or reinjection that involves drilling wells, some pumps, and connecting pipes to inject water into disposal wells. This program can be easily integrated with the expenditure package. Reinjection of produced water back into the oil reservoir, as an alternative to disposal, is also a component of the expenditure package. Still, it has the potential to generate significant revenue from the incremental oil recovery.

As mentioned earlier, the net cash flow is the difference between income and expenditure. The procedure to calculate the net present value of future cash flow is shown with the help of a simple exercise (Table 7.3) wherein

TABLE 7.3

An Example Illustrating the Techno-Economic Optimization Procedure

Assumptions for the exercise
1. Oil price US$ 50/bbl
2. Gas revenue ignored
3. Water production ignored
4. Discount Factor
DF=1/((1+i/n)^n*t)
i= Annualized rate of interest =10%
n= Number of compounding periods of a discount rate per year =1
t= Number of years

1	2	3	4	5	6	7	8	9	10	11
Option	O1	O2	O3		O1	O2	O3	O1	O2	O3
Year t	Ver Wells Cumulative	Dev Wells Cumulative	Hor Wells Cumulative	Capex Facilities MM$	Total Expenditure MM$ A1	Total Expenditure MM$ A2	Total Expenditure MM$ A3	Oil Rate bopd A4	Oil Rate bopd A5	Oil Rate bopd A6
1	2	2	2	50	70.0	73.8	93.1	0	0	0
2	4	4	3	75	95.0	98.8	118.1	0	0	0
3	6	5	5	50	70.0	73.8	93.1	0	0	0
4	14	13	11	50	110.0	121.3	162.3	8000	8800	10400
5	30	27	23		95.0	112.5	149.6	16000	17600	20800
6	40	36	31		70.0	83.1	125.4	22000	24000	28500
7	47	43	36		55.0	65.4	105.8	22000	24000	28500
8	50	45	38		35.0	41.7	79.6	22000	24000	28500
9	53	48	41		35.0	41.7	79.6	22000	24000	28500
10	56	51	43		35.0	41.7	79.6	22000	21120	23655
11	58	53	45		30.0	35.8	73.1	22000	18586	19634
12	60	55	46		30.0	35.8	73.1	19360	16355	16296
13	62	56	48		30.0	35.8	73.1	17037	14393	13526
14	64	58	49		30.0	35.8	73.1	14992	12666	11226
15	66	60	51		30.0	35.8	73.1	13193	11146	9318
16	68	62	52		30.0	35.8	73.1	11610	9808	7734
17	69	63	53		25.0	29.9	66.5	10217	8631	6419
18	70	64	54		25.0	29.9	66.5	8991	7595	5328
19	71	65	55		20.0	23.9	51.5	7912	6684	4422
20	72	65	55		20.0	23.9	51.5	6963	5882	3670
21	72	65	55		15.0	18.0	45.0	6127	5176	3046
22	72	65	55		10.0	12.0	30.0	5392	4555	2528
23	72	65	55		10.0	12.0	30.0	4745	4008	2099
24	72	65	55		5.0	6.0	15.0	4175	3527	1742
25	72	65	55		5.0	6.0	15.0	3674	3104	1446

	12	13	14	15	16	17	18	19	20	21
Option		O1	O2	O3	O1	O2	O3	O1	O2	O3
Year t	Oil Cashflow MM$ B1	Oil Cashflow MM$ B2	Oil Cashflow MM$ B3	Net Cash Flow MM$ B1-A1	Net Cash Flow MM$ B2-A2	Net Cash Flow MM$ B3-A3	Discount Factor fraction C	Discounted Net Cash Flow MM$ C*(B1-A1)	Discounted Net Cash Flow MM$ C*(B2-A2)	Discounted Net Cash Flow MM$ C*(B3-A3)
1	0.0	0.0	0.0	-70.0	-73.8	-93.1	0.9091	-63.636	-67.107	-84.615
2	0.0	0.0	0.0	-95.0	-98.8	-118.1	0.8264	-78.512	-81.668	-97.584
3	0.0	0.0	0.0	-70.0	-73.8	-93.1	0.7513	-52.592	-55.461	-69.930
4	146.1	160.7	189.9	36.1	39.4	27.6	0.6830	24.657	26.936	18.866
5	292.2	321.4	379.9	197.2	208.9	230.2	0.6209	122.446	129.695	142.964
6	401.8	438.3	520.5	331.8	355.2	395.1	0.5645	187.278	200.506	223.022
7	401.8	438.3	520.5	346.8	372.9	414.7	0.5132	177.950	191.375	212.813
8	401.8	438.3	520.5	366.8	396.6	440.9	0.4665	171.103	185.004	205.667
9	401.8	438.3	520.5	366.8	396.6	440.9	0.4241	155.548	168.186	186.970
10	401.8	385.7	432.0	366.8	344.0	352.4	0.3855	141.408	132.618	135.859
11	401.8	339.4	358.6	371.8	303.6	285.5	0.3505	130.305	106.410	100.060
12	353.6	298.7	297.6	323.6	262.9	224.5	0.3186	103.097	83.759	71.541
13	311.1	262.8	247.0	281.1	227.0	173.9	0.2897	81.435	65.762	50.383
14	273.8	231.3	205.0	243.8	195.5	131.9	0.2633	64.200	51.478	34.745
15	240.9	203.5	170.2	210.9	167.7	97.1	0.2394	50.498	40.153	23.242
16	212.0	179.1	141.2	182.0	143.3	68.2	0.2176	39.615	31.187	14.834
17	186.6	157.6	117.2	161.6	127.7	50.7	0.1978	31.969	25.268	10.029
18	164.2	138.7	97.3	139.2	108.8	30.8	0.1799	25.036	19.569	5.533
19	144.5	122.1	80.8	124.5	98.2	29.2	0.1635	20.355	16.050	4.778
20	127.2	107.4	67.0	107.2	83.5	15.5	0.1486	15.928	12.413	2.303
21	111.9	94.5	55.6	96.9	76.5	10.6	0.1351	13.093	10.341	1.437
22	98.5	83.2	46.2	88.5	71.2	16.2	0.1228	10.868	8.745	1.987
23	86.7	73.2	38.3	76.7	61.2	8.3	0.1117	8.560	6.835	0.930
24	76.3	64.4	31.8	71.3	58.4	16.8	0.1015	7.234	5.931	1.707
25	67.1	56.7	26.4	62.1	50.7	11.4	0.0923	5.732	4.678	1.052
						Total NPV →		1393.574	1318.664	1198.591

three development options, O1, O2, and O3, are considered to develop an oil reservoir with conventional vertical wells, deviated wells, and horizontal wells, respectively.

Techno-economic optimization is a standard oil industry practice that is used to rank and prioritize several development options. This practice strikes the right balance between technical and economic results and provides a sufficient ground for management's decision-making. Figure 7.9 presents a summary of techno-economic optimization results for the exercise shown in Table 7.3.

Table 7.3 is an abridged version of a detailed worksheet. It consists of a total of 21 columns split into two portions. The spreadsheet's top part includes columns 1 through 11, and the bottom section columns 12 to 21.

NPV of the future cash flows, IRR, and Payback Period for all three options, namely, development with vertical, deviated, and horizontal wells, are calculated. Based on the NPV and IRR, the development option "O1" with vertical wells is the most optimal for developing this reservoir.

Specialist accountants carry out a separate detailed economic evaluation that looks at depletion, depreciation, amortization, and several other factors. This evaluation is used to prepare the company's balance sheet, which goes to different stakeholders, including the government. Oil companies can use

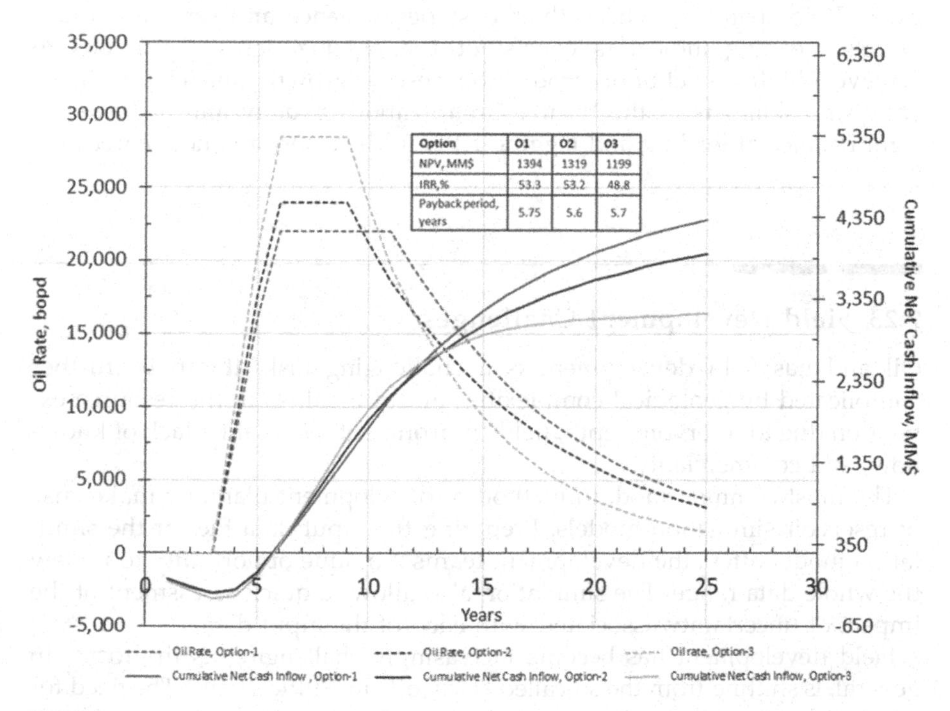

Option	O1	O2	O3
NPV, MM$	1394	1319	1199
IRR,%	53.3	53.2	48.8
Payback period, years	5.75	5.6	5.7

FIGURE 7.9
Summary of the Techno-Economic Optimization.

this evaluation or modify it based on the industry standards or their company practice to determine the development cost of a petroleum resource.

7.22 Alignment of Business Goals with the FDP

Almost all large oil and gas companies, particularly the NOCs and IOCs, are quite explicit in the proclamation of their mission, vision, goals, and strategy. It helps them to prepare their short- and long-term action plans. Field development planning has to be aligned with these fundamentals.

All organizations have a system to set and measure their performance. The KPIs are designed to cover the main business activities intrinsically linked to its strategic objectives. The responsible managers manage their staff members' performance to ensure the set timeframe's delivery of targets. KPIs have lagging and leading indicators. The purpose of establishing and monitoring the lagging indicators is to measure past performance. Accidents, injuries, and fatalities are some examples of the lagging indicators relating to HSE. Production or reserve growth achieved in the previous year are some other examples of lagging indicators that are used by E&P companies to check their past performance and set their future targets. Leading indicators consist of the organization's aspirations to achieve a higher level of oil production, reserve growth, financial profit, or employee skill sets in the future. Organizations usually aim to have the right balance of leading and lagging KPIs in their performance scorecard.

7.23 Field Development Challenges

Oil and gas field development is a challenging task. It can be further complicated by geological complexities and difficulties in the resource extraction due to reservoir depth, field environment, oil quality, lack of know-how, and commerciality.

The most common modern method of development planning makes use of reservoir simulation models. Preparing the input data file for the simulation model offers the development teams a unique opportunity to review the whole data range. The simulation also allows a quick assessment of the impact of uncertainty associated with each of the input data.

Field development has become increasingly challenging as the focus, in general, is shifting from the so-called "easy oil" to "difficult oil." The need for energy is continually pushing the technology to access hydrocarbon reserves from frontier areas such as deep waters offshore Brazil, India, and elsewhere.

As always, the new challenges are technical as well as commercial and environmental. The recovery of oil and gas from unconventional reservoirs and the harsh environment of the Arctic must not only be commercially attractive; it should also be environment-friendly. These challenges demand new ideas and technologies for the exploitation of oil and gas reserves.

Field development programs face a more significant challenge from the "have and have not" paradox. IOCs have the expertise to extract oil and gas reserves, but they do not have direct access to it. NOCs, on the contrary, may have the custody of oil and gas reserves, but they don't necessarily have the know-how. IOCs are generally reluctant to part with their expertise in oil and gas technologies and so are NOCs with their oil and gas resources. IOCs promote a culture of merit and excellence. NOCs intertwine it with the rule of equitable distribution among locals.

Because of these contrasting ideologies, the IOC-NOC coalition for exploiting oil and gas reserves is considered an exercise in conflict management. However, continually increasing technological, commercial, and environmental challenges offer ample opportunities for their harmonious coexistence. By working together, they have also reconciled with their newfound roles to complement each other.

7.24 Reservoir Monitoring

The maximization of economic oil recovery requires that the reservoir be developed and managed according to a well-devised FDP. The design and implementation of the FDP require extensive planning and efficient deployment of resources. Intermediate milestones are set to ensure that the field production is along the projected path to ultimate oil/gas recovery. Therefore, the reservoir performance must be continuously monitored and corrected if any deviation is observed from the charted path.

Reservoir performance is monitored and evaluated by measuring the production rate, water cut, GOR, and reservoir pressure of individual wells for each reservoir and its sub-layers. A chronological record of these parameters is maintained in tabular and graphical forms with annotations explaining unexpected values. The ability to measure production rate, water cut, GOR, and reservoir pressure may sometimes be constrained by the type of completion and use of the technology in some wells. In such cases, these data may be acquired from neighboring wells or individual critical wells that are "strategically" placed on the structure to accurately represent the overall reservoir behavior.

Reservoir behavior can be addressed by continually monitoring and correcting well-behavior issues. Measurements of production rate, GOR, water cut, and reservoir pressure with time are essential to understand the system's energy balance, which governs the well and reservoir behavior. Engineers must be conscious of the way these measurements are made and rationalized. Reservoir pressure coming from different sources such as Static Bottom Hole Pressure (SBHP) measurements, RFT, pressure build-up/drawdown/fall-off, or other types of tests must be referenced to standard date and datum for preparing an isobar map. Similarly, production rate, GOR, and water cut can be measured near the wellhead using portable test separators or gathering centers via test separators. Production rate, GOR, and water cut may be plotted in graphical form or bubble maps to show their variance with time. Bubble map is a common visual representation of the rate or cumulative volume of fluids, that is, oil, water, and gas, extracted by producing wells from a reservoir. A bubble is a circle of the radius corresponding to the volume of a fluid produced by an oil or gas well. The center of this bubble is the well-location on a suitable property map such as an isobar map, porosity, or thickness map. Bubble maps may be prepared to represent cumulative production or daily rates of oil/water/gas/liquid. Many bubble map variations are prevalent, including pie charts showing fractions of oil or water as a fraction of the total fluid production. Reservoir engineers use bubble maps to highlight the role of faults, direction, and effect of water encroachment and forecast the dry and wet oil production performance (Figure 7.10).

Performance monitoring is a crucial reservoir management function. It ensures that a consistent and definitive ascending, stable, or descending reservoir behavior trend is established based on reliable data. Any uncertainty in response can seriously affect the quality of decision-making on reservoir management issues. For example, an increase in GOR or water cut with time may require a zonal transfer, choke offending production, and injection wells for some time, or permanently. Likewise, the reservoir pressure approaching the bubble point indicates the need for early initiation of a pressure maintenance program.

These parameters' measurement frequency is of equal significance to capture the changes taking place in reservoir conditions. This ability is many times limited by the availability of resources. Specific values of these indicators set up practical steps to achieve reservoir management objectives in the short and long term. In case the situation is not corrected by such measures, a complete relook at the development plan may be necessary.

7.25 Role of Oil and Gas Processing Facilities

In their lifetime, oil and gas wells can produce a mixture of oil, water, and gas associated with many impurities. The presence of associated gas and water tends to exaggerate the volume of crude oil. Associated water can form an emulsion or be a corrosion source due to scale formation or salt precipitation. Oil and gas can also carry acid gases such as carbon dioxide (CO_2) and hydrogen sulfide (H_2S) that can pose severe HSE risks if present in higher concentrations. They can also be highly corrosive to the subsurface and surface equipment and material at lower concentrations. The other impurities can be dissolved salt, scale, corrosion material, and even sand, damaging facilities during separation and pipeline transportation. Therefore, the producer is obligated to bring the produced oil and gas to mutually agreed on quality specifications as laid down in the sales agreements.

The API gravity usually specifies the crude oil quality. However, the refineries may sometimes compromise with crude quality for a price and allow minor sulfur, salt, or water contents. This concession makes sense because the refinery infrastructure is more sophisticated and better equipped to deal with these impurities than the relatively simple oilfield facility, which primarily separates fluids based on small changes in pressure, temperature, and gravity. Such concessions are not uncommon between the domestic upstream and downstream oil companies.

Natural gas dissolved in crude oil or accumulated in an oil reservoir's gas cap is termed as associated natural gas. Natural gas that occurs as a separate gas pool with no association with crude oil is called free gas. Natural gas may also be produced by condensate wells and is rich in heavier hydrocarbons such as propane, butane, and pentanes.

Like crude oil, raw natural gas may contain impurities such as water vapor, hydrogen sulfide, carbon dioxide, helium, nitrogen, and other compounds. These impurities must be removed or diluted to bring the supply gas to the sale specifications or pipeline quality. N_2, CO_2, He, and CO having no energy content act as diluents and tend to reduce the heating value of the natural gas. Water vapor is also a diluent that can condense as liquid and burden the dehydration units at high pressure and low-temperature conditions. H_2S poses safety risks, while an excessive amount of sulfur can cause corrosion problems.

Natural gas is sold based on quantity and quality measured in a Thousand Cubic Feet (MCF) and British Thermal Units (BTU). The average thermal content of natural gas varies between 950 and 1150 Btu/scf. Historically, the produced associated natural gas was flared and was a source of carbon dioxide emission. Most companies now insist on zero gas flaring objectives. New environment protection regulations have forced the industry to invest in LNG (Liquefied Natural Gas) and GTL (Gas to Liquid) plants.

The produced water from oil wells must either be reinjected into the reservoir for pressure maintenance or underground shallow reservoirs devoid of hydrocarbons. If considered for utilization or disposal on the ground or into the sea, it must be treated to meet the regulatory requirements.

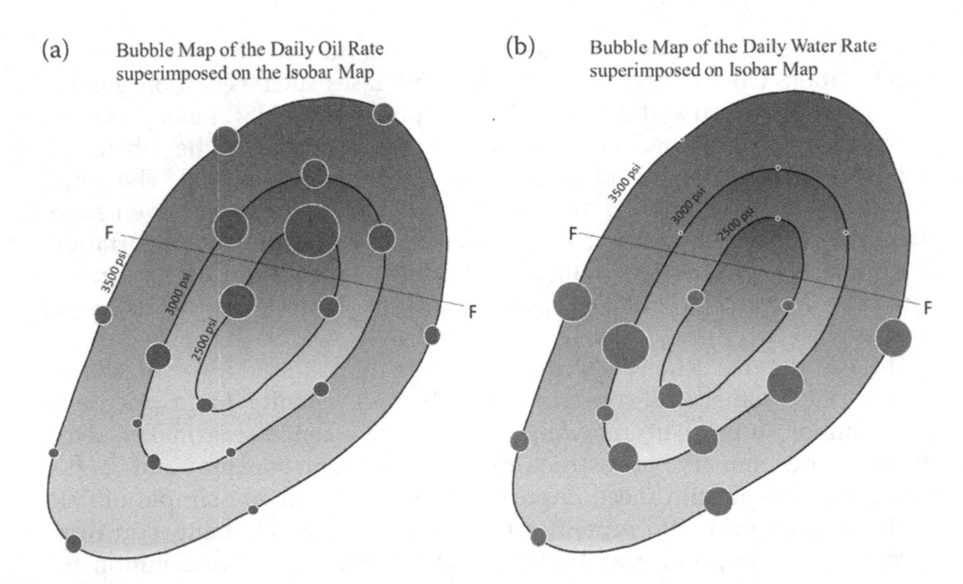

FIGURE 7.10
Daily Rate of Oil and Water Production Bubble Map.

An oilfield production facility represents a specific arrangement of necessary equipment between the wells and the sales terminal. It forms an essential link in the supply chain of crude oil from an underground reservoir to the customer. A facility's primary function is to gather and separate the wells' produced fluids into oil, water, and gas. It also provides necessary stabilization to the crude oil and makes it ready for sale or disposal in conformance with accepted standards.

While separated oil must meet sales requirements in terms of API, sulfur, BS&W, and salt content, the produced gas and water must also conform to sales or disposal specifications. The facility should even afford a system for sampling and measurement of crude oil to determine its value before it is delivered into the transportation system.

7.25.1 Oil and Gas Separation

A complex well-stream may consist of oil, gas, water, paraffin, sediments, and corrosion material. Separation of oil and gas from other undesirable constituents of the well-stream is necessary to make them suitable for supply to the refineries or other buyers. The most common sale criteria require that crude oil not contain more than 1% BS&W, and the sales gas should not have any free liquids. Some refineries or buyers may also specify a maximum limit for salt content.

Separation of gas from oil may start in the reservoir or the wellbore. It may progressively increase with a drop in pressure through the flow lines and surface handling equipment. Production stream from different wells is received at the gathering center. It is fed to oil and gas separators to separate oil and gas from other constituents present in the production stream. The separation of oil and gas is necessary to transform the oil and gas into a saleable product. It is also required because downstream equipment cannot handle the oil-gas mixture. Pumps require gas-free liquid, and the compressors and dehydration equipment require liquid-free gas. Oil and gas metering devices do not work correctly and are highly inaccurate in the presence of the other phase.

Three variables, fluid pressure, temperature, and composition, control the process of oil and gas separation. These variables are affected by physical processes like momentum, gravity, and coalescence. Fluids present in the production stream have different densities and, therefore, different momentum. Changes in fluid direction will separate fluids at different momentum. Gravity plays a role in separating oil, water, and gas phases due to the difference in their densities. Coalescence is a process that allows smaller droplets to aggregate into larger droplets. This process effectively helps recover small oil droplets from the gaseous phase and is promoted by coalescing devices fitted in the separators. Every sub-system inside the separator, such as gravity separators, mist extractors, cyclones, heated water baths, and so forth, promote these physical processes to separate the liquid from the gas.

A separator is a horizontal, vertical, or spherical pressure vessel used to separate liquids and gaseous phases. The horizontal separators facilitate improved phase separation due to their flow geometry than vertical separators. However, vertical separators are preferred in case of surface constraints. Irrespective of their geometry, separators aid in phase segregation, that is, oil, gas, and water, based on their density. Gas being the lightest stays on top, and water being the heaviest, settles at the bottom (Figures 7.11–7.13).

Oil and gas separators may be classified by their orientation, function, operating pressure, or application.[14] Oil and gas separators may have a horizontal, vertical, or spherical orientation. Horizontal separators (Figure 7.14) are well suited for the liquid–liquid separation and are invariably used for high GOR wells, foaming liquid well streams, and liquid–liquid separation.

A vertical separator (Figure 7.15) is ideal for high flow rate, low to moderate GOR wells, and large-sized liquid slugs. It provides better surge control and eliminates liquid carry over to gas inlet. Vertical separators may come as pre-fitted with accessories and can be fixed, skid, or trailer-mounted for quick and easy installation.

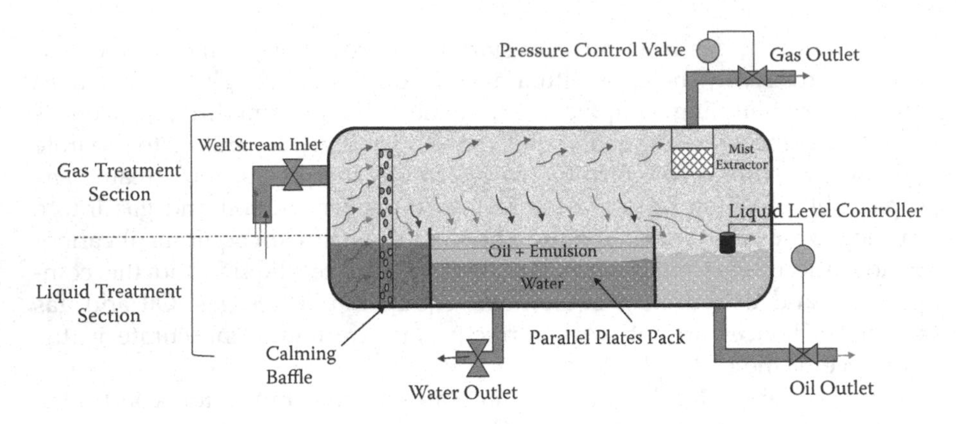

FIGURE 7.11
Schematic of a Horizontal Three-Phase Separator.

The spherical separator (Figure 7.15) though very efficient in terms of working pressure has limited surge control capability. Spherical separators are less efficient than horizontal or vertical separators due to their limited liquid settling section. They are also difficult to fabricate and are therefore rarely used.

Oil and gas separators must function in a wide range of pressures, from 5000 psi to near-vacuum. Based on their typical operating pressure range, they can also be classified as high (750 to 1,500 psi), medium (225 to 700 psi), or low pressure (15 to 225 psi) separators.

It is quite common to classify the oil and gas separators based on their application. They can be referred to as stage separators, production separators, test separator, or low-temperature separator. The term stage separator (first stage, second stage, and so forth) is used when a battery of separators is in operation for oil processing. Whether horizontal, vertical, or spherical, a production separator is used to process the daily production stream from a group of wells in the oilfield.

The other important criterion for the classification of the oil and gas separators is their ability to separate well fluids into two or three phases. Two-phase separators are good at separating well streams into two phases, liquid and gas. Three-phase separators can separate well fluids into three separate phases: oil, water, and gas.

Gathering centers usually have a test separator, which can test a new well after hook up to the gathering center for measuring its production rate, GOR, and water cut. Test separators are not required to have significant production handling capabilities like production separators. They can be used for periodic tests of oil and gas wells with dynamic changes in production rates. They can be vertical, horizontal, or spherical, portable or permanent, and two-phase or three-phase, depending on the need.

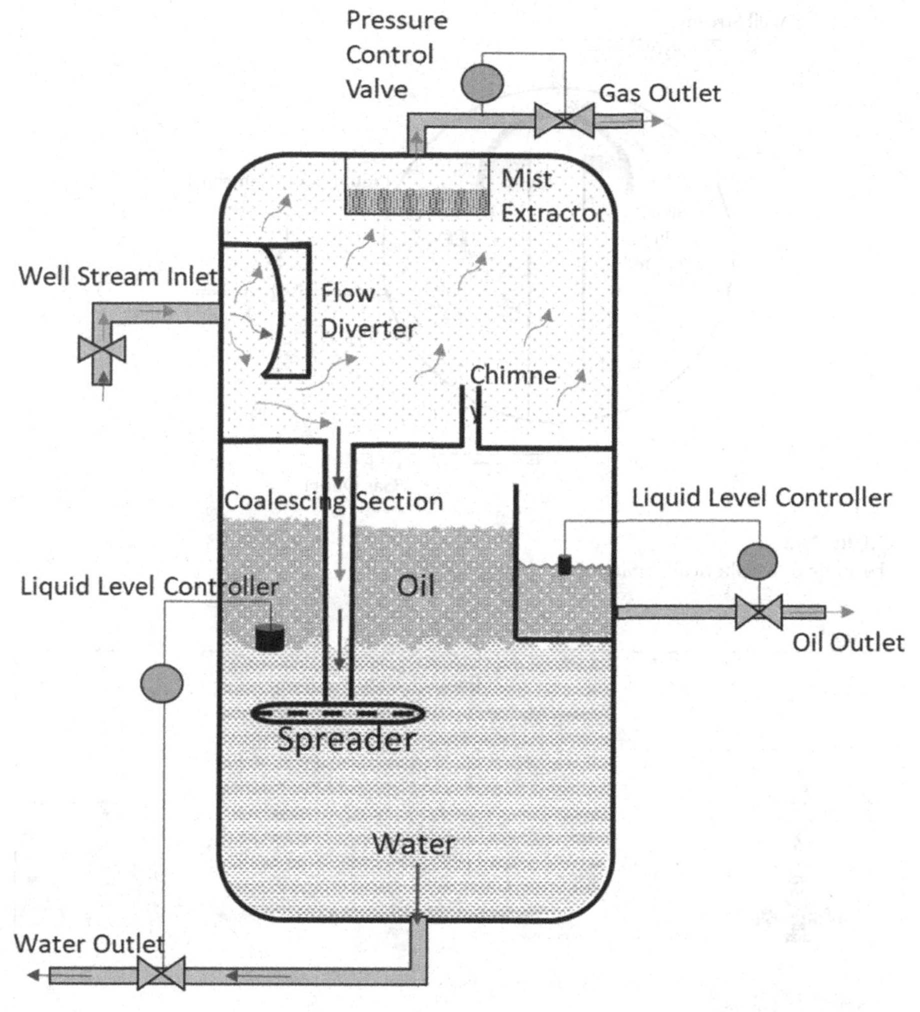

FIGURE 7.12
Schematic of a Vertical Three-Phase Separator.

Like pressure, the temperature also plays an important role in the separation of well-fluids. A low-temperature separator is one in which well fluids are suddenly allowed to expand through a pressure valve or choke to achieve a considerable reduction in temperature. This process is used to extract condensate and remove moisture from the gas at temperatures of 0°C to –15°C. Low-Temperature Separation (LTS) enables condensate recovery from the vapors that would otherwise leave the separator with condensate.

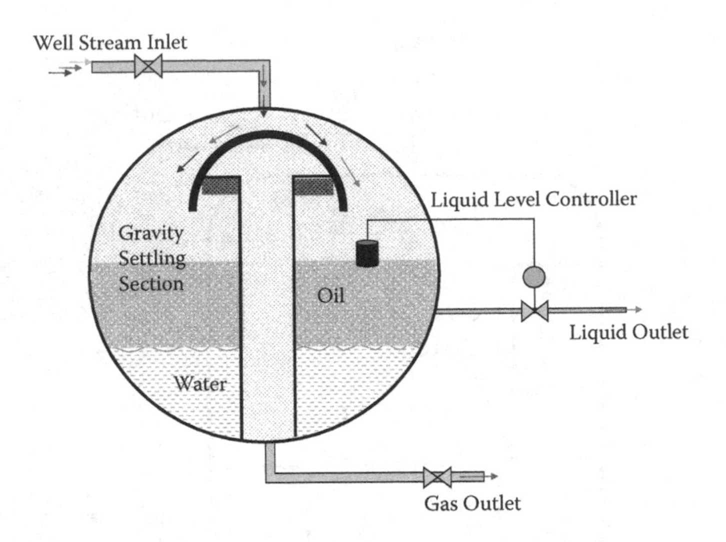

FIGURE 7.13
Schematic of a Spherical Separator.

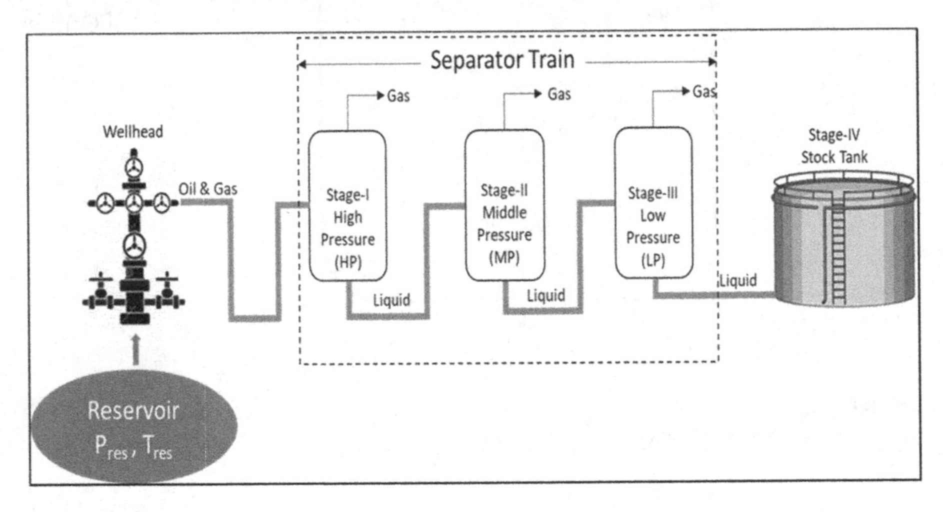

FIGURE 7.14
A Typical Multi-Stage (4-Stage) Oil and Gas Separation Process.

7.25.2 Functions of Oil and Gas Separators

The primary purpose of oil and gas separators is to handle and process the hydrocarbon mixture delivered by a group of wells into the saleable oil and gas quality. Recognizing that crude oil, natural gas, and formation or injected water are the main constituents of the production stream fed to the gathering center, the separators must efficiently achieve the following:

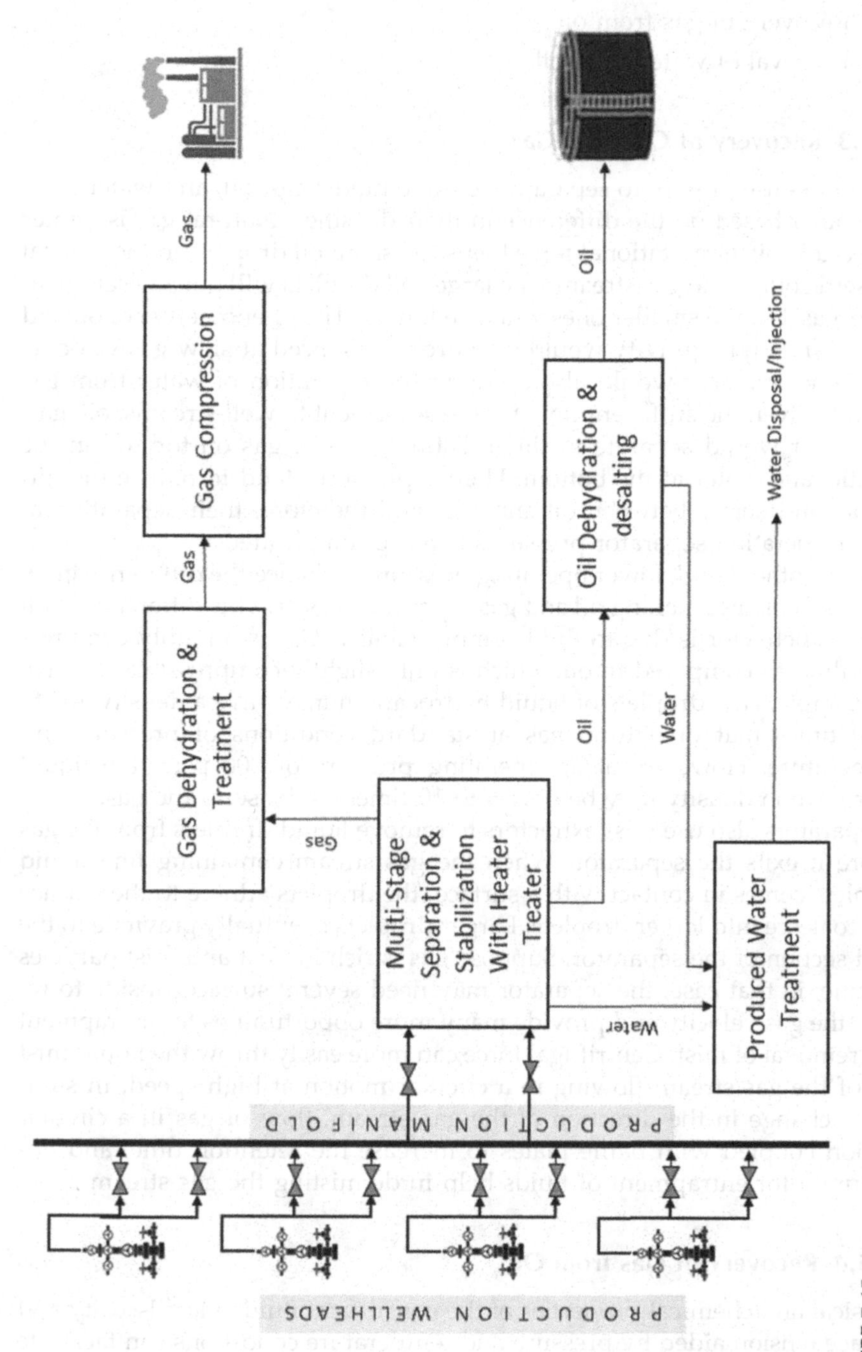

FIGURE 7.15
Simplified Oil and Gas Processing Diagram.

- Recovery of oil from gas
- Recovery of gas from oil
- Removal of water from oil

7.25.3 Recovery of Oil from Gas

Separators use gravity to separate the three fluids: gas, oil, and water from each other based on the difference in their densities. Natural gas is lighter than crude oil. Gravitational force helps the small oil droplets in the natural gas settle out of the gas stream. The larger oil droplets will quickly settle out of the gas, but the smaller ones will take longer. The phenomenon of oil and gas separation by gravity would be more pronounced at slow gas velocity. The same principle would also apply to the separation of water from gas and oil. Given the sufficient time for the settlement to well-streams, oil, gas, and water would separate in three distinct phases: gas on top, oil in the middle, and water at the bottom. Higher pressures tend to reduce the difference in density between oil and gas, and therefore, their separation at higher operating separator pressures is more complicated.

On the other hand, lower operating pressures enhance the difference in oil and gas densities, making oil and gas separation easier. This difference in oil and gas behavior is due to fluid compressibility. Gas is a highly compressible fluid as compared to oil, which is only slightly compressible. To give an example, "the droplets of liquid hydrocarbon may have a density 400 to 1,600 times that of natural gas at standard conditions of pressure and temperature. However, at an operating pressure of 800 psig, the liquid hydrocarbon density may be only 6 to 10 times as dense as the gas."[15]

Separators also use mist extractors to remove liquid droplets from the gas before it exits the separator. When the gas stream containing fine liquid droplets comes in contact with a surface, the droplets adhere to the surface and coalesce into larger droplets. Larger droplets eventually gravitate to the fluid section of the separator. Suppose gas is rich in mist and mist particles are fine; in that case, the separator may need several surfaces inside to reduce the gas velocity and provide many more opportunities for entrapment and removal of mist. Centrifugal force can more easily throw the liquid mist out of the gas stream flowing in a circular motion at high speed. In summary, change in the direction of the gas stream, flow of gas in a circular motion coupled with baffle plates to increase the retention time, and opportunity for entrapment of fluids help in de-misting the gas stream.

7.25.4 Recovery of Gas from Oil

Physical and chemical properties of the constituent fluids like viscosity and surface tension aided by pressure and temperature conditions can facilitate entrapment of non-solution gas and water particles in the oil. Several operating and design factors affect oil and gas separation; the most important of them are:

- Rate of throughput
- Design and operating pressures and temperatures
- Design for phase separation (gas-liquid or liquid–liquid)
- Design for the degree of separation (for example, remove 100% particles >10 microns)
- Size and configuration of the separator
- Fluid properties (density, compressibility, composition)
- Surging, slugging, or foaming tendencies of the feed streams
- Presence of impurities (paraffin, sand, scale, etc.)

The liberation of gas from oil is a function of the change in pressure and temperature. It can be further assisted by a combination of heating, agitation, baffling, coalescing packs, chemicals, and filters. Agitation causes the gas bubbles to coalesce for quicker separation.

Some of these features may be included in the separator design. Gas, oil, and water settle within the separator based on their gravity, gas on top and water at the bottom. The attached instrumentation controls the oil and water level optimally for separation. The separators usually achieve the bulk separation of water. A coalescer helps reduce water content by recovering the smaller water droplets that do not typically settle under gravity from an oil stream.

Oil and water can be separated in a three-phase separator by the use of chemicals and gravity separation. Sometimes the size of the separator may not be large enough for separation. In such a case, installing a free-water knockout vessel upstream or downstream of the separator can be planned.

7.25.5 Removal of Water from Oil

The separation of oil and water is never perfect, particularly when oil and water form tough emulsions. There is always some water left in the oil. Separators are known to be less efficient for lower API oils. In some cases, the separation may be so poor that 1% to 20% of water may not be removed. Buyer agreements specify the maximum water content and the maximum salt content in the crude oil. High water content can be corrosive to pipelines and cause problems with downstream processing. On the other hand, high salt content due to the salinity of the produced water associated with the oil may cause a problem with refinery equipment.

Some of the water associated with oil can be removed by heating the oil-water mixture to 80°C to 85°C-heating aids in reducing viscosity and increasing the density contrast between oil and water. The remaining mixture or emulsion then moves over to the next section of the treater, where the electrostatic force helps the small water droplets to coalesce, break free from oil, and settle at the bottom as a separate phase. The free gas is removed from the top of the treater.

A conventional treater with the provision of an electrostatic grid can reduce the water content in the oil to 0.3% to 0.5%. The oil from the treater flows into a dry oil tank. It is then pumped into a pipeline for transportation.

7.25.6 Multi-Stage Separation

Multi-stage separation is necessary to maximize the oil recovery and to stabilize the oil and gas phases exiting the final separator.

Reservoir oil volume shrinks as pressure is reduced, and vaporization occurs. This shrinkage is not only a function of the temperature and fluid composition, but it also depends on the process of separation of gas from oil, be it flash or differential. Sudden and drastic reduction in pressure of a single-phase fluid such as gas dissolved in oil can cause immediate separation of gas from oil, forming two distinct phases. The flash liberation most nearly approaches the situation in field separators or during the transportation of volatile petroleum products through pipelines.

Differential liberation is a process caused by a gradual decrease in pressure, which results in the separation of gas that is continually removed from the system. Multi-stage separation is a close representation of this process, subject to an optimum set of separator temperature and pressure, to achieve a larger oil volume in the stock tank per barrel produced.

Laboratories have reported that a differential process for oil and gas separation can increase oil recovery by 4% to 7.5% over a conventional two-stage flash separation process.

Field tests of the differential process yielded an increase of stock-tank oil of 5%. Field tests that deployed the differential separation process also corroborated a 5% increase in stock-tank oil recovery.[16]

Because of the above, oil and gas separation is carried out in several stages, with a series of separators operating at successively reduced pressures. The oil is finally collected in large stock tanks, which maintain atmospheric pressure and temperature. This oilfield practice is known as multi-stage separation. It mimics the differential liberation process to ensure the maximum recovery of stabilized oil from the production stream. Production wells may be operating at different wellhead pressures due to the variations in reservoir pressure and well-PIs. In a typical four-stage separation process, the wells with wellhead pressure higher than the HP separator are connected to the GC's HP separator. The same logic is applied to the wells that have their wellhead pressures higher than the MP or LP separators. Therefore, the fluids entering the HP separator are subjected to the four-stage separation process by HP, MP, and LP separators before delivery to the stock tank to maximize the yield of oil from the production stream.

On the other hand, the production streams from wells directly connected to the MP separator go through a three-stage separation process, namely, MP, LP, and the stock tank. In contrast, those connected to the LP separator have to pass only two separation stages, that is, LP and the stock tank. The multi-stage separation is similar to differential vaporization, which promotes maximum liquid recovery from the production stream and stabilized oil and gas.

Stabilization and separation are two different processes. Stabilization of crude helps increase the intermediate (C_3 to C_5) and heavy (C_6+) components in the liquid phase. Stabilization requires the vapor pressure of crude oil to be brought down to meet oil-pipeline specifications. The simplest form of stabilization is to produce a high-pressure well directly into the storage tank at atmospheric pressure. This method may get the true vapor pressure of oil down to atmospheric; however, many lighter components crucial for the crude oil's value addition are bound to escape with gas. Therefore, flash separation, in this case, results in stock tank oil with lower API. Multi-stage separation ensures that the crude oil is separated and stabilized in stages without losing valuable oil components.

In a typical four-stage separation process that includes the stock tank as the fourth and last stage, the liquid is first flashed at an initial pressure to the high-pressure stage separator. This separated liquid stream from the first stage, the high-pressure separator is directed to the middle-pressure and then to the low-pressure stage separator before entering the stock tank (Figures 7.14). Usually, three- or four-stage separation is considered sufficient for oilfield operations. However, due to the variations in the fluid properties and their behavior, detailed engineering studies and judgment are crucial to deciding the number of stages required for optimum separation.

Petroleum literature and industry professionals are not consistent in the way they report the number of separation stages. Some count the stock tank as a stage of separation, and some don't. Therefore, it is essential to qualify if the stage separation is inclusive or exclusive of the stock tank.

Operating pressures of the multi-stage separators hold the key to the liquid recovery from the hydrocarbon mixture. The following would typically constrain the separator pressure:

a. Minimum compression requirement for sales gas delivery to the consumers and

b. Minimum back pressure on the flowing wells to allow the full flow of fluids into the separator

The sales gas pressure is a constraining factor in the operating pressure of the high-pressure separator.

Table 7.4 provides an approximate guide to the number of stages in separation, which field experience indicates is somewhat near optimum.[17]

TABLE 7.4

Common Range of Operating Pressures for Multi-Stage Separation

Separator Pressure, psig	Number of Stages (Excluding Stock Tank)
25–125	1
125–300	1
300–500	2
500–700	2

More stages may be required for flowrates above 100,000 BOPD if economically justified by additional liquid recovery.

In contrast, the separators used for gas may have an operating pressure range of 1,000/2,000 psi.

7.25.7 Gas Dehydration

The produced gas is saturated with water, and it must be dehydrated before sending it for sale. Glycol dehydrators that use tri ethylene glycol (TEG) are very popular in the oil and gas industry. However, their downside is that they emit Methane, Volatile Organic Compounds (VOCs), and Hazardous Air Pollutants (HAPs) harmful to the environment. Optimizing glycol circulation rates reduces emissions and increases gas savings. New advancements in gas dehydration technology are capable of achieving zero-emission dehydration at lower costs.

7.26 Facility Constraints and Management

Constraints are a way of life. It is not possible to do everything due to money, time, or capacity constraints. Similarly, a facility has many constraints of design, construction, operation, and maintenance. These constraints commonly originate from cost and time of construction, limitations of available technology, and capacities of various equipment parts, such as a separator or dehydrator.

The constraints impose trade-offs on facility construction, performance, and results. Lower construction costs and time may reflect in loss of quality, high maintenance cost, increase in down-time, and risk to operations. In reality, a project must be executed in a world of constraints, both internal and external to the organization. Some examples of the internal constraints are the budget process and cut-off dates, project scoping and design, project award process, prescribed quality specifications, risk-taking, and project delivery time due to a commitment to the customer. HSE requirements also pose project constraints. External constraints can be in the forms of agreements, regulations, laws, or the state of technology that limit project options.

Economics is a universal driver for both the design of surface facilities and the estimated production profile. However, more often than not, the facility's maximum throughput rate becomes the overriding design consideration. This rate corresponds to the maximum plateau rate of production. However, field production is often lower than the estimated plateau rate or its duration. Therefore, the construction of a facility based on the plateau rate of production would often result in an oversized facility design. Studies have highlighted[18] that field complexity and uncertainties in the forecast have often led to the oversizing of the facility and its equipment. On top of this, the design also provides for the change in the production forecast. The other uncertain aspect of the performance forecast is the estimated water and solids production. In the absence of any direct testing, a history match is taken as evidence to represent the aquifer properties and behavior. This rationale may work well for forecasts with water injection where the rate of injection can be controlled as desired. But natural aquifers do not respond precisely similarly under the reservoir dynamics, often causing water production profiles to be underestimated.

At the GCs, oil and gas separators should not only be able to handle the maximum daily production from the field, but they should also process the fluids of variable composition and quality effectively to meet the sales specifications. The typical approach is to make the separators large enough to handle the maximum throughput but constrain the design to a maximum value of GOR or water cut based on the forecast. The fluid mix of the production stream during the plateau period is quite different from the fluid composition of the tail-end period and so is the efficiency of the separators. In summary, the size of the separators is a crucial requirement for handling field throughput. But processing fluids of the different properties and their variable mix to the required specifications is an equally important objective, which can only be achieved by focusing on the separator's internal design and quality elements. The real question is if oversizing the facilities to meet the short-time high-rate requirement at the cost of long-time low-rate conditions with ineffective fluid separation is acceptable.

The other examples of typical facility constraints are the facility's limited capacity, pressure ratings, or fluid-carrying capacity of the flow lines for transportation, the limited size of the separator, and the water treatment unit's limited ability to handle produced water. Unstable flow, excessive surging, foaming nature of crude, loading of solids, and corrosive H_2S and brine can also limit facilities' ability to perform effectively. In gas lift wells, the gas injection rate based on the wells and reservoir properties should be in a specific range to escape unstable flow in the tubing.

Upstream oil and gas companies function in a challenging national and international environment. Local governments maintain tight control over their activities and expect strict compliance with regulations and standard operating procedures. But the oil and gas price does not necessarily consider the problematic logistics, the complexity of operations, and associated risks. The companies have to continually invest in new technologies and training and development of its workforce to reduce the cost of their services and improve their facilities, assets, and employees' performance. Focus is shifting from "reactive" to "proactive" solutions enabled by digital technologies that have improved human access and insight into reservoirs. Experts can look at the wells and complex operations remotely and suggest solutions. Digital Oilfields have provided a platform to collect data at the desired time and frequency for collective decision-making. Companies are investing in Centers of Excellence/Research Institutes to focus on future research and capacity building. Advanced Condition Monitoring (ACM) or Condition Based Monitoring (CBM) is the latest technology with predictive capabilities that can help avoid a possible operational failure or facility accident based on equipment and material performance. The regular practice of a pre-set maintenance schedule for a facility cannot address malfunction/poor performance until the next scheduled maintenance. Predictive maintenance tools can evaluate maintenance requirements in advance to prevent malfunction and achieve optimum performance with minimum downtime.

Facility maintenance directly impacts facility operations and can paralyze oil and gas production if not appropriately managed. Oil and gas facilities are costly and require expert attention for replacement or repair. Being a critical link in the supply chain, any inadequacy in their maintenance could result in a significant financial loss to the company. Timely and comprehensive support is the key to avoid these losses and must be preferred over maintenance short-cuts with costly long-term implications. Some of these implications are substantial depreciation of poorly maintained assets, loss of production, work hours, and revenue apart from missed production targets.

The safety of the facilities is a significant issue in all oil and gas companies. Many fires, explosions, sabotage, and spill accidents have been reported, with considerable losses to production, property, human life, and environmental damage. Oil and gas are flammable and, therefore, susceptible to catching fire and explosion. HSE is generally a top priority in all oil and gas companies; the gaps in the organization's safety culture and management of change procedures are common factors for the accidents. Equipment failure, human negligence, inadequate maintenance, not following safety norms/industry standards, and tight schedules can lead to avoidable accidents.

7.27 Project Management Slippages

In E&P companies, the project management department is responsible for engineering design and construction of the surface facilities for handling, processing, and transportation of oil and gas. This department may be known by other names such as Engineering & Construction, Major Projects, Facility Engineering, or Construction & Maintenance in other companies.

The project management department is generally responsible for creating new facilities and upgrades/revamp of existing facilities, offshore or onshore. Planning and executing the projects from conceptualization to commissioning and turning over to the asset is a part of their responsibility. The department provides total support throughout the project life cycle. The timely completion of these facilities and their trouble-free operation is essential for the oil company to achieve its production objectives.

Poor project management is a common source of slippages in the projects. Despite early knowledge of a project's criticality, it is usually behind schedule, over budget, and compromises with standard and quality. Project management literature is full of such case histories where more projects reported slippages than in-time completion. General conclusions about why so many projects fail to conform to the budget and schedule are weak assumptions, changes in the project's original scope, and lack of compliance with established policies, procedures, and project-specific quality assurance plans. The general tendency is to blame the processes and procedures for these lapses. One wonders how the system and processes created to facilitate organizational workflow can be a bottleneck and very excuse for its delay or failure!

Later, the companies may usually carry out a *post mortem* intending to remove any lacunae and improve things in the future. Lessons learned are recorded and shared with all concerned. The next time around, it is another vital project, the company wants it executed to perfection, but history repeats itself!

At this juncture, a comment about the processes and people is in order. Many organizations place too much emphasis on the process than people. The downside of this is that processes are often blamed for project delays. Incompetent staff can take shelter from the procedures to shield themselves if the processes are not time-bound and the timeline is not strictly followed. On the contrary, emphasizing people makes them feel important and responsible for project delivery.

Because the production and data acquisition are vital for new fields, the company might consider fast-tracking production with a makeshift EPF until the full-fledged facility is operational. EPF enables early cash flow and reservoir data collection from extended production tests during the initial production period. However, this crude processing method is suboptimal and only temporary though it can give project management the breathing space and opportunity to cover the slippages in the facility project completion.

But EPF is not an option to make up for the delays in the additional oil/gas processing or water injection facilities. The project management department seldom announces a delay of 1 year or more in one go. Instead, it uses the tactic to inform about the postponement in piecemeal, in several minor, inconsequential looking time-steps. These "seemingly harmless" short time intervals are used as the pretext to continue to produce oil/gas/water at the old rate rather than regulating it based on reservoir management principles. Notwithstanding the delays in the project commissioning, the company continues to drill new wells in the meantime. However, the new wells may not be hooked to any production facility (because there is no room to process more oil) or be connected to an existing facility with appreciable overload. None of these situations is desirable. The result is that:

- The production is not correctly processed at the GCs
- An excessive amount of associated gas/water loads the system
- The oil rate is significantly reduced
- Excessive withdrawal of reservoir fluids without proper pressure maintenance, in case of postponement of injection facility

These conditions can cause a substantial loss in income and impact reservoir health. It is not difficult to recognize that the delay not only implies extra cost and time; it also means missed opportunities for production and revenue collection. In real life, this "opportunity cost" may be enormously higher than the sum of its parts.

7.28 Urban Planning for Surface and Subsurface Management

The development of oilfields is similar to the development of townships. Oilfield development activities increase after the discovery of oil/gas is confirmed. First development wells are usually drilled on the structure where the risk of finding oil is low, and the expected oil rate is high. Further, development proceeds gradually from around these wells towards flanks with new information about the structure, rock quality, role of faults and their patterns, and the reservoir limits. As the number of wells grows, oil production from the field is maximized. Over time, field production declines, and more infill wells are required to maintain the production. When infill drilling is rendered inadequate to sustain the field production, secondary recovery methods are applied to improve oil production and recovery. Additional injection and production wells are drilled for secondary recovery operations. These operations may continue for long until a reasonable value of S_{orw} is achieved, and EOR methods are introduced for additional oil recovery. The rapid growth of wells resulting from the urgent need for oil production presents formidable challenges for properly managing available space in the subsurface and surface.

The oilfield development continues with drilling and hooking up of a large number of wells. Building production handling/processing facilities, booster stations, oil storage and pumping stations for exports/supply to consumers, water treatment and injection facilities, and constructing roads to access drill sites and wells and offices. At the beginning of the development, both surface and subsurface have ample room, and therefore, oil companies can lavishly use the available space. Most of the decisions about routing the flowlines and GC locations are based on hydraulics and cost. Therefore, the entire production paraphernalia is located directly above the reservoir's footprint, congesting the surface to the extent that drilling wells at a closer spacing in the secondary recovery phase may become difficult. The companies usually pay little attention to surface and subsurface planning in the early period of field development because they do not have a full and final view of the field development stage. The problems of surface congestion dawn as the activities reach the mid-secondary recovery stage, and the first view of EOR starts emerging. Unplanned and haphazard mushrooming of flowline clusters crisscrossing the field can be particularly problematic and pose problems for drilling wells at closer spacing.

Surface congestion due to production infrastructure and high well-density can pose difficulties for movement or placement of drilling rigs and force changes in well design and trajectory. Vertical, directional, and horizontal well profiles, which can sometimes be crooked, extend over a long-range, and can swarm the subsurface, increasing the risk of well-collision. It complicates the drilling and placement of well trajectories in the desired place.

Many mature fields in the US, Europe, and the Middle East are facing this problem. The problem arises because the focus of development activities, mainly in the early and middle stages, is on cost and oil production and not on proper space utilization on the surface or aesthetics. Urban Planning is a term used to describe reorganizing existing surface infrastructure in the way of oil and gas field development. It takes the name after the concept, process, and philosophy of town planning. A master town plan controls the design and construction of infrastructure for public utilities such as water, electricity, sewerage, telephone services, natural gas, car parks, playgrounds, road network, schools, hospitals, shopping areas, and so forth, for the convenience of the residents. Modern urban planning aims to allow sustainable development grounded in engineering, architectural, social, political, and economic considerations.

In oil and gas field development, urban planning aims to meet the surface requirements over the entire life cycle until abandonment. It may involve rationalizing flowline routes in the form of corridors and re-organization/relocation of electrical lines, power stations, roads, offices, moveable equipment, machinery, and so forth, that obstruct drilling wells or other oilfield activities.

No matter how good the planning, it is challenging to keep an oilfield master plan/layout unchanged beyond a time horizon of 20 years. Possible extensions of oilfields, the discovery of new reservoirs in the same or adjoining area, deployment of new technologies, and new ways of doing business can usually force an overhaul of a longer-term business plan and road map to ultimate oil recovery.

Urban planning and architectural engineering concepts are applicable and helpful towards efficient utilization of space, housekeeping, and general aesthetics in congested oilfields. The task is quite ingenious and can challenge the best of minds when accomplishing these objectives with minimal oil production loss and at a low cost.

7.29 Reservoir Management Execution

The present-day significance of oil and gas is at the root of reservoir management considerations and practices. National governments and oil companies are conscious that oil and gas are a precious national resource and their proper extraction and management are crucial for value creation. The governments exercise their control by way of policymaking and regulations. Or else they may choose to explore and develop the resource through their NOC. The government's direction and input are essential to decide the pace of development, formulation of the strategy, and capital and collaboration arrangement.

The most fundamental questions to a company's operating philosophy are whether it will work in a domestic or foreign setting, and for a short or long period. Local companies have the advantage of knowing the native culture. On the other hand, international companies take time to familiarize themselves with the environment and be careful about local sensibilities. Accordingly, a NOC's focus is primarily on long-term value creation, which includes the company's financial growth, the national economy's progress, and society's development. A POC or IOC's may not be so generous towards social obligations.

A company's vision, mission, and value statements enshrine a broad reservoir management goal (Figure 7.16). It is essential to break down this rather "abstract" goal into SMART (specific, measurable, attainable, relevant, and time-bound) performance measures. Smartly devised metrics can help bring focus to measure and evaluate the results of a particular activity, process, or project for improvement. Scientific investigation and

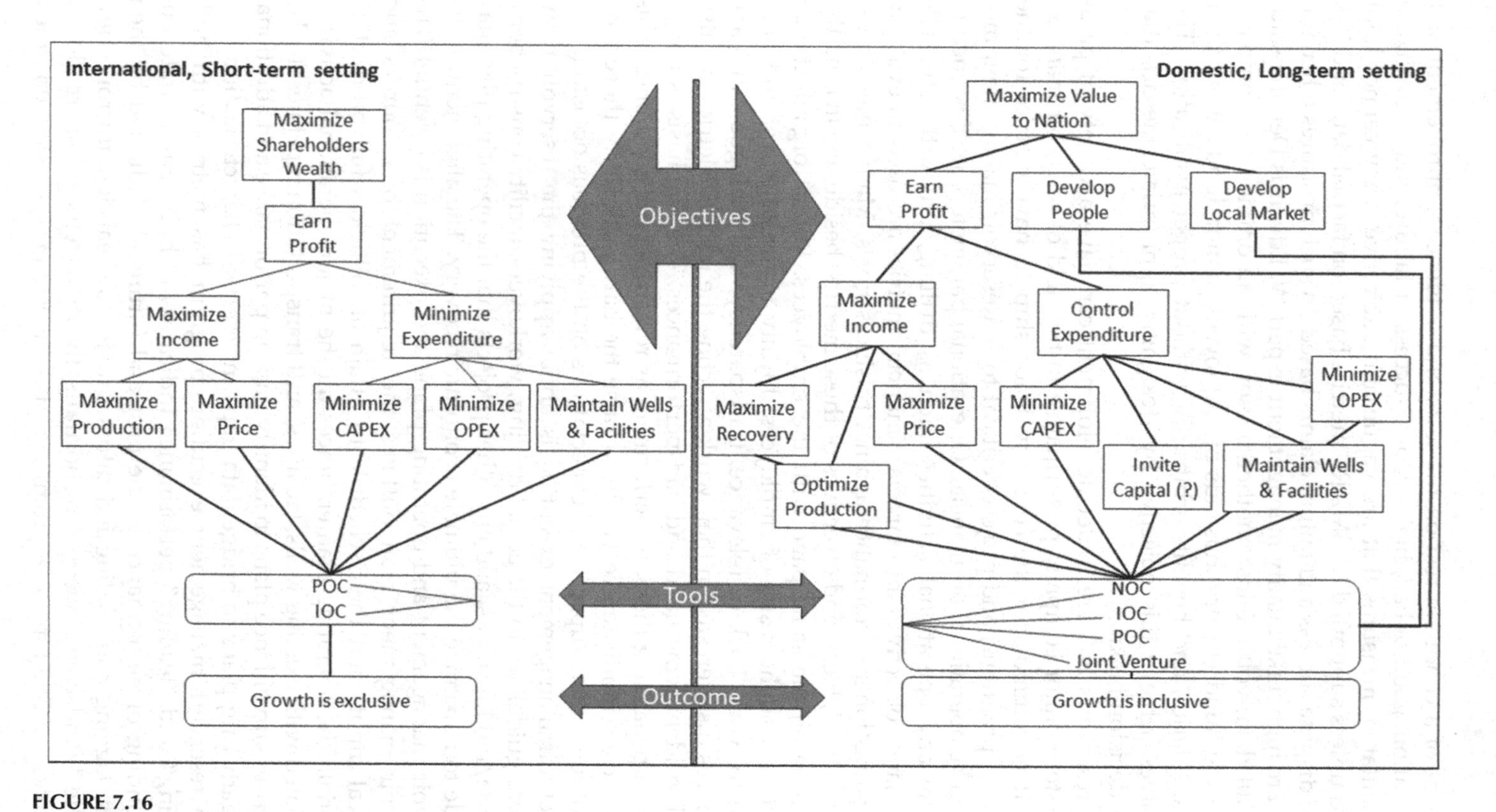

FIGURE 7.16
Basic Reservoir Management Structure-Strategic Objectives, Tools, and Outcome.

treatment of a condition rest on the tests and evidence. Improvement and deterioration, good or bad, high or low, success or failure, and so forth, are the qualitative measures that are vague and subjective. They cannot hold ground unless supported by SMART metrics based on factual data, and it is easy to dispute success and failure claims based on such measures when the stakes are high. Like quality management expert W. Edwards Deming said, "Without data, you are just another person with an opinion." A company comprising hundreds or thousands of employees cannot decide about their careers, salaries, and benefits arbitrarily without proper justification. The companies achieve this objective by designing appropriate metrics that justify decision-making.

The reservoirs that are under development can have a variety of parameters to monitor company performance in the field of reservoir management. It is common for the company leadership to have metrics for the processes, programs, and projects critical to it. Reservoir management metrics can be operational or financial. Operational parameters are designed to translate any operational activities such as drilling, production, or HR performance objectives into targets. Financial metrics are used to evaluate an organization's performance against business goals. Management must share the need, basis, and objectives of these metrics, besides evaluating the data to be on the same page with the employees. Robust organizational performance is the result of employees' loyalty to the employer and dedication to work. Lack of relevance, transparency, and dialogue can leave employees disinterested in their work affecting the results. Metrics are also useful in identifying good and bad performances. The emphasis is not so much on penalizing the poor performers as rewarding the good ones. This practice does motivate everybody to raise the bar, particularly those who did not meet their supervisors' expectations on the previous occasion.

Reservoir management execution is about applying the reservoir management guidelines to the actual drilling, workovers, facility construction/upgrade/modifications, and any other projects that the operating plan may include to maximize the ultimate economic recovery. The alignment of the reservoir management and operating plans and execution is essential. The reservoir management plan outlines the requirement of resources and physical inputs and their schedule to attain an oilfield's full production potential. The operating plan embraces all the major operational activities associated with existing wells, facilities, and transportation pipelines. It also comprises surveillance plans to identify gaps in performance and additional data gathering plans to bridge data gaps in reservoir characterization.

The reservoir management execution program has metrics with both "lagging" and "leading" performance indicators to build an accurate understanding of performance in the past and future. As indicated by their names, lagging and leading indicators provide a "rearview mirror" view and the "windscreen" view[19]. In shortfalls, the management may reallocate resources or fine-tune the programs and investments to stay on track.

TABLE 7.5

Examples of Operational and Financial Reservoir Management Execution Metrics

Performance Measure		Unit	Lagging Indicator (Previous Year)			Leading Indicator (Current Year)	
			Min	**Max**	**Actual**	**Min**	**Max**
Operational Metrics	New Well Drilling	#	20	23	22	23	27
	Average Oil Rate	MBOPD	75	77	77	77	85
	Water Injection	MBWPD	95	102	98	98	120
	Water cut	%	8	10	10	10	12
	Add Reserves (Cumulative)	MMSTB	30	50	50	50	75
	Construction progress of new GC	%	60	62	62	97	100
Financial Metrics	Revenue generation	MM US$	1350	1405	1405	1405	1551
	Development cost of oil and gas	$/bbl	5.0	5.3	5.1	5.1	5.3

Table 7.5 presents a simple example of reservoir management execution metrics.

It is tempting to measure everything, but it should be avoided. The data collection, analysis, and feedback loop could be quite cumbersome and even unproductive in some conditions. Selecting the right metrics is the key because it improves focus and channelizes its energy in the right areas. Each of the parameters listed in Table 7.5 requires collecting, processing, and reconciling data that the functional teams maintain in their repository. For accuracy and consistency in calculations and reporting of the results, the management must identify and entrust this responsibility to a champion who functions as a single point of contact. The champion's role is:

- To hold periodic reviews of reservoir management activities and accomplishments with management and between groups
- To liaise with the reserves management team for charting out future work plans based on the management's input
- To analyze the performance of the allocated resources to ensure their optimal deployment
- To share the experiences and lessons learned among the participating teams, and
- To identify and implement the best practices to improve the performance and results

The reservoir management execution champion has a crucial role to play between the technical teams and the management. Champion's integrity, expertise, and continuity have a bearing on the quality of the reservoir management program, its execution, and its results. However, due to rotational policies, promotions, and resignations, it is often difficult to retain the champions in their position for long. Therefore, the organizations are interested in calculating and reporting KPIs independent of the individuals.

The scale of operations, volume, and diversity of data necessitate that the champion deploys the computerized business intelligence tools to automate this process. These tools, known as dashboards, are like specialized apps capable of extracting and merging data from disparate sources into a single database to reveal patterns of complex data sets. Usually built in Microsoft Excel, the dashboards display the latest status of the reservoir management metrics on a single panel in charts, tables, or views, just like the motor car's dashboard. Their strength lies in their immediate visual impact due to simplicity, clarity, and a crisp pictorial presentation. The dashboard panels are extremely popular with busy managers and business leaders as they afford the means to grasp the critical gaps in the KPIs at a glance. Modern dashboards have moved away from Excel to combine data visualization capabilities with business intelligence and analytics software to improve the user experience.

7.29.1 Quantitative Appraisal of Reservoir Management Program

Reservoir management is a fundamental requirement for managing oilfield activities. It has been in place in all E&P companies in subtle or obvious forms. The general approach to the reservoir management plan can be compared with the daily consumption of food unmindful of its nutritional value. Modern companies do not favor such an unstructured use of scarce resources. Their plan must be comparable with a balanced diet program that provides all the organs and tissues with proper nutrition. It must contain enough protein, fat, carbohydrate, minerals, and vitamins that allow the body to work effectively without any disease, infection, fatigue, and poor performance.

A quantitative appraisal of the reservoir management program is essential to know the gaps in performance and focus on improvement. The maxim "if you can't measure it, you can't manage it" aptly describes the need for a quantitative appraisal. Carefully planned metrics such as shown in Table 7.6 can gauge the quality and effectiveness. These metrics aim to put together various KPIs to ascertain the key projects' progress. KPIs may depend on the maturity of the reservoir, stage of oil recovery, and the importance of any specific project that the company wants to evaluate at a given time. The company may set realistic or ideal targets in line with its capability and check its performance quarterly, half-yearly, or annually.

The quantitative appraisal helps in the following:

- To establish the current reality against the plan

TABLE 7.6

Examples of Quantitative Measures of a Reservoir Management Execution Program

#	KPI & Measure	Attribute, Definition & Remarks	Ideal Target
1	Drilling Management • Drilling Efficiency • Well Quality	$DE = \frac{\text{Total no. of wells drilled}}{\text{Total no. of wells planned}}$ $WQ = \frac{\text{Total oil rate achieved}}{\text{Total oil rate planned}}$	100% 100%
2	Well hook up to production facility (Time interval between completion of drilling and hook up to facility)	An oil well connected to the production facility as soon as completed	Zero delay
3	Production Management • Oil rate per reservoir • GOR per reservoir • Water Cut per reservoir	$Qo = \frac{\text{Actual oil rate}}{\text{Predicted oil rate}}$ $GOR = \frac{\text{Actual GOR}}{\text{Predicted GOR}}$ $WC = \frac{\text{Actual water cut}}{\text{Predicted water cut}}$	100% 100% 100%
4	Oil Rate from Rigless and Rig Work Over Activities	$\text{WOR Qo} = \frac{\text{Actual oil rate}}{\text{Target oil rate}}$	100%
5	Oil Well Inventory Utilization (Number of wells online)	$OWIU = \frac{\text{No. of oil wells online}}{\text{Total no. of oil wells in the reservoir}}$	100%
6	Reservoir Pressure Management	$Pm = \frac{\text{Actual reservoir pressure}}{\text{Target reservoir pressure}}$	100%
7	Oil Recovery Management for Primary and Secondary recovery phases: • Voidage Replacement Ratio (VRR) • Volumetric Sweep Efficiency(VSE) • Estimated Ultimate Recovery (EUR)	$VRR = \frac{\text{The total volume of fluids produced at reservoir conditions}}{\text{The volume of water injected at reservoir conditions}}$ $VSE = \frac{\text{Actual Estimated VSE}}{\text{Predicted VSE at the same time}}$ $EUOR = \frac{\text{Actual EUR based on reservoir performance}}{\text{Planned EUR for the design of the development plan}}$	1.0 to 1.2 100% 100%
8	Produced Water Utilization (Utilization of produced water for injection and other oilfield operations)	$PWU = \frac{\text{Amount of produced water utilized}}{\text{Total water produced from the reservoir}}$	100%
9	Gas Utilization (Utilization of produced associated gas)	$GU = \frac{\text{Amount of associated gas utilized}}{\text{Total Amount of associated gas produced}}$	100%

(Continued)

TABLE 7.6 (Continued)

Examples of Quantitative Measures of a Reservoir Management Execution Program

#	KPI & Measure	Attribute, Definition & Remarks	Ideal Target
10	Produced Water Disposal (PWD)	$PWD = \frac{\text{Amount of surface disposal of produced water}}{\text{Total water produced from the reservoir}}$	100%
11	Accuracy of Overall Reservoir Performance Prediction	1. Quality of match of historic oil, gas & water production/injection rates & cuts 2. Quality of match of reservoir pressures 3. Quality of match of fluid saturations 4. Quality of match of original and current in-place, recoverable and mobile oil and gas volumes	High
12	Production & Injection Facility Uptime	Production and injection facilities up and running time	98%
13	Budget Availability	Finance available on time to support field operations, material procurement, etc.	100%
14	Project Delivery on Schedule	Completion of projects to support/ enhance production and injection as planned	100%
15	Oil Production Cost/ Barrel	Consistency in the calculation to be maintained for comparison with industry standards	US$ X

- To identify the key obstacles in achieving production targets, oil recovery
- To identify areas for continual improvement
- To ensure accountability at all levels
- To reallocate resources according to reservoir management needs
- To benchmark performance against competitors or industry standards
- To generate or reorient action plans

Table 7.6 presents elementary and straightforward examples of reservoir management measures and KPIs. Asset teams can select and design their own KPIs the way they work best for them.

7.30 The Link between Reservoir Management and Organization

There is a direct link between reservoir management and the organization. Oil and gas companies with many employees follow a hierarchical setup with a chain of command. In such a scenario, the organization's structure and culture are two subtle factors that influence employee performance. The hierarchical structure is slow and bureaucratic by nature. It limits personal creativity and reduces an individual's role to that of a robot. However, there is no confusion about the roles and responsibilities of different departments and the control line.

As mentioned in Chapter 5, reservoir management is a comprehensive E&P organization system to maximize the benefits of the rate, recovery, and profit/value. BSC is a management tool that aids in measuring and monitoring the organization's progress towards these objectives. It links the organization's long-term objectives with short-term goals and lists the key deliverables in line with its business plan.

As mentioned in Chapter 5, reservoir management is a comprehensive E&P organization system to maximize the benefits of the rate, recovery, and profit/value. BSC is a management tool that aids in measuring and monitoring the organization's progress towards these objectives. It links the organization's long-term objectives with short-term goals and lists the key deliverables in line with its business plan.

Figure 7.17 presents a short version of a BSC of a typical NOC in the E&P sector. In reality, a company's BSC is much more extensive. It lists various projects/programs critical for the business, along with the groups responsible for delivery. It is a standard practice to assign a priority and weightage to these deliverables based on their significance to stay focused. The process of linking the reservoir management objectives with the organization and its multiple organs starts at the top and moves down the ladder. BSC allows various department and individuals to design their goals and KPIs. They can work with their group managers/supervisors to set their SMART performance targets in sync with the BSC.

The BSC helps to define, set, and institutionalize organization-centered goals. It promotes cross-team communication and the delivery of shared targets for the success of the organizations. The digital age has powered the web-based tools that can integrate the BSC with the Performance Management System to track and monitor ongoing efforts in real-time and intervene when necessary.

An organization's success depends on its leader and the people who are part of it. A leader is the most crucial link between the employees and the organization. He/she can convince and motivate people to accomplish specific goals, even in adverse circumstances. He/she must be result oriented with a proven record of accomplishments. When required, he/she should have the ability to bring people with conflicting opinions to a common ground and receive their whole-hearted willing cooperation. He/she must be a role model with exceptional communication and professional skills that draw unequivocal appreciation from colleagues and naked admiration from competitors. Clear vision, strategic thinking, and a strong desire for constant improvement are the hallmarks of a good leader.

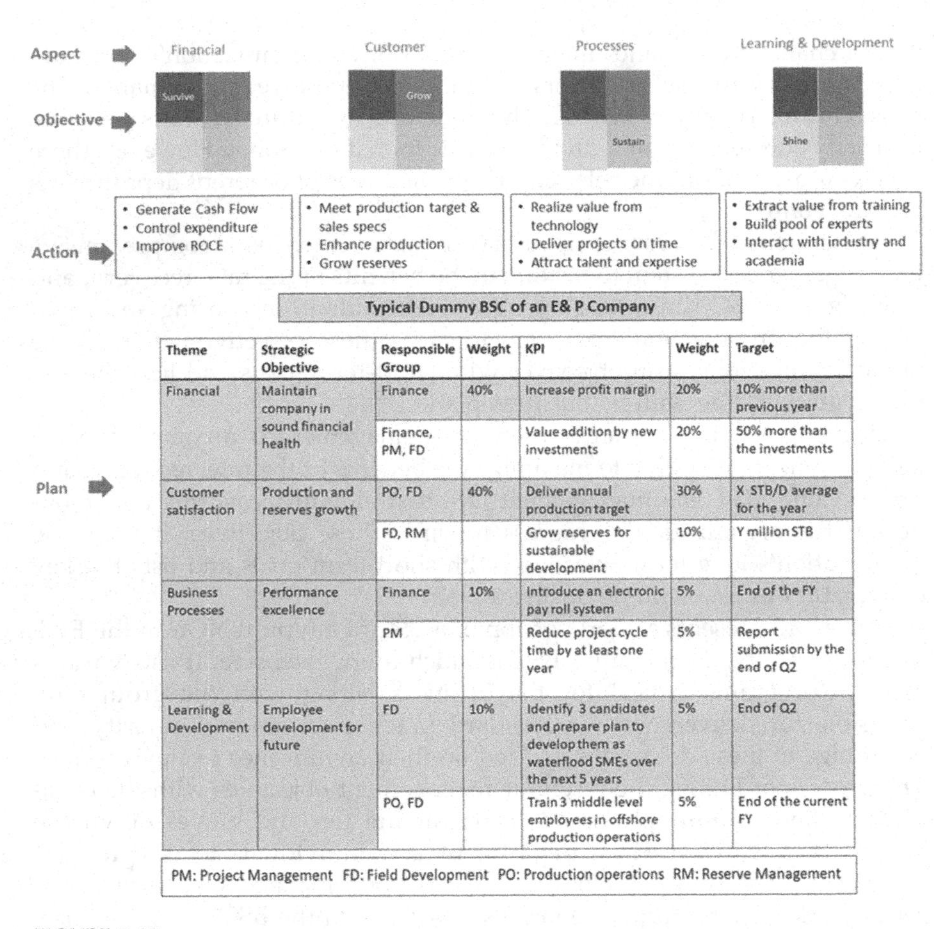

Theme	Strategic Objective	Responsible Group	Weight	KPI	Weight	Target
Financial	Maintain company in sound financial health	Finance	40%	Increase profit margin	20%	10% more than previous year
		Finance, PM, FD		Value addition by new investments	20%	50% more than the investments
Customer satisfaction	Production and reserves growth	PO, FD	40%	Deliver annual production target	30%	X STB/D average for the year
		FD, RM		Grow reserves for sustainable development	10%	Y million STB
Business Processes	Performance excellence	Finance	10%	Introduce an electronic pay roll system	5%	End of the FY
		PM		Reduce project cycle time by at least one year	5%	Report submission by the end of Q2
Learning & Development	Employee development for future	FD	10%	Identify 3 candidates and prepare plan to develop them as waterflood SMEs over the next 5 years	5%	End of Q2
		PO, FD		Train 3 middle level employees in offshore production operations	5%	End of the current FY

PM: Project Management FD: Field Development PO: Production operations RM: Reserve Management

FIGURE 7.17
Link between Reservoir Management and Organization through the BSC.

The leader should have the courage of conviction and making hard decisions in the face of difficulty. As a leader of modern oil and gas corporations, he/she must be particularly adept at dealing with uncertainties and managing risks.

Uncertainty management is a particular skill that is required to avoid risks to oilfield development operations and investments. The organization needs a proper risk mitigation strategy to manage financial, strategic, and political risks. Risk is described as the possibility of undesirable results. It can originate from "uncertainty" and end in a "consequence." Therefore, the magnitude of risk can be minimized by adequately managing uncertainty and consequence.

Risk is unavoidable in the oil business and is closely associated with rewards and performance. E&P organizations' "no risk, no reward" strategy is based on calculated risk-taking that is planned after careful consideration of all probable outcomes. History of past failures may tend to cause risk-aversion on the slightest pretext. However, fear of failure is the enemy of performance. Deployment of the right expertise can help overcome this fear and achieve success.

An organization that can quickly reorient its operating tactics and strategies and organizes its resources to take advantage of the market conditions can extract the full mileage from reservoir management activities. This organization will be bound by mutual trust between the leader known to walk the talk and the workforce. He/she will not hesitate to grant autonomy to the deserving and empower the achievers to enhance the sense of belonging and output quality. He or she knows how to turn good ideas into actions and actions into success stories. Learning from the past and be optimistic about the future will be his strengths. He or she will respect knowledge and work to build a professional workforce. His conduct will be exemplary and in line with company policies and national statutes. Rewarding the deserving and punishment to defaulters will come naturally to him or her.

On the other hand, a professional workforce is expected to be disciplined, reliable, efficient, ethical, proficient, and willing to upgrade its skills. It will respond to its duties and challenges to introduce a positive change. Their active engagement is a crucial requirement for the sound economic health and well-being of the enterprise. Experience has demonstrated that a higher level of employee engagement results in overall job satisfaction paving the way for enhanced professionalism, productivity, and organization profitability.

Notes

1 Austin Boyd, Harold Darling, Jacques Tabanou, Bob Davis, Bruce Lyon, Charles Flaum, James Klein, Robert M. Sneider, Alan Sibbit, Julian Singer; The Lowdown on Low-Resistivity Pay; Autumn 1995 Oilfield Review
2 Li, Ming & Yang, Hai'en & Lu, Hongjun & Wu, Tianjiang & Zhou, Desheng & Liu, Yafei. (2018). Investigation into the Classification of Tight Sandstone Reservoirs via Imbibition Characteristics. Energies. 11. 2619. 10.3390/en11102619; https://www.mdpi.com/journal/energies
3 Special Core Analysis Challenges, Pitfalls and Solutions by Colin McPhee SPE London May 26 2015; www.lr-senergy.com
4 Mirzaalian Dastjerdi, A., Eyvazi Farab, A. & Sharifi, M. Possible pitfalls in pressure transient analysis: Effect of adjacent wells. *J Petrol Exploration and Prod Technology* **9,** 3023–3038 (2019). https://doi.org/10.1007/s13202-019-0701-2
5 Iraj, E., & Woodbury, J. J. (1985, February 1). Examples of Pitfalls in Well Test Analysis. Society of Petroleum Engineers. doi:10.2118/12305-PA; https://www.onepetro.org/journal-paper/SPE-12305-PA
6 Livescience website; https://www.livescience.com/2493-mind-limit-4.html

7 Reservoir Simulation; Calvin C. Mattax & Robert L. Dalton SPE Monograph Volume 13, Henry L. Doherty Series

8 MA Thompson, "Conference paper (SPE-99639-MS) Best practices for fields producing below the bubble point" SPE Europec/EAGE Annual Conference and Exhibition, 12-15 June, 2006 Vienna, Austria

9 Dyes, A. B. (1954, October 1). Production of Water-Driven Reservoirs below Their Bubblepoint. Society of Petroleum Engineers. doi:10.2118/417-G

10 EPA United States Environmental Protection Agency website; https://www.epa.gov/ghgemissions/understanding-global-warming-potentials

11 Working Document of the NPC North American Resource Development Study Made Available 15-Sep-2011; https://www.npc.org/Prudent_Development-Topic_Papers/2-11_Subsea_Drilling-Well_Ops-Completions_Paper.pdf

12 *Wikipedia, The Free Encyclopedia;* https://en.wikipedia.org/wiki/Oil_platform

13 Investopedia, Profitability Index; https://www.investopedia.com/terms/p/profitability.asp

14 From Wikipedia, the Free Encyclopedia Separator (oil production); https://en.wikipedia.org/wiki/Separator_(oil_production)#Classification_by_application

15 Wikipedia, the Free Encyclopedia Separator (oil production); https://en.wikipedia.org/wiki/Separator_(oil_production)#Density_difference_(gravity_separation)

16 Kalish, P J, and EnDean, H J. Flash-differential separation increases stock-tank recovery. United States: N. p., 1965. Web.OSTI.GOV, U.S. Department of Energy, Scientific and Technical Information; Flash-differential separation increases stock-tank recovery: https://www.osti.gov/biblio/6626005

17 Oil and Gas Separator website; www.oilngasseparator.info/oil-handling-surfacefacilities/stages-separation-selection.html

18 Deaver, J., "A History of Too Much Capacity and Too Much Complexity in the Deep Water Gulf of Mexico;" Proceedings of AIChE/SPE Joint Workshop: Next Generation Deepwater Facilities

19 Utilizing Windshield and Rearview Metrics; https://www.ceoreport.com/utilizing-windshield-and-rearview-metrics/

8

Reservoir Development and Management: Important Lessons Learned

8.1 Introduction

Everyone makes mistakes in life, some are inconsequential, but some can have severe consequences. To err may be human, but to go repeatedly wrong is not acceptable to conventional wisdom. An error of judgment or a miscalculation in business can cost the company a substantial financial loss and reputation. Therefore, the organizations have a structured process to ensure that lessons learned from past mistakes are absorbed in their workflow and decision-making process.

The *Lessons Learned* process is a crucial element of knowledge management. The principal objective of this exercise is continuous improvement in the quality of a job/service with a cost reduction. It ensures the incident-free accomplishment of a task within the regular schedule and budget and incorporates the new knowledge and learnings into future work practices.

A common complaint about the lessons learned process is that organizations seem to benefit very little despite their best efforts. The answer, it seems, lies in their ability to execute the process with discipline and precision. A process alone is not the end in itself; it is the execution, which is the most critical piece of an endeavor to achieve results.

Lessons Learned is an integral part of all modern E&P organizations. What value they derive out of this process depends on how meticulously they implement it. We discuss hereafter some of the most important reservoir management lessons the industry has learned the hard way.

8.2 Reservoir Development and Management

Resource development is a natural extension of successful oil discovery and appraisal programs. It requires an investment of funds and human effort to

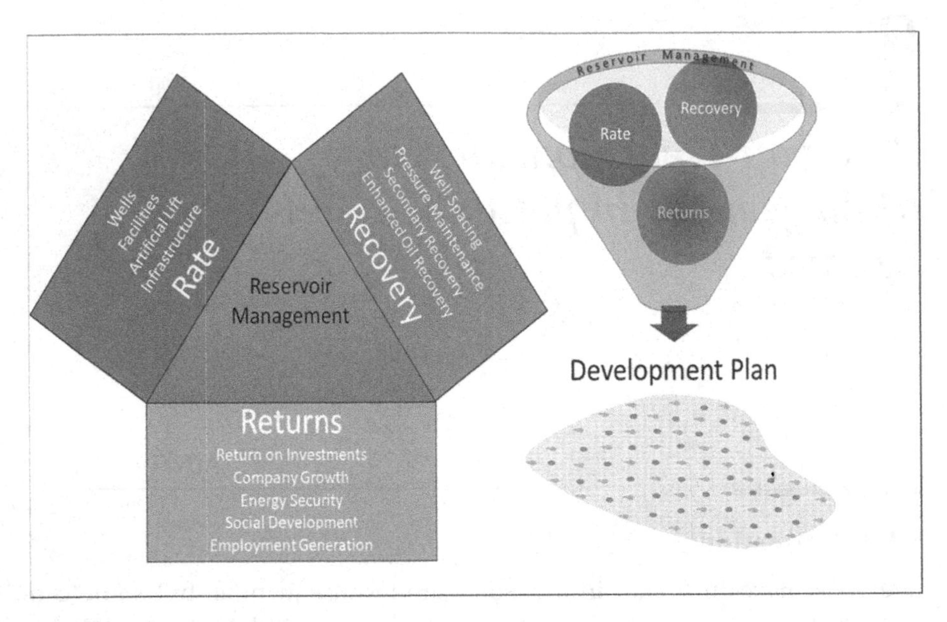

FIGURE 8.1
Three Rs of Reservoir Management.

convert dormant oil and gas resources into active cash flow. Field development teams design the most suitable development plan that meets their technical and economic objectives. These objectives emanate from the reservoir management philosophy built around the maximization of the three Rs: rate, recovery, and returns within the development plan (Figure 8.1) period. As discussed in Chapter 7, it is impossible to maximize all three Rs due to the usual conflict between the rate and recovery goals. Therefore, a compromise, which in mathematical terms is called "optimization," is made to achieve the most rational value of all three Rs in a single development plan.

Company priority between rate, recovery, and returns can change from time to time, depending on the business environment. For a new company, the oil rate may be the priority because it needs to generate a quick return on the money. A company in charge of multiple assets with no urgent need for high cash returns from a new field may maximize its oil recovery and not necessarily the rate.

The question of returns is very specific. In preceding chapters, we have differentiated between the private, international, and national oil companies based on their objectives of wealth and value creation. It is essential to revisit this issue from a different perspective to get a more objective answer.

The real question is, what is the difference between "wealth" and "value" and when we say that a POC/IOC creates wealth, then what we mean is that they create wealth for themselves, not for the host nation. And when we say NOCs create value, they make money and drive the host country's socio-economic growth. The same NOC's agenda changes when it participates in foreign E&P ventures. So, when we analyze the question of "returns," the right question to ask is "returns for whom?"

Indeed, POCs/IOCs don't only create wealth for their stakeholders; they also create significant value for their employees and local people. The employees gain experience in new oil and gas fields. They get to see new places, traditions, and culture, learn new languages, travel to new destinations, and make new friends. Higher salaries and perquisites help them have lavish lifestyles, fulfill family obligations, and pursue growth and development opportunities. They rent or buy local property, hire local employees to help at home and in offices, and purchase local products. The same applies to their company too. It looks for suitable avenues to help the host nations besides paying royalty and taxes.

In summary, despite their primary objective of wealth creation for their shareholders, their presence in the host country leads to significant value creation for the local society. Nevertheless, their contribution is limited compared to takeaway. Therefore, many countries prefer IOC's participation in their national E&P program on a fee basis rather than as an operator.

Rate and recovery are, in reality, two sides of the same coin differentiated by the span of time. Both are achieved by investment of capital and human effort. The word "rate" has a short-term connotation, while "recovery" implies long-term sustenance of oil production rate. In contrast, returns can be in cash income, employment generation, growth of the economy, and social service.

Therefore, proper reservoir management involves managing the following three elements in line with company policy and strategy.

- **Rate Management**

The oil rate is a good indicator of the size of an asset. A higher production rate allows a quicker recovery of the invested capital. For a new company, the oil rate may be a priority because it needs to generate an immediate cash flow to sustain its operation and activities.

Rate management requires successful drilling and completion of new wells, artificial lift methods, proper handling and processing of produced fluids by facilities, and transportation infrastructure. Identification and release of needed well locations, rigless and rig workover activities to maintain production rate, assigning production and injection well-allowables, production rate measurements of oil and gas for sale to consumers are essential steps to achieve this objective.

- **Recovery Management**

A company in charge of multiple assets of a mixed level of maturity may maximize oil recovery. Rate and recovery management is usually an exercise in conflict management. The oil companies that decide to accept higher oil recovery may have to be content with a somewhat lower production rate. But they can probably enjoy the benefits of higher cumulative oil production and financial returns in the longer term.

Recovery management requires that subsurface and surface development plans be coupled seamlessly to gain the maximum liquid recovery in the surface tanks. Proper design and timely execution of primary, secondary, and tertiary recovery programs are key to maximizing oil recovery.

- **Return Management**

Financial analysis is a standard method used to optimize the rate and recovery objectives of a development plan. It allows the first view of the project economics to the company management and professionals. All other benefits/value to the company, society, and nation flow from the profit that a company makes. Higher the profits or returns to the operating company, the higher the benefits to the host country.

The question of "returns" must also be examined in the context of the host government's "expectations." NOC, as an operator, is expected to manage the field production and recovery in a professional way to secure an acceptable rate of return on investments. It must also commit to employee training and development programs, providing health, recreation, and education facilities. As a domestic operating company, NOC offers direct and indirect support to various social and government-sponsored human welfare programs for poverty alleviation and economic upliftment of the masses. It generates employment opportunities and connects the oil and gas company with the local market.

But this broad-brush painting of NOCs as good Samaritans is countered by allegations of inefficiencies, corruption, mismanagement, and mediocrity. The employee development and other programs they offer for their employees hardly meet their objectives. Many NOCs did not live up to their mandate of reserve replenishment, a general trend with most oil companies, showing their reluctance to take exploration risks.

The subject of returns needs to be evaluated in terms of the expectations of the other party. A POC/IOC can match or exceed the NOCs standards of oilfield operations and management, ensuring improved production rate and recovery management. However, what they can probably not match is the expectation of their endless support to the host government's social programs. Nor should they be expected to do as much. A business contract is based on the deal between the host government and the POC/IOC, and returns to the host government are based on this thoroughly pre-negotiated document.

8.3 Safety First

An E&P organization is expected to carry out the complex operations of extracting highly inflammable hydrocarbons. Oil and gas occur at high pressures and temperatures under challenging terrains and harsh environments. The material used to complete the wells and facilities/pipes is impaired over the reservoir lifecycle due to material fatigue, aging, and corrosion. Material fatigue causes structural damage to the metallurgy and weakens the well-completion equipment, facilities, and pipelines. Corrosion and high-pressure and high-temperature conditions can aggravate this phenomenon. Pressurized wells, vessels, pipes, and flammable liquids can combine with human error to cause explosions and fires. Such incidents cause loss of production and mandays; their consequences can be devastating to individuals and the organization.

The most common and catastrophic disasters in oil and gas wells are blowouts, explosions, and fires. Workers operating heavy-duty machines, equipment, and drilling rigs for long hours are vulnerable to tragic accidents and injuries. Offshore workers engaged in drilling and production platforms activities are exposed to graver risks due to evacuation limitations. The release of hydrogen sulfide from wells in higher concentrations is a significant risk and can cause grievous harm to the workers.

History has witnessed the worst ever man-inflicted disasters of mass-scale oilwell blowouts and fires in Kuwait. These fires burned for more than 8 months, causing massive damage to the national economy and environment. The crude oil and high salinity formation water discharged by oil wells during blowouts, in addition to the seawater that was used to extinguish fires, formed the so-called "oil lakes" in the open desert area. The crude oil from these lakes contaminated the soil and possibly the shallow aquifers in the north Kuwait area.

HSE takes the top priority in the KPIs of all E&P companies today. Their HSE policy is all-inclusive. They aim to have their core business operations and allied activities at well sites, on the road, and in offices to be accident-free. All organizations have the tools and systems to map and audit the HSE performance. The results speak of the positive HSE performance many organizations have brought about in a relatively short time. Fatalities or disabilities due to vehicular accidents, ergonomic hazards, and working conditions in engineering or construction projects have registered a steep decline. Following the Occupational Safety and Health Administration (OSHA) and the National Institute for Occupational Safety and Health (NIOSH) guidelines help the organizations control the possible hazards for oil and gas workers' safety.

Yet, we heard several stories, almost regularly, of oilfield disasters: an offshore platform capsizing, deepwater horizon oil spill, burning of an oil gathering facility, a 30″ natural gas pipeline explosion, tank batteries catching fire by lightning, collision of a multipurpose support vessel with the gas exporting risers of a production platform, and so forth. It only proves that *Lessons Learned* is an eternal program to improve current processes, practices, and procedures, avoid future mistakes, and reduce risks.

8.4 Project Management

The projects, more importantly, and their timely execution are critical for the growth and expansion of oil and gas production. Companies usually bundle drilling, production, processing, transportation, and construction activities to create a stand-alone project. This methodology helps to track the progress of an essential program during the planning and construction phase. Later on, it aids in ascertaining the cost-effectiveness of individual projects by comparing the costs versus benefits.

More often than not, operating oil companies usually rely on outsourcing their projects to international engineering and construction companies. The project is awarded to a company based on the project cost estimate, the experience of doing similar projects, and the track record of project delivery. However, despite modern sophistication in project management tools, processes, and controls, coupled with the knowledge base available with the owner and contractor teams, projects fail to achieve specific goals and meet their success criteria. The reasons are mostly organizational and behavioral. A patronizing attitude, command and control mindset, and bullish reporting on the owner team's part prevent true collaboration with the contractor team. It can ultimately contribute to poor outcomes of the project. Generally, the larger the company, the more processes, procedures, practices, and paperwork can slow down the project execution. Agile leadership; the delegation of authority; and respect for competence, capability, and personal integrity of the owner and contracting team members improve project delivery.

Based on their analysis of 58 major oil and gas development projects executed between 2011 and 2016, the UK's Oil & Gas Authority found a strong relationship between project execution efficiency, the quality of people engaged in the project delivery, and their job continuity. The lessons learned from various oil and gas projects commissioned in the UK and other parts of the world can be summarized as follows-

- Keep the project design simple
- Precisely define the project scope at an early stage
- Assign accountability for the project delivery
- Improve collaboration among stakeholders

FIGURE 8.2
Triple Constraints of Project Quality.

Successful project execution must overcome the triple constraints of time, cost, and scope represented by a triangle (Figure 8.2). In many companies, project definition and scope are changed until the last moment. People spend much time accommodating these changes within the original budget and schedule, not realizing that it is impossible to change one side of the quality triangle without impacting the other two sides.

Therefore, the time and effort to prepare a robust FEL at the early stage of the project definition are well spent. Simultaneously, the ability to incorporate design changes is relatively high, and the cost to make those changes is relatively low. An understanding of the uncertainties associated with project design helps develop probabilistic estimates of project schedules and costs. These estimates help avoid personal biases and unreasonable expectations.

The *Lessons Learned* exercise is usually undertaken in the form of a survey or a workshop. The idea of this exercise is to have honest feedback from participants about what went well with the project and what didn't. The participants must have the tact and courage for plain-speaking to realize the value of such sessions.

8.5 Business Contracts

Contracts are a way of life in the E&P business. The host country may not have the capability, time, or resources to explore and extract oil and gas reserves indigenously. IOCs and contractors with their reputation for knowledge, services, operations, technology, and trained staff can be their ideal partners to provide expert support in exchange for money. IOCs and contractors do not limit their operations to a specific country. They go where the opportunities are to make money. They may prefer countries rich in oil and gas and are politically stable, but in reality, they may go where the money is. Contracts are the instruments that guarantee the safety and security of their life, property, and business.

Contracting is a mechanism between two or more parties to facilitate their business interests over a pre-defined period. In case of any disputes or eventualities, the contract being a legal document protects their commercial interests in the court of law. Therefore, the agreements must anticipate situations that may arise from changes in the organizational, social, political, or economic environment and hurt the business. The provisions of long-term contracts may have to be revised several times to make room for new local regulations. More often than not, the contract a service provider starts with is not necessarily the contract it ends up with.

Typically, the governments have a range of controls over E&P activities. These controls take a different dimension if the participating company is of a foreign origin. The internationalization of the business might offer exciting growth opportunities, but it also brings tremendous responsibilities and increased risks for contracting parties.

Knowledge and technology are the two main drivers for international contracts in the E&P sector. Some of the common problems that complicate the designing of these cross-border contracts are the rights, duties, and obligations of participating co-venturers because of the following:

- Exposure of oil and gas operations to HSSE risks both onshore and offshore
- Possibility of the lower success rate of exploratory drilling efforts
- Significant operational risks in development projects
- Uncertainties in estimates of oil and natural gas reserves
- Volatility in crude oil and natural gas prices
- Multilateral legal and fiscal framework for cross-border oil and gas pipeline
- Application of new technology
- Challenges of managing a JV or a Consortium
- Interdependence between reserve estimates and oil price under PSAs
- Modalities for extraction of oil and gas from border fields
- Royalty and tax issues

In international contracts, the force majeure clauses assume increased importance and consider the geopolitical and environmental context. Long-term agreements should not rule out the possibility of natural calamities and conflicts. Such instances can impose an unforeseen burden of direct or indirect liabilities on the parties bound by the contract.

Most contract disputes could be avoided through open discussions about the expectations at the outset. Litigation is an expensive, time-consuming, and energy-draining process that goes by the contract's verbatim requirements. Careful drafting and correct use of the contract language, including punctuation marks, can avoid major headaches in the long run.

8.6 Reservoir Assessment

Many oil and gas reservoirs fail to realize their full production potential. According to a Westwood Global Energy Group study, "half of oil and gas fields are not producing to expectations when on stream"[1] due to inaccurate assessment resulting from the inadequate or incomplete appraisal of the subsurface asset. The primary objective of an appraisal phase is to reduce the uncertainties associated with the reservoir's size and configuration. The appraisal is also the most critical phase that enables a clear understanding of the hydrocarbon-in-place, well-productivities, and production potential.

There were instances when oil companies, driven by their urgent need for hydrocarbons, reduced the appraisal programs to accelerate oil and gas production. Bypassing the conventional wisdom of developing oil and gas fields in a sequential "exploration-appraisal-development-abandonment" process can prove risky and can put ambitious management and staff members at a disadvantage. They may proceed with development planning based on over-optimistic estimates of oil-in-place, reserves, and production capacity without a clear understanding of uncertainties. These figures are the foundation for the production facility's design, pressure maintenance program, and injection system that require a massive financial investment.

In the Westwood study conducted over 70 producing oilfields, 84% of the assets failed to match the reservoir performance, while 51% did not match the reservoir volume. Both performance and volume issues are interrelated and can cause significant errors in production forecasts and estimating the time a reservoir takes to reach the bubble point. Reservoir health can deteriorate very quickly below the bubble point pressure. Therefore, most companies are keen to implement pressure maintenance projects well before the bubble point is reached. These are long lead-time projects, and compressing their delivery schedule is not easy. A wrong assessment of time for a reservoir to reach the bubble point can negatively affect the project size, oil recovery, and cost. Optimistic estimates imply that the operator will probably exploit a reservoir that has lower oil-in-place rather aggressively. It hastens the arrival of bubble point sooner than planned. If the pressure maintenance project cannot be expedited, production below bubble point results in excessive gas production, sharply declining reservoir pressure, and significant oil recovery loss. The size and cost of an injection facility also have to change due to alteration in the base design parameters. It means that more water for pressure maintenance needs to be injected into the reservoir, which may require larger pumps and more injection wells. Drilling more injection wells requires a change in the development plan, and so on.

Delays in project execution are universal, whether for genuine or trivial reasons. Projects can suffer from a bureaucratic mindset and widespread system inefficiencies costing the organizations a fortune. The people responsible for delays are more critical for organizations than the project. The people responsible for delays are more critical for organizations than the project and therefore, there is a tendency to sugarcoat failures under the guise of "deferred success" and move on by "thinking positive."

8.7 Production Forecasting

All organizations have a process around setting their annual targets reflected in their KPIs. The KPIs may be composed of the core and financial objectives apart from CSR. For an oil company that is engaged in E&P activities with an inventory of oilfields at different stages of maturity, setting production targets is a tightrope walk. Due to their direct or indirect embedment in national energy and economic plans, production targets are always under media and government focus. As a result, production targets need to be technically as well as politically correct.

Production forecasts constitute the rationale for making investment decisions in E&P activities. Comprehensive, transparent, and data-driven predictions have become increasingly valuable for long-term planning of projects, improvement in recovery factors, and development of the sale contracts. The short-term, particularly the annual production forecasts, can be a subset of long-term production forecasts and tweaked appropriately to accommodate specific events and schedules. Alternatively, the annual estimates can also be generated independently of the long-range projections, based on the intimate knowledge of resource allocation, specific inputs, and schedules. Accuracy of short-term production forecasts is crucial because of their use for the annual target setting.

Reservoir engineers use various classical reservoir engineering and model-based tools for forecasting the reservoir performance under possible scenarios of constraints and the corresponding oil and gas production. A reservoir may use multiple models/tools because a single model/tool may not adequately represent the reservoir flow process correctly. Knowledgeable reservoir engineers do exactly know which forecasting method to use to produce a unified and realistic range of outcomes. They are also well versed in developing and customizing their tools to process forecasting results in a specific format.

Deterministic methods contain significant uncertainty forcing only a single view of the production trend and reserves that may nowhere be close to reality. The companies that generally use deterministic production profiles representing the "most likely" scenario may miss out on the upside (P10) opportunities and the downside (P90) potential. Probabilistic methods, on the contrary, quantify the uncertainty and provide an improved EUR with associated risks.

Ideally speaking, the most likely production forecast must be based on EUR's P50 case and not the P50 case of OOIP. The compartmentalized reservoirs may have uneven recovery factors, and a P50 case of OOIP may not always be equivalent to P50 EUR. Not all companies may choose to maximize the EUR; some depending on their business goals, may seek to maximize NPV as an objective function in EUR. Probabilistic forecasts help rationalize budget allocations in line with the resource estimates and business strategy. Generating high (P10), most likely (P50), and low (P90) forecasts also present an opportunity to sense check if the forecasts are in the correct order and representative of actual reservoir performance.

Nexus, Avocet, PetroVR, and IPSM are a few of the advanced software systems that facilitate end-to-end asset modeling. Some of these applications are better than others in identifying the system constraints and bottlenecks that prevent the realization of the asset's full production potential over the whole supply chain from the reservoir to the sales terminal.

The most common reasons for inaccuracies in the production forecast originate from the following:

- Inaccurate rate measurements
- Inadequate understanding of the fluid flow in the reservoir
- Use of the old/unrepresentative well-test data to predict the new well productivity
- Success rate, restoration time, and production rate after workover
- Deviations in agreed schedules for the availability of drilling or workover rigs
- Uptime estimate of wells and facilities
- Mismatch in the base production rate (the production rate a day before the start of the forecast) because the plan is usually made at least a month in advance

Despite all the micro-detailing of events and schedules, short-term forecasting is a daunting task. Accuracy of the forecast and underlying assumptions come under scrutiny from the first day of the forecast period. Each time a company fails to achieve the production target, the forecasting team is criticized for making wrong assumptions, though the real problem may lie elsewhere.

The production forecasts based on classical reservoir engineering methods or reservoir simulation usually suffer from inconsistent and inaccurate datasets. DCA is an extensively used method to provide deterministic estimates for future performance and remaining reserves. Deterministic production forecasts based on "noisy" production data are susceptible to subjectivity and personal bias. Drawing the correct decline trend on a statistically representative data sample and associating it with the correct type of actual reservoir decline (linear, exponential, hyperbolic) can be quite challenging in a large and erratic dataset. Extrapolation of incorrect decline trends results in wrong forecasts of production and reserves.

Similarly, in reservoir simulation, the history match process can inspire confidence in the forecast but does not guarantee accuracy. A history match is a time-consuming process and is never perfect due to model complexity or data issues. Many wells in the model may have an excellent pressure match but not necessarily an oil, gas, and water rate match. Therefore, modelers working on large models have to be content with about 80% solution. The remaining 20% of the wells are not correctly history-matched because they may have serious data issues or are less understood. These wells can sometimes be the primary source of error in the production forecast and reserves. Unfortunately, the accuracy of the production forecast and reserves can only be tested in hindsight. The ultimate proof of reserve estimation accuracy comes with the receipt of oil volume into the tanks in the abandonment phase. For these reasons, final production forecasts may sometimes have to be hand-edited to smoothen anomalous data or trends.

8.8 Role of Best Practices

The petroleum industry has realized the importance of the use of "Best Practices" the hard way. Many operational accidents and environmental disasters that caused immense loss of life, company property, and reputation could have been averted if the best practices were followed in letter and spirit. Many national and international conferences have emphasized the role of best practices in E&P operations over the last decade or so. Yet, their format and content only highlight the inconsistencies in the concept and execution. There is an increasing trend among geoscientists and engineers to claim that their operational activities are based on the industry's best practices. Many such assertions appear to gain an advantage of self-promotion and visibility among the community, customers, and management in the absence of any official evidence of organizational backup.

Despite an apparent disconnect between the individualistic proclamations and organizational approach towards developing best practices, there is a consensus about their need and application. The convergence of these tactics is creating opportunities for in-depth reviews that improve the quality and results of complex E&P operations. Participation in national and international conferences is a modern and powerful way of information sharing across the professional community network. Exposure to such events and industry experts opens new avenues of learning by observation, introspection, self-assessment, and benchmarking.

The use of the word "best" might sound like an overstatement, but the best practice concept is certainly rooted in positivity and improvement. In the context of organizational activity, it must reflect the highest standards of performance, cost optimization, and professionalism. The best practice cannot be a sporadic event to deliver the desired outcome by fluke or chance. It has to be a systematic framework of the policy and procedures that ascertain success at a low cost with minimum risk. The best practice may also not be everlasting and unique for the entire industry. Innovation, technology, and local factors can play a significant role in their constant evolution, development, and customization. The organization being the ultimate beneficiary, its support and sponsorship are the main drivers for designing and implementing a best practice on a sustained basis. Lack of quantifiable measures makes it challenging to classify a particular method as the best practice. Therefore, organizations are conscious that their best practices must focus on optimizing time, cost, and quality. Their methods must also rank high on the criteria of the deployment of appropriate technology, repeatability, customer satisfaction, legal, ethical, and HSE compliance, and the scrutiny of other peer organizations.

Many organizations commit to using best practices by adopting international quality standards to ensure quality, safety, security, and reliability in the petroleum, petrochemical, and natural gas industries. Several domain-specific standards and codes are prevalent in the industry; some of the key standards applicable to E&P operations/activities are discussed here. The associated personnel are expected to be well conversant with these codes to avoid a performance glitch or incident.

The ISO 9000 series is a family of quality management standards that continuously helps organizations achieve, maintain, and continuously improve service/product quality.

The goal of the ISO 14000 family of standards helps organizations manage their operations and processes in an environmentally compliant manner.

ISO 15926 standards' objective is data integration, sharing, exchange, and hand-over between computer systems.

ISO 14224 standards apply to collecting and exchanging reliability and maintenance data for equipment, materials, equipment, and offshore structures.

ISO 20815 standards are used for oil and gas production assurance, maintenance, and reliability management of upstream (including subsea), midstream, and downstream sector facilities/activities.

WITSML, PRODML, and RESQML are a family of open, non-proprietary, digital technology-centric industry initiatives/standards to ensure that all the electronics and software technologies work together properly and do not suffer because of incompatible data architecture and designs. The evolution and increasing popularity of digital fields are also due to a standard and consistent real-time dataset transmission to all the stakeholders at the same time. The efforts to develop, manage, and adopt these Extensible Markup Language (XML) and web-based data transmission and exchange standards are led by Energistics, which is a global consortium of about 115 partners and Special Interest Groups (SIG) drawing experts from participating companies and member organizations.[2]

WITSML (Well-site Information Transfer Standard Markup Language) industry standard for transmitting, exchanging, or receiving data from the drilling rig or offshore platform to the oil company representatives to monitor and manage wells, completions, and workovers. Its present capabilities include wellsite-to-office data transfers, completions (equipment, events, and flows), wireline and LWD well logs, and calculated and planned well trajectories.[3]

PRODML standards facilitate automated production data gathering to monitor, optimize, and manage hydrocarbon production from the reservoir to the point of sales by invoking necessary surveillance applications, model analysis, simulation applications, and so forth. Present versions of PRODML standards support fiber-optic distributed temperature surveys and distributed acoustic sensing, fluid and PVT analysis, Wireline Formation Testing, production volume reporting, and so forth.

RESQML is a series of standards for exchanging data on reservoir characterization and facilitating the construction of static and dynamic models. The standardization includes capturing the representative property values and navigating through the complex scenarios of multiple geological realizations and stochastic modeling. It also supports waterflood operations with streamline modeling capability to improve waterflood management. Modeling is a multi-disciplinary task involving multiple software packages and data formats that are not necessarily compatible and conducive to productivity and quality.

IEEE std. 730.1 Software Reliability standards are spearheaded by the Institute of Electrical and Electronics Engineers (IEEE) and apply to the quality assurance and reliability of software[4] used in E&P operations.

The Professional Petroleum Data Management (PPDM) 3.9 Data Model is a specific module of the International Petroleum Data Standards (IPDS) relational data model prevalent in Master Data Management strategies. It is designed and developed collaboratively by all the stakeholders in the industry.[5]

Reservoir management has hundreds of planning, operations, laboratory, modeling, and decision-making-related activities where the concept of best practice can be applied to improve the team and organizational performance. Oilfield operations are complex, risky, and costly. Best practices and good teamwork can reduce time, cost, and risk. Small improvements in how things are done, managed, and their scheduling can make a big difference to the final results. Proper drilling, completion, production, injection, well logging, and disposal methods can contribute to significant cost savings. Timely interventions for maintaining well integrity, controlling GOR and water cut, modification in waterfloods, production optimization, water management, and other similar activities can significantly impact the costs and development plans. In summary, the mandatory use of best practices can improve oil and gas production, recovery, and economics and reduce risks.

8.9 Data to Decision-Making

Data are the essential fundamental requirement for all businesses. But it is even more critical to the E&P business because the technology to acquire data from invisible target reservoirs that occur in the subsurface is limited, and the cost is very high. The mechanics of E&P operations can be compared with the medical procedures where initial diagnostics are run based on symptoms (interpretation). Then tests are conducted to confirm the suspicion (validation). The knowledge gained from the experiments is utilized to recommend further treatment or surgery (application). The final decision to use medication or surgery is based on compiled knowledge or second opinion (wisdom). The design and workflow from data acquisition to decision-making are quite similar irrespective of the nature and complexity of the business (Figure 8.3). Let us take a basic example to demonstrate how computers and humans interact to process raw data into decision-making.

Every decision is primarily grounded in data. Therefore, data quality is fundamental but not the only requirement for sound decision-making. A computer may store data in a UTF 8 (8-bit Unicode Transformation Format) binary code shown in Column 1. It must be first correctly interpreted to make sense to an ordinary human being and supplemented with additional data or information. The level and quality of information exponentially improve as data become information, and the information develops into knowledge after going through the process of validation. The application of verified information can transform knowledge/experience into wisdom.

The journey from data to decision-making is supported by the processes of interpretation, validation, and application. Each of these processes upgrades the level of understanding from data to information, knowledge, and

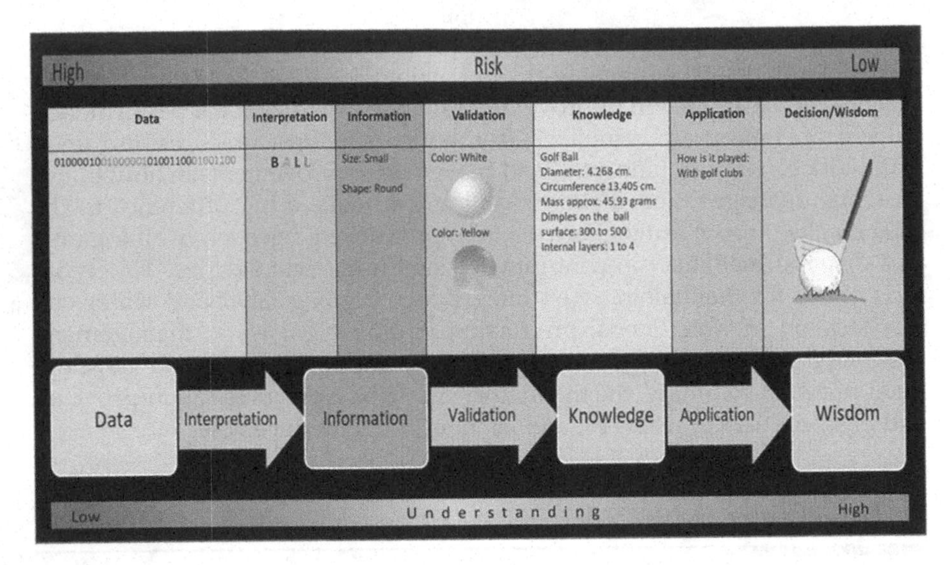

FIGURE 8.3
Transformation of Data into Wisdom; Computer and Human Interaction.

wisdom. The last step in this learning journey is the application of cumulative expertise to produce the desired outcome. Of course, the secret of the progressive advancement in education lies in the meticulous use of other intermediate processes in this input–output chain.

The preceding discussion explains why the data is so essential and data quality so critical for decision-making. A decision maker has the discretion to decide anywhere along the way, but in the absence of sufficient experience (knowledge), the decision has higher risk exposure. The suboptimal performance of some waterflood projects is a good example of an inadequate understanding of reservoir behavior.

Reservoir management faces a variety of problems due to technical uncertainties forcing a shift in decision-making processes. Geological complexity gives rise to many technical issues that include lower oil-in-place, loss in oil recovery, and poor reservoir performance. These problems are quite complicated and are managed on a routine basis by invoking probabilistic methods. Decision makers in this paradigm can no longer deal with a single assured value of hydrocarbon volumes, production profiles, and recovery factors with 100% confidence. Instead, they have to deal with a range of estimates associated with varying levels of confidence (P10, P50, and P90).

There are other forms of uncertainties that impact reservoir management decisions and introduce an element of risk of one of the different kinds. Economic risks can arise from unpredictable oil prices, loss of demand, and unexpectedly high operating costs. The organization must be mindful of the financial risks on account of environmental damage or pollution caused by its operations. And, of course, no organization can absolve itself from political, contractual, reputational, and HSE risks that go with E&P operations and activities.

E&P organizations bring about significant improvement in this process by collecting more and more data to validate suitable realizations and hypotheses. More data usually translate into less uncertainty, provided that the information is of good quality and all of it is properly analyzed. However, this is not always true. More data can also imply more "noise" and more uncertainty that can be counter-productive.

IT is generally an isolated department in many oil and gas organizations, seldom hardwired into its mainstream. The result is that IT professionals cannot fully appreciate the core business teams' need and functional requirements. Their hardware and software acquisition/development plans are also not necessarily aligned with their computing requirements.

Data management is a real issue in many E&P organizations. The size and complexity of the company database can assume alarming proportions with ever-increasing data flow resulting from the expansion of business and operations. Many employees prefer to maintain their personal customized databases because they may not be clean and suffer from quality issues. The organizational structure that promotes closer interaction of the IT department with core business groups can reduce these inefficiencies.

The power of Information Visualization lies in making terabytes of data visible in the form of graphics and images so it can be more accessible and understandable. Visualization technologies condense extensive data inventories into crisp representative pictures of objects that can be grasped quickly and retained by human minds. Scatter-plots, bar/column/pie charts, maps, cross-sections, and 3-D model images improve busy professionals' ability to check data quickly and recognize the issues quickly.

Computer graphics is a unique technology aided by advanced hardware and software that can generate still or animated images. Reservoir architecture and fluid flow process simulated with computers' aid may not make much sense to a layperson unless presented in the form of graphs and 2-D or 3-D images. In a fast-paced business world, where industry experts and decision-makers struggle to manage their time, it is challenging to sift through huge models and large spreadsheets every time. Computer graphics in the form of plots and pictures make up for a quick and reliable tool for improved and transparent decision-making.

8.10 Water Management

Water is a valuable resource for the entire ecosystem, and its constant depletion is a matter of great concern for society. Many countries and their states where water is in short supply do not allow wastage of water or disposal into open grounds or subsurface without proper treatment. Untreated wastewater is contaminated with several impurities that are harmful to the ecology.

Water is a prerequisite for oilfield operations. It is the prime source of energy and operating drive-mechanism in a petroleum reservoir's life cycle, whether natural or injected. Water acts as the displacing fluid to expel the oil out of the porous network. The efficiency of natural water drive depends on the aquifer's size and activity, whereas the injected water's effectiveness depends on the location, quality, and quantity of injection.

Proper subsurface water management is fundamental to maximizing oil recovery, but surface water management is crucial for overall reservoir management. The use of horizontal wells and hydraulic fracturing further intensifies water management challenges. The value of subsurface water management strategy lies in maximizing injection water efficiency to displace the oil represented by the ratio of oil production per barrel of injected water. Failure in proper subsurface water management does not only result in lower oil recoveries, but it also escalates the volumes of produced water that cannot be directly disposed of or reinjected. Water quality is the focal point of the surface water management strategy. It ensures that the water used for injection and the discarded water for disposal meet their separate specifications to achieve their objectives. Maintaining the desired quality of injection or disposal water is a subject that encompasses the cost, and more importantly, the deployment of appropriate technology and the best practices.

Oil recovery from reservoirs operating under depletion drive can be significantly improved by initiating the water injection before bubble point pressure. The preparation for water injection can take about 5 to 7 years. Many reservoirs may not allow this much time for development because of a small cushion between bubble point pressure and initial reservoir pressure. Therefore, planning for water injection must start in the appraisal phase itself. Maintaining water quality specifications and injecting a correct water volume at the right location are equally critical for maximizing the areal sweep efficiency.

Another good practice is to prevent/control the production of bad water. Undesired water production can be attributed to many sources such as downhole tubular or packer leaks, poor cementation, moving oil-water contact, watered out layers and fractures, or faults conducting water to the surface. It must be addressed as soon as it is detected.

In the new business paradigm, the oil industry is expected to carry out E&P operations without any risk to the environment. Any misuse of water or environmental damage is questionable by regulators, shareholders, and community members and is also punishable. The governments encourage operations that require less fresh water and utilize water-recycling technologies. Therefore, water management strategies must abide by the local or federal regulations and consider the additional cost for regional ecology safety.

The US has one of the mature and leading oil industries in the world. Consequently, it is one of the first few to experience the good and bad of a particular technology. The amount of wastewater from mature oilfields in the US has significantly increased. Unconventional oil and gas recovery operations, mainly fracking, are also responsible for large volumes of produced water. The issue of growing wastewater in the US is addressed based on regulatory, geologic, and technological considerations. The practice of wastewater disposal into EWD wells is not only expensive; it is also laden with the risk of polluting freshwater aquifers. Therefore, many oil and gas companies are moving away from wastewater disposal into the subsurface. For example, oil companies resort to recycling wastewater for reuse in states like Pennsylvania[6] that permit wastewater injection into the EWD wells. Pennsylvania extracts significant quantities of methane-rich natural gas from tight shale. The geologic formations in this area are predominantly tight with poor injectivity and do not justify wastewater disposal into EWD wells. Wastewater is also a matter of grave concern for the states like New Mexico that, despite being water-stressed, use enormous volumes of freshwater to recover oil and gas.

8.11 Reserve Estimation

Oil and gas reserves are the single most crucial indicator of the wealth of an E&P company. Therefore, accurate assessment of oil and gas reserves is a primary requirement for optimal reservoir development and management. The process of reserve evaluation continues through the reservoir lifecycle. The use of the word "estimate" reflects that the magnitude of oil and gas reserves is subject to uncertainties arising from the interpretation of available data. Therefore, the petroleum industry follows the practice of estimating a range of oil and gas reserves, rather than a single fixed value associated with uncertainties and confidence in the estimation. It enables the operating company to understand and mitigate risks through a focused risk management program. The magnitude of unpredictability and risk progressively diminish as the field advances through its lifecycle from exploration towards abandonment (Figure 8.4).

The degree of uncertainty in reserve estimation can be attributed to various factors such as the type of reservoir, operating drive mechanism, quality and quantity of G&G and engineering data, and various development plans. The development plan can use different strategies, options, technologies, and assumptions that affect reserve estimates. Quality of reserve estimation also depends on the team's knowledge and experience responsible for the task.

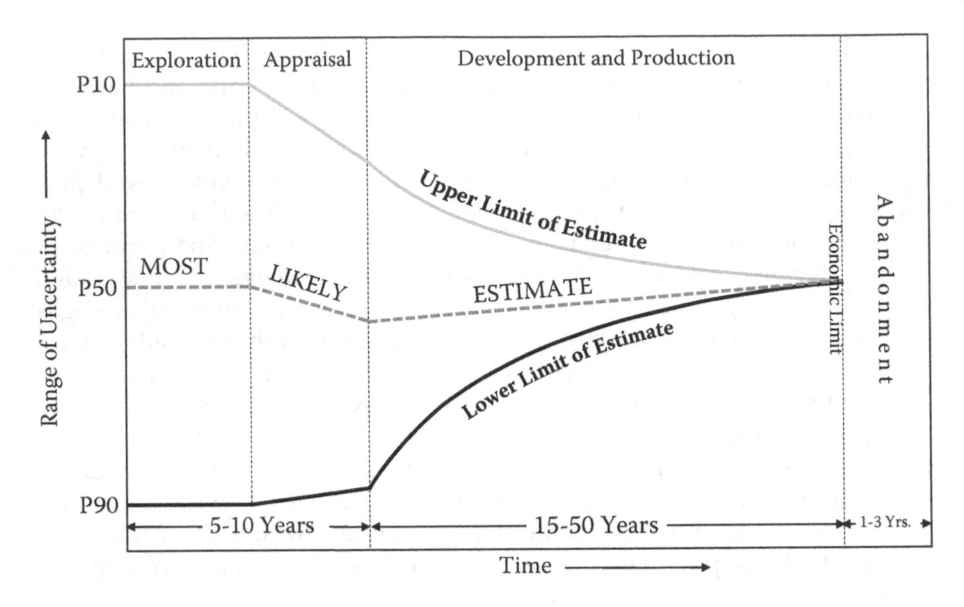

FIGURE 8.4
Schematic Representation of the Accuracy of Reserve Estimates with Time.

The reserve forecast methods discussed in Table 8.1 may use a deterministic or probabilistic approach. E&P companies know the value of each of these methods and use the results for comparison. A close agreement between the reserves estimated by the two approaches increases confidence in the calculated reserves.

Based on the experience gained by various companies in reserve estimation, the following recommendations can be useful:

a. **Multi-Disciplinary Team Approach:** Reserve estimation must be led by a qualified professional with adequate inline experience. The Lead Estimator must be well versed with the procedures, best practices, and guidelines of oil and gas reserve estimation. He/she will need the support of a multi-disciplinary team consisting of specialists in geology, geophysics, petrophysics, reservoir, and petroleum engineering to accomplish the objectives. Engaging the specialists' team benefits the process with additional insights, consensus building, and community ownership.

b. **One Goal, Many Methods:** Reserve estimation is an ongoing process. In the early stages of the lifecycle, there is insufficient data to validate the reserves. These estimates are, therefore, subject to higher levels of uncertainty and lower levels of confidence. However, this changes with time as the reservoir matures and performance data become available for validating the results.

TABLE 8.1

Uncertainty in Reserve Estimation

Reserve Estimation Method	Stage	Comments	Uncertainty
Analogy	Exploration, Appraisal	This method assumes that the performance of two reservoirs with similar geological settings will also be comparable. It involves comparing two reservoirs, one under exploration/in an early stage of development and a relatively mature reservoir, based on their lithology, rock, fluid properties, drive mechanism, etc.	High
Volumetric	Exploration, Appraisal, Development	The volumetric method is primarily a method to estimate the in-place volumes of hydrocarbons based on reservoir extent and rock and fluid properties. These estimates have inherent uncertainties owing to reservoir maturity, data quality, interpretation, and assumptions. Reserve estimates are obtained by multiplying the volumetric estimates by a suitable recovery factor.	High
DCA Method	Development, Production	It is a reservoir performance-based method and is ideally applicable after a reservoir/field is past its plateau production rate. The historical trend of the decline rate is used to forecast reserves.	Low
Material Balance Equation (MBE) Method	Development, Production	MBE method is another performance-based tool for estimating oil and gas reserves. Long pressure-production history, coupled with good quality of pressure, production, and PVT data, can improve reserve estimates' accuracy.	Low
Reservoir Simulation	Appraisal, Development, Production	This method attempts to mimic the reservoir processes of fluid flow and mass transfer by partial differential equations. The model considers the reservoir geometry, heterogeneity, gravity, capillary, and viscous forces acting on fluids in the pore spaces. The quality of the history match decides the accuracy of reserve estimates.	Low

Validation helps to reduce uncertainty and increase confidence in estimates. Different approaches improve perspective, provide assurance, and reduce risk to the field development plan.

c. **Range of Estimates:** It is a standard industry practice to present estimates as a range associated with a probability of occurrence and confidence in forecasts. More straightforward empirical forecasting methodologies (e.g., decline analysis or analytical models) are subject to probabilistic improvisations via Monte Carlo add-in applications. Proficient engineers can employ efficient programmable spreadsheets such as Crystal Ball to determine the range of reserve estimates. However, more than the tools, input data, and data ranges may have an individual bias, affecting the ultimate result. The idea to use a variety of methods and approaches to reserve estimates is not for the sake of ticking the box; it is to ensure that the outcome is valid and consistent in range.
Modern reservoir simulation methods employ an "Assisted History Matching" (AHM) technique that can reconcile several plausible geological realizations with production/injection history in a relatively quick time. This process aims to eliminate human bias and intervention by automating the history match procedure. The AHM leverages the prowess of simulation hardware and software for model optimization with probabilistic forecasting. It can examine the full range of probabilistic cases made up of tens of deterministic instances to decide the best fit in the space of uncertainty.
The lower range of estimates relates to higher confidence and vice versa. The exuberance and strategic interests can sometimes push the owners and reserve estimators to go overboard by projecting numbers outside the acceptable standards. It is an industry best practice to relate probabilistic reserves to P10 and P90.

d. **Review, Review, and Review:** The reserve estimation process must be repeated as additional geological data, petrophysical data, core analysis data, PVT analysis data, well test data becomes available from the newly drilled wells or new analysis. This data aid in narrowing the range of uncertainty in reserve estimates. Review opportunities must be used to make the forecast more scientific, unbiased, and data-driven based on hard evidence.

e. **Don't Overscience:** Reserve estimation is not an exact science; it has many pitfalls. Factors such as the quality of input data, gaps in reservoir characterization, erroneous probability ranges, and lack of understanding of the recovery mechanism can affect the estimates. Quantifying uncertainties in input data is not an easy task in the first place. Inadequately defined input ranges can further complicate it. All reserve estimation methods have their limitations and are applicable under a set of given assumptions.

In heterogeneous reservoirs, rock and oil properties may have high uncertainty. The data may never be sufficient for precise reservoir description, no matter how much money is spent on data acquisition. Many reservoir properties that go as input to the reserve estimation are dynamic, related to time and stage of depletion of the reservoir. Spending money on the acquisition of these data and its timing must be carefully planned. There is no point in over-sciencing the estimates. The focus should be on the proper integration of multi-disciplinary data and reconciling the gaps in understanding.

f. **Avoid reducing reserves:** Reserve addition to the books of the company is, in general, an upbeat and smooth process. However, the department responsible for maintaining the reserve books is not so generous when removing the reserves. Therefore, it is recommended that the reserves be linked with specific projects and made in smaller increments based on these projects' actual progress. It brings transparency to the process and minimizes the resistance to reduce reserves when necessary. However, there will be situations when canceling reserves may be required, which should be acceptable based on a solid rationale.

8.12 Production Target Setting

Oil and gas production is a crucial measure of an E&P organization's performance. It is one of the main drivers of success and a direct indicator of its potential earnings. Oil and gas production grows with time as resources are directed to perform related activities. The rate of production growth is directly proportional to resource allocation.

The business owner and management are always keen to promote growth and maximize profit out of investments. Their business and investment plans for tomorrow are built on today's success. The process of wealth and value creation starts with the way the organization manages its performance. Measuring success by achieving production targets is an age-long standard industry practice, yet setting a production target is always a tricky business.

The concept of setting and achieving a target is quite sporting, and it requires full integration and commitment of the individual players in the team. A sport is more enjoyable when the goal is tight and competition stiff. Also, the reward tastes sweeter after a hard-fought match.

There is no consensus on one single approach that should be used to set production targets. High targets can motivate some teams and individuals who like to be challenged. Achieving stringent production targets comes as an assurance to the management that their organization can deliver high performance under odds. The leadership can then pitch the company to a higher pedestal of growth. A constant pressure to perform under stressful conditions can cause behavior dysfunctionalities in managers and employees who take their job seriously. They may try to reach the end by any means and engage in unethical and dishonest practices to achieve the production target. The consequences of the failure to complete high objectives can be demotivating and detrimental to team morale. Persistent failures to hit achievable goals can play havoc with the psyche and careers of some individuals. It is also bad news for the image and credit rating of the organization.

Low production targets, on the other hand, do not test the true team potential and capability. Some companies that link employee performance with incentives that include bonuses and even promotion to a higher grade end up paying out hefty sums to employees for average effort. Therefore, the production target is a make-or-break affair both for the employer and employees.

The target setting is central to performance management in all organizations. Production targets must be SMART and in sync with organizational capability. The target setting must be finalized by the end of Q4 and announced at the beginning of the target period. Any shortfall in production on the zero date (start of the target period) can push the asking rate of output to unmanageably high levels. Delays in the production target announcement only add to the challenges and can only be mitigated by providing additional resources. The production target is the outcome of a rigorous forecasting exercise based on past performance and incremental production planned from scheduled activities during the target period. Management's commitment is necessary to ensure the availability of all the planned resources in the given timeframe.

Most companies follow a structured, transparent, and collaborative process to set production targets. Despite doing everything right, companies make mistakes in setting production targets and achieving set goals. It may be due to an incorrect assessment of the realistic target rate, inadequate understanding of the reservoir performance, or lack of organizational capability. Facility performance issues, diversion of critical resources such as drilling rigs to other assets, the shortfall in tubular or downhole equipment can pose unexpected challenges to production targets. The companies that do not make mid-course adjustments to the goals due to such issues can miss their mark. The old wisdom of "high but achievable" and management's preferred option to "fall short of an ambitious" rather than "exceed the average" can also be responsible for missed targets. Some large company setups may experience roadblocks arising from the diversity of authority, communication gaps, and misalignments in the participating teams' priorities. These unhealthy situations can result in protracted discussions without ever a consensus on production targets. Under such circumstances, management may have to impose a production target that becomes a bone of contention from day one.

Some federal governments of the net oil-importing countries follow the practice of signing a Memorandum of Understanding (MOU) with their oil and gas companies to deliver a pre-negotiated quantity of hydrocarbons annually. Based on such MOUs, governments can decide how much oil or gas to buy to meet their domestic demand. Quite clearly, the oil companies' failure to comply with a commitment made in the MOU has financial and political implications. Company management may sometimes use the tactic of setting an "internal" production target for the company, which is higher than the one committed in the MOU. This approach works well for reservoirs with good health and higher production potential, maybe with additional resources. However, the final production target is usually a challenging number that goes in such MOUs for a nobler cause of reducing the national oil-import bill.

In large companies, particularly the NOCs, the chain of command is lengthy and also weak. The staff has a mix of excellent, average, and non-performing employees who are more conscious about their rights than duties. Only senior management, managers, and conscientious employees feel the heat and pressure of targets in most cases. Others are least affected by the proceedings around them; their only interest is in the size of incentive at the end of the annual business cycle.

In his introductory speech, a new CEO to the staff and management of an E&P organization congratulated them for achieving the 10 years' production target in a row. However, he added in the same breath that "this makes me suspect that your targets were deliberately set low."

A similar argument is valid for the organizations that fail to achieve their production targets year after year yet end up making huge profits. These organizations must review their organization's capability vis-à-vis the process of target setting. Repeated failure to achieve production targets is a chronic problem that should not be allowed to persist for long unless the management deliberately sets "high target" as a strategy and is content with somewhat lower results.

The critique here has no aim to present the good or bad consequences of the target setting. Goal setting and seeking, within or outside the context of organizational performance, is an exercise that tests one's character. An organization is a collection of individuals with different ideas, skills, work styles, and temperaments. Holding them together and channelizing their talents to achieve organizational goals is quite a task in itself. Real leaders take pride in how their organization, teams, and individuals rise above petty issues to overcome a challenge. They learn from the failures, which have many positive outcomes; one of them is experience. They also learn from success, which can have many negative fallouts such as complacency, overconfidence, and arrogance, which can destroy an organization's value system bit by bit. They believe there is no shame in losing a hard-fought battle, and there is not much pride in an accidental win.

8.13 Production Optimization

Almost all producing assets have the opportunity to optimize production. Significantly huge assets with hundreds or even thousands of prolific producers do not have enough time to focus on individual wells. Therefore, many wells with suboptimal performance can go unnoticed. An article "Analytical approach to maximizing reservoir production[7]" claims that their approach had the potential to "improve the global average underground recovery factor by up to 10%, equivalent to unlocking an additional 1 trillion barrel of oil equivalents (boe)." Some of us might dispute this figure per se. Still, most of us will agree in principle that there is a significant production potential that oil companies could tap from production optimization efforts and avoid unnecessary drilling of new wells.

A systematic analysis of wells utilizing the nodal system analysis approach can help keep the production flowing at its optimum and lower lifting costs. The software can reduce the physical and mental effort to design a well completion program for a natural flow or artificial lift. It can also quickly integrate the inflow performance, vertical lift performance, and bean performance to perform a quick analysis of oil gain from additional perforation and well stimulation. All analyses and design that engineers did manually before; can be done using the software more efficiently.

All oil companies have an ongoing process of systematically reviewing the existing well stock of poor performers. These wells are taken up for suitable remedial measures such as zone transfers, sidetracking, replacement of corroded tubing, change of pump, redesigning the artificial lift completion, waterflood optimization, and so forth. Such efforts may be reasonable to feed the routine rigless and rig workover programs; they soon become mundane, lacking innovation and fresh ideas. No wonder they don't necessarily bring the kind of value the technical paper[5] refers to above. The teams that dare to break away from the old beaten path can hope to meet exciting results. That is why changing the composition and inducting new members into the team can be so productive sometimes.

Production wells may be suffering from a variety of problems that need to be addressed by different specialists. The team working for incremental oil gain from production optimization must be passionate about their work. It must be ready to try new ways of increasing production and should be protected against failures, just in case. Open-minded discussions can overcome human-conservatism and dated mindsets to remove any bottlenecks in the production system, be it subsurface or surface.

Mature fields pose a more significant challenge to increasing production. Ideally, an engineer should not be in charge of more than 50 wells. He should be responsible for maximizing production from these wells and fully supporting the multi-disciplinary team consisting of specialists from geophysics, geology, petrophysics, reservoir engineering, and facility/process engineering. The engineer in charge should lead a properly designed, structured, well-review process for his set of wells to identify opportunities for increasing production. His team should have a holistic review of each well from the reservoir to the facilities. It includes going over geology, petrophysics, drilling, completion, and production history to understand the current flow capacity. From this point onward, deeper involvement of operations and process/facility engineers is mandatory. Each well is checked for the possible bottlenecks and constraints in the wellbore, wellhead, flowline network, and routed separators. This team must have the authority to draw SMEs/specialists as necessary to find a cost-effective solution to realize the full production potential of individual wells.

Sometimes drilling of additional wells may not help in increasing production because surface facilities have constraints. It is essential to have a correct assessment of the subsurface potential and production handling capacity of the surface infrastructure through a "Produce to Limit" exercise to identify the system bottlenecks. The debottlenecking process addresses specific areas and equipment that turn out to be a constraint in oil and gas facilities' performance. Comparing current with design dry/wet crude handling capacities of the GCs and operating ranges of the main pieces of equipment, such as separators, desalters, pumps, etc., facilitates an understanding of the bottlenecks. In some situations, facility debottlenecking may involve simple actions such as changing system parameters to match recommended design values. However, on other occasions, the solution may be more elaborate, requiring replacing an entire piece of equipment to match the facility's needs. Like well review, it is essential to have an estimate of oil gain from specific debottlenecking programs.

The end product of such an overall well and facility review is an action plan which includes oil gain opportunities both on subsurface and surface sides. These opportunities can be ranked and prioritized based on the expected incremental oil gain. This process being transparent has the concurrence and ownership of all the stakeholders.

In many productive assets, the number of dedicated drilling and workover rigs may dictate the frequency of such reviews. The asset must continuously feed the locations to these rigs; otherwise, they can be diverted to other needy assets, and getting them back may be difficult. It is not a healthy practice yet a practical problem. The teams usually maintain an inventory of wells that come out of well reviews conducted some time back. One after another, these wells are passed on to the rigs to keep them engaged. The same process would be more effective if the asset gets a rig when it needs it rather than the rig asking for locations.

8.14 Value of Peer Reviews

Peer review is an integral part of critical projects, planning, or studies with high stakes or long-term implications. It is a comprehensive and successful process to assess the quality and standard of a study, project, or research work based on infield experts' insights. These reviews are often necessary when technical teams want to double-check if their project work moves in the right direction or if their study/strategy results are in order. The management may also request peer reviews before making a crucial investment decision or deploying a potential game-changing strategy.

The idea of such discussions/reviews is to address the apparent pitfalls in a project to improve the chances of success. They can be conducted on more than one occasion by people held in high esteem both by the project team and the management to facilitate trouble-free follow-up actions. Peer reviews for high value-projects held near the start of a project can help the project team orient and focus on the right items.

To avoid the peer reviews turn into an ugly "power game" or "tick the box" eye-wash, all the stakeholders, that is, the project team, reviewers, and the management must be on the same page about their objectives. The project team must not get the impression that the reviewers are out to catch faults on their own or at the behest of the management. Such a threat might force the project team to win at any cost, even by concealing vital information. By the way, the project team is probably the most knowledgeable among all the concerned parties because it stays with the project on a 24 × 7 basis. Its cooperation and support are the most critical success factors for the project review and completion. Some peer reviews might be conducted to show that a due/rigorous process was followed to arrive at an important decision. While a complete alignment between the project team and the reviewers is not an unwelcome situation, it is essential that it follows as an outcome of the due diligence and not as a mere premeditated conclusion. All stakeholders, project team, reviewers, and management must be in the collaboration mode to remove the impediments for improving the probability of the project's success.

The selection of reviewers is the single-most-important success factor of the peer reviews. The reviewers must be knowledgeable, well-rounded, unbiased, and respectable professionals who can quickly go over details. Depending on the subject of discussion, they may be internal or external to the organization and may come from industry or academia.

Peer reviews' success also depends on the quality of study material provided to the reviewers and the time allowed to go through it. Reviewers are usually busy people, and they need sufficient time to go through the study material about a new project and understand it correctly before issuing comments.

Preparing the study material for peer reviews is time-consuming and burdensome for the project team. It must organize all critical data sets, files, maps, and models to present and substantiate its case before the reviewers who act as the jury. Any vital missing data or piece of information may be questioned and decision withheld, something not appreciated by concerned parties. Any delay in sending the study material to the reviewers is not in the interest of the review. It is seen as an old and deliberate ploy used by the project team that does not augur professionalism.

A useful peer review is like a second opinion, which provides peace of mind to the decision makers. It highlights the points of vulnerability and possible obstacles in the success of a project/study to the stakeholders. In-time completion of high-value projects is critical for business organizations. Project timelines are usually tight, and the organizations cannot afford any slip-ups that affect the product's quality and final specifications. If conducted properly, it can bring much-needed clarity and assurance to the decision-making process. Based on the conclusions of the reviews, the organization can decide to move forward decisively or make suitable adjustments to the project. It can even completely change course when there is still enough time left to do it safely. Peer reviews conducted in a collaborative environment with a positive attitude can be powerful tools for a project/study's success.

8.15 Keep It Simple and Measured

It is an expert's job to keep things simple. An amateur who doesn't fully understand an undertaking's intricacies is likely to produce something complicated and costly to manage. Such designs are destined to fail, especially in critical times.

Field development plans are not perfect because they are based on assumptions and uncertainties in data. The objective of maximizing the profit by exploiting hydrocarbons from Mother Earth's depths by drilling hundreds of wells with the help of hundreds or thousands of employees without risking their life is no mean task. There is no need to complicate it any further by over-sophisticated well design, completion, and technology unless utmost necessary. Simplicity offers many direct and indirect benefits. Simple plans are not dependent on specific people or experts. They improve efficiency, quality, and reliability, reduce cost, and minimize the need for supervision. It is a different matter that, with practice, even complicated tasks can be accomplished without a glitch. Conventional and straightforward designs also have the advantage of assurance about performance and are easy to fix when necessary.

Reservoir management team should work to develop SMART criteria that will promote the use of best practices. Limiting values of Reservoir Health Indicators (RHIs) and Well Health Indicators (WHIs) must be established by sensitivity studies and communicated to the responsible asset teams to protect the reservoir health. Everybody understands the language of gain and loss in terms of barrels and dollars. Therefore, the consequences of violating the limiting parameters must be emphasized with the loss in oil recovery and revenue. The issue of the production below the bubble point may be dealt with in the same manner.

The qualitative approach to reservoir management must be replaced by quantitative assessment. Specialists know how to relate quality with quantitative measures in their area of expertise. Quality of oil wells, water injection, stimulation, and other treatments can be converted into incremental oil gain/loss and value/cost. This approach is essential to improve the quality and delivery of results. It simplifies performance assessments of groups and individuals. Management also needs this information to fix responsibility and reward merit.

8.16 Teamwork

Teamwork is the essence of group tasks that require various skills, sharing of workload, and delivery of targets. Organizations manage their oil and gas assets with the help of teams. A team is a basic unit that consists of many individuals with complementary skills to achieve a common goal. But its size and scope increase with objectives.

E&P activities are varied and vast in range and scale. They require the engagement of several teams with clearly defined roles to manage exploration, development, production, and transportation phases. A single team lacks the skills to achieve the ultimate goal of profit-making from oil and gas extraction. Still, when they work together, they can deliver exceptional value to the organization.

Teamwork is one of the most sought-after qualities in time-bound group activities. It requires a clear understanding of each other's roles, responsibilities, and capabilities. Honest and open communication, whether written, verbal, and non-verbal, are the foundation of teamwork. Respecting colleagues' views and public acknowledgment of their contributions help build the bridges of personal rapport with coworkers.

A team draws its strength from individual personal qualities, such as problem-solving, conflict-resolution, and determination to deliver against odds and deadlines. Real team players have team vibes. They revel in each other's success and don't let personal ambitions rule over the team's interest. They are reliable, willing to take responsibility, share the workload, and always ready to help. They act as gelling agents setting examples for selfless giving, their sole motivation being the improvement in character and unity of the group. Their goal is not to be the best on the team. They want their team to be the best.

Working with such teams where performance, not the hierarchy, empowers people, growth possibilities are limitless. Collaboration, knowledge management, and innovation are the key elements of their work culture.

Workplace camaraderie in such settings can sometimes reach a degree where the lines between family and colleagues, home and office, and weekdays and weekends are blurred. Driven by their love for work and colleagues, people put aside their convenience and end up spending more time in offices than with their families without any complaints.

In the divided opinion situations, good teams capitalize on interpersonal relationships, mutual respect, and personal rapport among coworkers. They resolve conflicts by persuasion, not by authoritarian diktats, and make decisions in the interest of the common goal without causing ill will or dissent. Such teams are the real assets of the organization. With their unselfish attitude, they prepare the ground where success comes to the organization as a habit and not by chance.

Notes

1 Westwood Global Energy Group; "Why so many oil and gas projects fail to produce as planned" 5-Feb-2018, Extract from a research report published in Westwood's online Wildcat service entitled: "Appraisal efficiency – can it be improved?"; https://www.westwoodenergy.com/news/westwood-insight-many-oil-gas-projects-fail-produce-planned
2 Energistics; https://energistics.org/category/members/
3 WITSML Data Standards - Energistics; https://energistics.org/portfolio/witsml-data-standards
4 IEEE Std 730.1–1989, vol., no., pp.0_1-, 1989, doi: 10.1109/IEEESTD.1989.114461; IEEE Standard for Software Quality Assurance Plans; https://ieeexplore.ieee.org/document/213705
5 PPDM 3.9 Data Model; https://ppdm.org/ppdm/PPDM/IPDS/PPDM_Data_Model/PPDM_3_9_Data_Model/PPDM/PPDM_3.9_Data_Model.aspx?hkey=fed7573b-c57d-4909-b15a-a61880fb8d2b
6 Adrian Hedden, Carlsbad Current Argus; Waste to Water: 4 lessons learned from our national tour of oil and gas operations; 12 Sep 2019; https://www.currentargus.com/story/news/local/2019/08/28/natural-gas-oil-fracking-produced-waste-water-solutions-takeaways/2121119001/
7 Anton Maximenko, Otto van der Molen, and Francesco Verre; McKinsey & Company; Oil & Gas Our Insights; Article "Analytical approach to maximizing reservoir production"; 21-Sep-2017

9

Proactive Reservoir Management

9.1 Introduction

It is clear that geological complexities are not the only hurdle; the reservoir management plans are subject to errors of judgment due to uncertainties, operational failures, accidents, and poor decisions. The frustrating life experience voiced in Murphy's law, "if anything can go wrong, chances are it will," can deliver a deadly blow to the company's plans and sometimes even to their image. Most organizations, based on industry experience, prefer to manage their programs and projects proactively than reactively. Being proactive is a matter of attitude which involves planning and preparing for possibilities in the future. The proactive approach centers on controlling the results of the outcome. Forward-looking enterprises that plan for the upcoming events/operations have better chances to avoid pitfalls and achieve their goals. Reactive management usually refers to "firefighting." This approach focuses on dealing with the consequences of the outcomes or, in other words, fixing what is broken or failed. "make things happen" versus "let things happen" are the distinguishing features between the underlying philosophy of the proactive and reactive management approaches, respectively.

Following the preventive maintenance schedules of machinery and equipment is an excellent example of proactive maintenance because it results in uninterrupted service. The cost of reactive maintenance can sometimes be too high to afford due to loss of production, mandays, and possible injuries. Therefore, E&P organizations' only way to protect their business from unplanned breakdowns is to adopt a proactive maintenance strategy. Servicing transport vehicles, drilling rigs, pumps, engines, an inspection of pipelines for corrosion monitoring, and other activities of this nature are all part of preventive maintenance.

The implications of reservoir management activities affecting oil and gas production and supply are far too important to be attended on a reactive "run-to-failure" basis. This chapter is dedicated to the new emerging trends and technologies that will take center stage in the coming years to protect the company's interests.

9.2 Use of Drones in the Oil and Gas Industry

In aviation technology, a drone is referred to as an "unmanned aerial vehicle" (UAV) that can fly without a pilot as long as its fuel lasts. The technology has shown great promise in remote, harsh weather, and military operations where human life faces higher safety risks. Various industries are interested in deploying this technology for their business operations due to the clear advantages of speed, cost, and efficiency. The most exciting application of drone technology is under testing to allow online shoppers to receive their consignments in less than a day.[1] Today, drones can deliver precise aerial intelligence, which is extremely useful for planning and monitoring a wide range of infrastructure projects or regulatory compliance activities inland or offshore.

9.2.1 Seismic Surveys

Possible drone technology applications in the oil and gas industry are quite promising because its reach is extended to remote deserts, deep offshore waters, and the sites of floods and fires. Conventional seismic surveys, pipeline inspections for leaks, and routine surveillance operations for remote wells and facilities under challenging terrains can now be efficiently conducted without risking personnel and equipment safety.

Drones can carry out safe and quick aerial inspections to detect any damages/abnormalities in the wellheads, piping, pressure vessels, tanks, and miscellaneous components of the facilities enabling the company to take preventive action. It provides the project managers with a fast and cost-effective means of predicting and managing the risks.

Today companies can acquire high-quality seismic data by employing aerial drone mapping.[2] They can plan the surveys by strategically positioning the seismic wave generators and receivers in the area of interest to avoid surface features and soil types that can affect data quality. The seismic wave source can be an explosive or a mechanical vibrator that can produce seismic waves of controlled frequency and transmit them through the earth. These seismic waves are captured by geophones that are set at predetermined intervals from the source point.

It is possible to equip large drones with a range of cameras and advanced apparatus necessary for high-resolution ground imagery. In conjunction with the Geographic Information System (GIS), these drone images are used for optimizing the source and receiver's positions while planning the surveys.

The frontiers of technology are continually expanding. Successful deployment of darts and drones in a pilot seismic study in one of the most formidable terrains of Papua New Guinea by the supermajor Total is an indication that seismic hardware can be placed where required without sending any person to the location of the survey.[3] This technology

FIGURE 9.1
Seismic Surveys by Drones and Darts (Courtesy Total)

spearheaded by Total will undoubtedly revolutionize how seismic surveys are conducted in the future, particularly in rainforests and highly vegetated areas, without raising any environmental concerns for the governments and local inhabitants. It permits the placement of receivers known as DARTs (Downfall Air Receiver Technology) due to their resemblance with a dart by drones. Once released by the drones from a height, the biodegradable DARTS can be embedded in the ground at specific locations. The DARTS sensors can transmit seismic data via a high-speed, real-time radio telemetry system to a central site where the data can be processed to create subsurface imagery of the ground (Figure 9.1).

9.2.2 Oil Well and Drilling Rig Inspection

Drones also facilitate frequent photographic inspection of oil wells, storage tanks, and drilling rigs located in remote areas. Still, photography and videography provide the much-needed data and evidence to control and manage the HSE incidents such as oil spills, blowouts, and fire.

9.2.3 Pipeline Inspection

The pipelines are a safe and standard mode of transportation of petroleum and its products across the globe. To keep things in perspective, the total length of pipes in various countries of the world adds up to 6.23 million kilometers, approximately 16 times the distance between the earth and the moon (3,84,400 km). The world's longest natural gas pipeline West-East Gas Pipeline to be operated by PetroChina will be 8,707 km long. In contrast, the longest crude oil pipeline operated by Russian oil transportation company Transneft is Eastern Siberia-Pacific Ocean Oil Pipeline, which is 4,857 km long.[4]

Quite clearly, oil and gas pipelines can be very long and wind through varied, hostile, and ecologically sensitive terrains. They are susceptible to different kinds of impairment from internal and external corrosion, cracking, and third-party damage. Any leak, rupture, or physical harm due to corrosion, pressure loading, and drastic temperature changes along the route, sabotage, and terrorist attacks can jeopardize the integrity of the asset resulting in loss of revenue, explosion, casualties, property damage, the release of hazardous materials, and long-lasting damage to the environment and groundwater. The company responsible for the operation of the pipelines must follow a proactive approach for their maintenance. Such an approach ensures uninterrupted and smooth functioning for efficient use of the asset over its life and safety (Figure 9.2).

That is why periodic and regular inspections are so essential to maintain these expensive assets in good health. The most prevalent Non-Destructive Testing (NDT) methods of pipeline inspection to assess the condition of the material without causing any damage to it are as follows[6]:

- Visual inspection
- Ultrasonic techniques
- Thermography,
- Radiography
- Acoustic emissions

Visual inspection is carried out by a trained professional to detect superficial cracks, welding defects, corrosion growth, and structural failure.

Ultrasonic testing with frequencies 50 kHz–50 MHz beyond the audible range of human beings can identify the welds, cracks, delamination, thickness changes, and other types of material or structural defects.

Electromagnetic radiation or radiographic methods are useful for checking the strengths of welds on pipelines. These methods typically

FIGURE 9.2
Pipeline Inspections by Drones[5]

employ short-wavelength radiations viz., X-rays or gamma rays that can pierce the pipeline and pressure vessels to evaluate their strength and integrity.

Thermographic inspections focus on measuring the difference between the temperature of a pipeline and the surrounding environment to locate the point of leakage of oil/gas and determine the quality of pipeline insulation.

Acoustic emissions technique can capture the rarefaction waves that are caused by leaks in pipes. Rarefaction is the opposite phenomenon of compression. It comes into play when oil and gas moving through pipelines under pressure release through leaks.

Visual inspections and X-rays are the most commonly used methods to inspect pipelines that are above the ground. However, the underground/inaccessible pipes use "pigs" that scrape through the pipeline with the flow of air or liquid from one end to the other to perform the inspection. These pigs can record valuable data by ultrasonic, electromagnetic, thermographic, or acoustic emission methods, which can be analyzed to detect any problems or flaws in the pipeline.

Aerial surveys by drones are proving an effective way to inspect oil and gas pipelines. These UAVs enable coverage of large and difficult terrains by recording high-resolution photos and videos of above-ground pipelines for a preliminary examination. They can also capture thermal images of the hotspots with infrared cameras to detect leaks in the pipeline that are not visible to the naked eye. UAVs can also provide information about the areas of concern, and technicians can be dispatched to the site to address the problem. Inspections by drones can also reveal unauthorized intrusions, which could pose a danger to the pipeline network. The benefits gained in terms of cost, value, and efficiency of the UAVs attract more and more companies to deploy this technology for oil and gas pipeline inspection.

9.3 Maintenance Policies

The Industrial Internet of Things (IIoT), driven by artificial intelligence and UAVs, enables the maintenance of wells, pipelines, facilities, and infrastructure to a new level to increase their productivity and reduce unplanned downtime. The connectivity of instruments, sensors, and computers in real-time has allowed the companies to visualize the economic benefits that can be reaped from timely data collection and decision-making. Their age-old maintenance policies are progressively transforming from reactive (run-to-failure) and preventive (scheduled maintenance) to proactive/predictive methods, which can usher in a sea change in equipment and organizational productivity.

In the 1940s, companies practiced the reactive/corrective maintenance policy. The term reactive policy refers to repairs after the failures occur. Machines and facilities are allowed to run until they stopped to perform. This approach was taken over by preventive maintenance sometime in the 1970s when companies realized that the operating company could ill afford the consequences[7] of the reactive approach as both the governments and local population became more vigilant about HSE aspects of the business. They introduced the policy of scheduled maintenance whereby all critical pieces of the facilities, machinery, equipment, and material were checked and overhauled/serviced during planned shutdowns at a predefined time. This approach reduced the number of incidents and performance issues, yet it was not good enough for the industry. The biggest challenge that faced the preventive maintenance policy was the timing and interval between successive planned shutdowns for various asset components (Table 9.1).

Advancements in technology and know-how have allowed the oil and gas companies to resort to a proactive/predictive maintenance model. This approach to maintenance aims to eliminate system breakdowns and incidents by addressing the problem before it develops. This approach allows maintenance decisions to be made based on the continuously monitored asset; it is also referred to as condition-based maintenance (CBM). Based on the dynamic understanding of the running equipment, baseline performance of machines/assets/equipment can be used to diagnose the faults and make a prognosis about the failures. A proactive maintenance program is the result of the fusion of predictive and preventive maintenance strategies.

TABLE 9.1

Highlights of Maintenance Policies

Reactive (Maintenance after Failure)	Preventive (Scheduled Maintenance to Avoid Failure)	Proactive (Maintenance Only When Necessary, as Predicted)
High cost, unplanned shutdowns	High cost, planned shutdowns	Cost-effective
Staff and time Required minimal	Increased asset life, reduced risk	Increased asset life, reduced risk
	Enhances productivity	Visibility of asset health Predicts time of asset failure
		Minimum downtime, maintenance carried out before unavoidable failure
		Need for condition-based monitoring equipment/models

Matured oil and gas fields where E&P infrastructure is more than 15 to 20 years old can significantly leverage this new approach. This infrastructure includes drilling rigs, electrical submersible pumps, oil and gas separators, condensers, pressure valves, heat exchangers, storage tanks, flare stacks, pumps, compressors, turbines, crude oil and gas treatment system, transportation pipelines, and a variety of other kinds of machinery and equipment. IIoT led predictive maintenance involves installing acoustic, ultrasonic, temperature, and different types of sensors. The sensors are installed at vulnerable points to measure and detect anomalies in pressure, temperature, torque, vibrations, flow rate, and so forth, vis-a-vis the predictive models and send alerts to the maintenance teams impending performance issues/failure. This approach aids in improving asset integrity, reliability, and cost savings.

Similar technological advancements are in progress for inspecting offshore and other arduous sites. The world's first autonomous offshore robot ANYmal designed by ANYbotics can check offshore sites and generate 3-D maps of its surroundings to carry out inspections and operations more efficiently. The day is not far when manual inspections will be obsolete and replaced by robotic inspections of facilities, including inside tanks and pipes. These robots will share data and images of the inspected area and carry out delicate operations such as painting the substructures and intricate access points.[8]

9.4 Smart Oil and Gas Wells

Production optimization and recovery improvement objectives require that reservoir energy in dissolved gas, free gas, or water drive be utilized most optimally. It implies that oil and gas production and water injection must be controlled to overcome the effects of reservoir heterogeneity, something challenging to achieve with conventional well completions and traditional methods of operations.

Driven by technological advancements in electronics and information technology, smart completions offer the benefits of monitoring and controlling the production/injection rates in real time for optimizing the overall productivity of the well. These completions may be tailored to avoid or delay undesired water or gas production, thereby conserving the reservoir energy, leading to an increase in the well-productivity, improved oil recovery, and better reservoir management. Since the first successful installation of downhole pressure, temperature, and flowrate control valve in the North Sea[9] in 1997, the popularity of such completions grew in the offshore wells as E&P activities advanced to deep waters and subsea completions became frequent. The increasing complexity of well designs such as horizontal and multi-lateral adds to the cost and makes the surveillance and intervention highly expensive and risky. Smart/intelligent completions gained broader acceptance in the industry because they enabled better control and management of various productive zones with the permanent installation of pressure and temperature sensors, multi-phase flow meters, and flow control devices.[10] The new technology facilitated a significant reduction in CAPEX because a smart/intelligent well completed across multiple zones could very well match the production of several conventional wells. Not only this, but it could also be operated remotely without any worries of weather conditions or sending a technician to the well (Figure 9.3).

The smart/intelligent wells can be designed to exercise reactive and proactive controls on oil/gas wells. inflow control valves (ICVs) are set to control GOR or water cut below the predefined target value in the reactive control mode. In case installed ICVs are unable to control the GOR or water cut, they are replaced. The proactive control pivots around the presetting of ICVs to limit the GOR and water cut values within the range specified by the tested simulation models.

In summary, intelligent-well systems achieve the vital objective of improved reservoir management in monitoring reservoirs to optimize production rate, maximize oil recovery, and minimize capital and operating expenditures. The technology application can significantly improve the sweep efficiency in heterogeneous and multilayered reservoirs undergoing water/gas floods.

9.5 Digital Fields

Digital, smart, or intelligent oilfields are the same thing that takes full advantage of the integration and interaction of people, technology, and processes. They were primarily born out of the necessity to extract oil and gas from challenging terrains and hostile environments with minimum safety risks. The story of their evolution is closely linked with the growth of electronics, information, telecommunication, and instrumentation technologies and their successful application in various disciplines of oil and gas. Oil and gas technology has followed the maxim "necessity is the mother of invention." It has been unraveling solutions necessary to overcome the challenges posed by the field location and environment even before the digital revolution. SCADA systems were already used to control and monitor large parts of the facilities and perform automatic controls on flow, pressure, current, or voltage by the remote terminal units (RTUs) or programmable logic controllers (PLCs). Numerous technologies were being tested to support deepwater drilling and information collection for remote

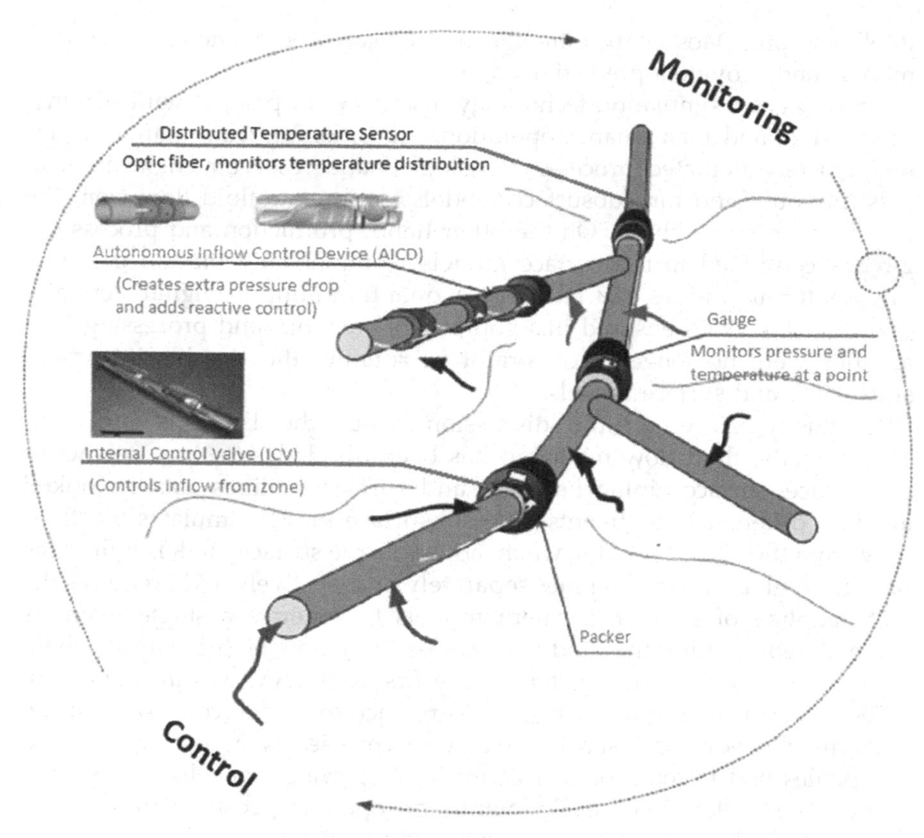

FIGURE 9.3
Intelligent Well Completions for Improved Reservoir Management (Source- Rafiei, Yousef; Improved Oil Production and Waterflood Performance by Water Allocation Management)

operations and disaster response. However, their application was limited and arbitrary.

The digital versions of the new business operations are more intelligent, efficient, and cost-conscious. Digitalization in drilling enables automated well design and execution,[11] optimized well planning and reduced drilling time. Field development planning acquires a considerable wealth of data in digital form for analytics, development drilling, modeling, visualization, and predictive purposes. IAMs can replicate the behavior of fluid flow from the reservoir to the well and through the flowline network. The use of intelligent completions for optimizing oil and gas production from vertical and NCWs is rising. Hard-to-access production facilities and pipelines can be monitored by drones, transmitting data in real-time to carry out predictive maintenance. Pipeline integrity can be assessed and restored with intelligent pigs. Most of the oilfield processes can be simulated by computer models and provide a predictive edge.

The reservoir simulation technology also grew in parallel with drilling, production, and maintenance operations transitioning from manual to intelligent or automated processes. Geologists and reservoir engineers successfully built and ran subsurface models to simulate fluid flow from the reservoir to the wellhead. On the other hand, production and process engineers continued to use surface models to understand oil and gas flow physics through pipes and facilities. A quantum jump in digital technologies after the 1990s ensured that computers' memory and processing capabilities were no longer a constraint in running the sizeable integrated subsurface and surface models.

At this juncture, a little discussion about the IAMs is in order. Traditionally, fluid flow modeling has been divided into three segments: subsurface, surface piping network, and surface facilities usually looked after by different departments. The subsurface model simulates the fluid flow from the reservoir to the wellhead, while the surface models mimic the flow through pipes and facilities separately or collectively. IAM requires the soft coupling of all such segment models to generate a single seamless "clone," representing the fluid flow across the whole oil/gas supply chain from the reservoir to storage tanks. The first such IAM was introduced in 1976[12] when the industry struggled to replace the produced hydrocarbon reserves. Onshore and shallow water discoveries were dwindling, and companies had to look for petroleum in deep waters and harsh environments. The exploration, development, and operating costs of these assets were very high due to the requirement of advanced technologies for drilling, completion, surveillance, and maintenance under challenging logistics due to the associated safety risks.

IAMs are a vital component of any digital field because they cover the entire reservoir fluid flow and facilities domain. They help identify the piping network's constraints and debottleneck the system for improved oil production and recovery. The size of the model and computational requirements usually require powerful machines to make end-to-end simulation runs. Nevertheless, the cost and effort of digital fields are worth it due to the following value additions:

- A clear understanding of the physics of the fluid flow and pressure losses across the production chain from the reservoir to the storage tanks/export terminal
- Resolution of any gap in production between the wellhead and in-tank volumes
- Reduction in the undesirable production of gas or water to achieve the model forecasts

- Making operational decisions on problematic wells/piece of assets in real-time to avoid any interruption in production
- Receiving specialists/experts' input, regardless of their location, for oilfield problem-solving to improve the quality of decision-making
- Harsh weather, difficult terrains, and remote locations are no longer a constraint for data collection. Technology enables data collection at any time and frequency without any technician visiting the site.
- Frequent validation of asset models with the field data
- Improved field production, oil recovery, and project economics
- Elimination of non-productive time and increase in productivity
- Safer, well-managed operations

Digital transformation has touched almost all aspects of oil and gas operations. When the focus is on the value in the modern business environment, all digital fields need not be entirely digital or equally smart. An oilfield's intelligence level is closely related to the business objectives and the investment plan for digitalization. Accordingly, some smart fields may have only basic intelligence with limited functionalities. In contrast, their more sophisticated versions may have capabilities to remotely operate and control the wells, flowlines, and facilities from the safety of a collaboration center.

Digital oilfields, iFields, eFields, Smart Fields, and so forth are the oil and gas fields that are information technology-centric and where the attendant instrumentation and software facilitate oil and gas operations. Digital fields are unique because they utilize cutting-edge technologies to support operational and production excellence, remote operations and management, Collaborative Working Environments (CWEs), and real-time data collection and visualization. A combination of oilfield know-how with these technologies enhances the speed and quality of decision-making for excellent reservoir management and control via automated processes and efficient workflows.

In the modern business environment, smart fields can serve as an instrument to change the business model and operating strategies. Riding the wave of these successful disruptive processes and technologies, the companies visualize a future where they can extract oil and gas from challenging environments without much risk to company assets and human safety. Most of the operations that earlier required the on-site physical presence of staff can now be accomplished via automated processes from the safety of a remote control/collaboration room set up in company offices. The intelligent field technologies have practically defuncted the traditional organizational structures along the functional lines. The new ways of working have demonstrated all-round improvements in drilling and completions, production optimization, surveillance and control of subsea production

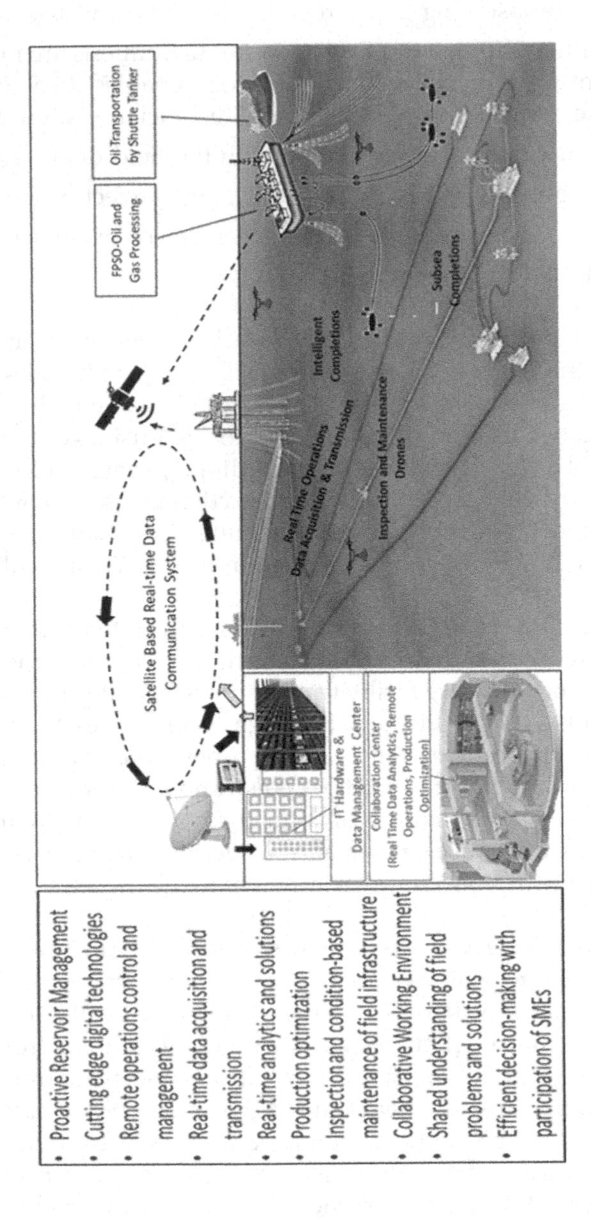

FIGURE 9.4
Digital Fields Achieve Corporate Objectives via Technology Integration

systems, reliability and maintenance, environmental monitoring, asset integrity management, and decision-support (Figure 9.4).

Integrated intelligent operations have given a big boost to the oil and gas production from offshore fields. Engineers prefer digital fields because they can secure real-time data and specialists' virtual presence to support the complex decision-making under constraints. To give an example of what happens to the overall field production if a wet oil well is routed to a GC with limited water handling capacity, the IAM needs to be run with given capacity constraints to find an answer. Digital fields allow quick simulations to be performed in the background before the results can be presented to make a final decision. Managers prefer digital fields because they can see both the reasons and consequences of their actions without delays. And of course, the management is happy because they see the sprouting of a new work culture based on the rules of teamwork, the spirit of collaboration, and the practice of workflows that promote the corporate objectives of increased operational efficiency, transparent decision-making, and cost reduction in business operations.

The industry is looking at the future of the digital fields with interest. It is likely that with further advancement of technology and cost reduction, the integrated digital fields will be a standard both for onshore and offshore areas.

Notes

1 Dezeen website; https://www.dezeen.com/2019/06/06/amazon-prime-air-drone-news/
2 Terrex Seismic, Drone Aerial Mapping; https://www.terrexseismic.com/advanced-technology/drone-aerial-mapping/
3 Darts and Drones-the future of onshore seismic; Seismic study in one of the most formidable terrains of Papua New Guinea by Total; https://wirelessseismic.com/darts-drones-future-onshore-seismic/
4 Offshore Technology, Transporting oil and gas-world's longest pipelines; https://www.offshore-technology.com/features/worlds-longest-pipelines/
5 Canadian Aviator Magazine; New UAS Rules Published; https://canadianaviator.com/new-uas-rules-published/
6 IS website, 5 most popular inspection techniques for the oil and gas industry; https://info.industrialskyworks.com/blog/5-most-popular-inspection-techniques-for-the-oil-and-gas-industry
7 Iqbal, Hassan & Tesfamariam, Solomon & Haider, Husnain & Sadiq, Rehan. (2016). Inspection and maintenance of oil & gas pipelines: a review of policies. Structure and Infrastructure Engineering. 10.1080/15732479.2016.1187632.
8 Offshore Technology, Robot revolution: five robotics developments in offshore oil and gas; https://www.offshore-technology.com/features/robotics-oil-gas/
9 Gustavo Carvajal, Marko Maucec, Stan Cullick; Chapter Seven; Smart Wells and Techniques for Reservoir Monitoring, Intelligent Digital Oil and Gas Fields, 2018;

ISBN 978-0-12–804642-5, Gulf Professional Publishing; https://doi.org/10.1016/B978-0-12–804642-5.00007-4

10 Rafiei, Yousef. (2014); "Improved Oil Production and Waterflood Performance by Water Allocation Management." 10.13140/RG.2.2.30524.33925.

11 Oil and Gas Journal-"Digital Transformation: Powering The Oil & Gas Industry;" https://www.ogj.com/home/article/17297879/digital-transformation-powering-the-oil-gas-industry

12 Petrowiki, "Digital Oilfields;" https://petrowiki.org/Digital_oilfields#Overview

10

Pricing of Crude Oil

10.1 Introduction

Petroleum is the most important source of energy in the world today. It powers national economies, drives modern societies, and facilitates contemporary lifestyles. It is the primary source to provide energy for personal and public travel, manufacturing goods, agriculture, mining, and construction. The other major sectors that make generous use of oil and natural gas are power, commercial and residential sectors. With such massive implications on businesses and people, petroleum is undoubtedly the most sought-after resource. The narrative of petroleum reservoir management has changed significantly over time in changing national and international perspectives and policies. Therefore, modern oil corporations cannot afford to be content with only the rate and recovery improvement programs. They must walk the extra mile to ensure that they get the real price of their product.

NOC's presence on the center stage has added new dimensions of energy security and pricing. NOC is not just another oil company; it is a vehicle for state participation in managing and controlling oil and gas resources. The national government's backing and patronage allow it access to formal and informal communication channels with other governments and international agencies. Therefore, E&P operations under NOCs are not for mere extraction of energy fuels; they serve as strategic tools to achieve the national commercial and non-commercial objectives.

The oil and gas touch the life of a common man as much as the economy of a nation. Volatile oil prices can upset the household budget of ordinary citizens as severely as the government treasury. Aspiring industry leaders must appreciate the problems and related issues of the oil markets, market structures, and pricing mechanisms, even if they cannot fully comprehend the solution.

Figure 10.1 presents the top 20 countries' crude oil reserves and their percent share. Quite clearly, Mother Nature has been kind to endow these countries with a bounty of reserves. Out of the world's total proven oil reserves of about 1.5 trillion barrels, approximately 90% are held by 20 countries. The world's oil consumption is of the order of 90 million BOPD. Interestingly, most countries rich in oil reserves do not consume much oil

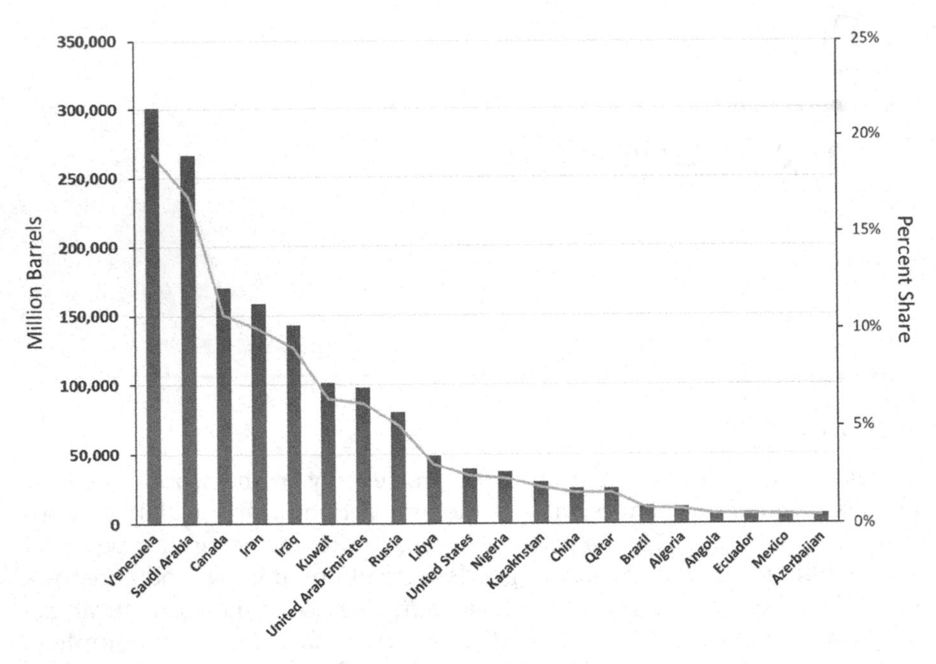

FIGURE 10.1
Countries with Major Proven Crude Oil Reserves[1].

except for the US, China, Saudi Arabia, Canada, Brazil, and Mexico. Therefore, the countries with a high net surplus of crude oil availability have high export potential.

10.2 The Global Oil Markets

The major oil-consuming nations are either the most populous or those with massive industrialization programs (Figure 10.2). The difference in the geographical location of crude oil reserves and its consumption creates opportunities for trading. Accordingly, the oil-rich countries with low crude oil requirements have high export potential and significant export opportunities. Likewise, some oil companies that are entirely vertically integrated over upstream (E&P) and downstream (Refining and Marketing) sectors of the business can refine the crude oil they produce and market products through their retail networks. The companies that focus exclusively on upstream operations have to find buyers for their crude oil. The companies engaged in refining crude oil are also interested in securing crude to their facilities on a sustained basis. Industry sources agree that a significant share of the oil and its products,

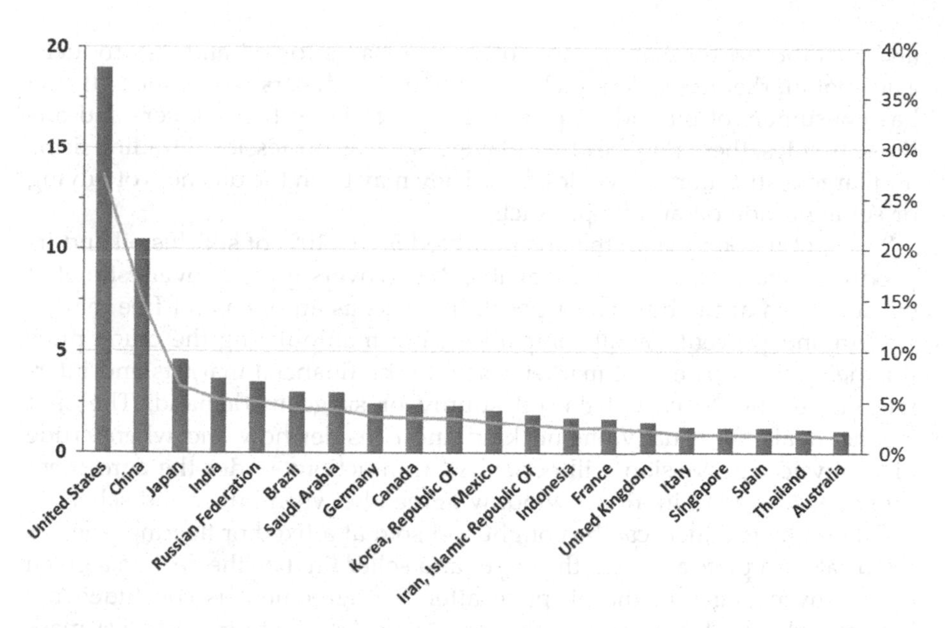

FIGURE 10.2
Countries with Major Crude Oil Consumption[3].

to the extent of 90%-95%, is sold through the pre-negotiated term contracts[2] as spot markets are deemed unreliable to guarantee the quality, quantity, and price of the crude. The sale of crude oil under term contracts removes any uncertainties about the quality and amount of crude oil. Volatility in the oil price is taken care of by building some tolerance into the contracts.

Like other commodities, the movement of crude oil is also governed by the universal law of demand and supply. It is affected by global oil production and the need and economic condition of the buying countries. Oversupply and shrinking demand cause the oil price to go down while increasing demand and decline in production tend to push the oil price up. At higher oil prices, buyers demand less, but sellers can supply more.

Most of the oil is procured under term contracts. Such contracts provide clear long-term visibility and reassurance to both the buyer and seller as their objectives from the sale or purchase of oil will be met. These contracts rely on the crude price prevailing in the spot market. National governments of the crude oil buying and selling countries may sometimes agree to settle payments in their currency or even use the barter option.

10.2.1 Spot Market

The spot market is where any buyer can buy oil or its products at the market price for near-term delivery. It features physical transactions of oil cargoes and sets the price basis for most other purchases for oil and its products. The spot market has sellers such as crude oil producers and refiners. It also has consumers of oil and oil products such as large fuel retailers and airlines. Besides, there may be other players, such as professional trading firms and large distributors or wholesalers. They may be in the business of buying or selling crude oil and its products.

The spot market places the uncommitted 5% to 10% of surplus oil and its products, which may become available due to oversupply or over-estimated consumption at the buyers' disposal. It works as an open and free market system and prevents small companies from manipulating the crude price for their profits. The spot market reacts to the financial markets and international events to curtail the oil supply or surge in demand. The spot market deals are usually one-off kind and consider how and where crude oil's physical possession will occur. All transactions under the agreement must be concluded in a time window agreed between buyer and seller.

Oil and its products can be bought and sold at a fixed or floating price. A fixed rate is a price at which the buyer and seller finalize the deal on a given date. Any changes in the oil price after the agreement is concluded are immaterial. The fixed-price deals are not flexible in terms of price management. There is a potential of possible financial loss to the buyer/seller owing to the price change when the oil cargo is in transit. Floating price deals are dynamic and open to changes in price with the market. In this kind of transaction, the buyer and seller agree on the deal's broader framework, such as grade, product volume, delivery schedule, mode of transport, the port of delivery, and so forth. The price of oil is set after loading it onto a tanker or its delivery at the port. Powered by modern communication technology, floating price sales have become the preferred mode of transaction and purchase over the fixed price sales. Floating price also allows a higher flexibility in managing risk and in optimizing the deal value.

10.2.2 Financial Markets

Crude oil and its products are actively traded commodities. Refineries are always interested in buying the light, sweet crudes which can be refined to extract high-value products such as gasoline, diesel fuel, and jet fuel.

Financial or Paper Markets can buy and sell oil on paper without any physical transactions. These markets are highly speculative as they trade oil for delivery in the future, commonly known as "futures" or "forward." Both futures and forward are bilateral agreements between the buyer and seller, not necessarily involving the physical delivery of the commodity (Table 10.1). They are used to buy and sell oil or its products at a set price at a future date irrespective of its current market price. Futures are also used by investors hoping to profit from transactions based on the accuracy of their "speculation" about the change in oil prices. This speculation may be

TABLE 10.1

Difference between Futures and Forward Contracts[4]

	Futures	Forward
1	Futures contracts are traded on an exchange and have full system support to guarantee the transactions.	Forward contracts are private agreements between a buyer and a seller. The possibility of defaulting by the buyer or seller is not ruled out.
2	Futures trade oil contracts for future delivery. Changes in asset prices are settled daily until the end of the contract.	Forwards trade oil contracts for future delivery. The price change is settled only when the deal ends.
3	Futures contracts are preferred by speculators who bet on the direction of the price movement.	Forward contracts are primarily used by hedgers who are interested in moderating the volatility in price.
4	The market for futures contracts is highly liquid. Investors can enter or exit the market when they like.	Forward contracts are not readily available to retail investors.
5	Futures have price transparency, the certainty of transactions, and practically no possibility of defaulting	The market for forward contracts is unpredictable and prone to risks against the possibilities of windfall profits.

driven by their knowledge of the market or purely intuition. The function of both these contracts is the same though their details may vary.

Airlines or refiners invest in futures to protect their operations against price increases in the future. Oil producers may also choose to sell oil futures to reserve a price for a specific period. Because the futures are traded on an exchange, they have price transparency, the certainty of transactions, and practically no possibility of defaulting. The futures market has ample liquidity/cash, which gives investors the freedom to enter or exit the market when they decide.

Unlike futures, forward contracts do not trade on an exchange. These are private agreements, details of which are kept between the buyer and seller only. Their terms and conditions are more flexible and known only to the buyer and seller. Like futures, forward contracts facilitate the buying and selling of an asset at a future date. The price of the asset is set at the time of negotiating the terms of the agreement. The contract is settled once its term comes to an end; the contracted price holds, regardless of whether the real price increases or decreases.

Forward contracts are private agreements that are not freely available to retail investors. These contracts are vulnerable to high risks on two counts. First, it is difficult to predict market behavior. Second, one of the contract parties may default if the price difference between the actual and contracted price is significant.

Oil supplies can also be secured through options and swaps that allow the investor to lock in a price. Like futures, options can be traded on exchanges as well as on over-the-counter (OTC). Options give the buyer the right to take and the seller the right to deliver oil.

Swaps are purely financial instruments. They do not involve physical delivery of oil and are usually traded OTC.[4]

E-trading has gained popularity in modern times. It offers the companies/investors the flexibility to trade the oil/products at a posted price or through auction, tender, or negotiations.

10.3 Shipping Agreements

An important feature associated with the sale and purchase of crude oil is its transportation and safe custody transfer. A large volume of oil needs to be transported long distances across international borders by routes that may be unfriendly or prone to piracy. Long transit times and changing weather conditions may also pose additional risks to the consignment. Crude oil shipping agreements are finalized considering these aspects. Spot transactions are usually subject to the exchange of physical crude or product volumes at a specific location. It is typical to have crude transactions at the point of loading onto a tanker or into a pipeline near the production region. Product transactions, on the other hand, can occur at a major refining or transportation hub.

The role of transportation or logistics hubs is critical for regional crude oil and product distribution, supply, and trading. Transportation or logistics hubs have:

- substantial storage capacity
- provision of alternative modes of transport such as railhead, seaport, and motorable road network with the inventory of tankers, barges, and trucks for local transport
- interconnections among many pipelines for supply and distribution

These hubs serve as regional gateways and enable the market to respond to supply and demand changes quickly.

The terms of the shipping agreements are explicit whether a cargo will be dispatched by a tanker, barge, or pipeline and its cost basis-CIF (cost, insurance, and freight) or FOB (free on board). CIF and FOB are two commonly used terms used in the crude oil shipping agreements between the buyer and seller. Under CIF, the shipment is the seller's responsibility until it is received by the buyer, implying that the seller bears the cost of loading, insurance, and freight. FOB terms, on the other hand, make the buyer responsible for all expenses and liabilities once oil is shipped (Figure 10.3).

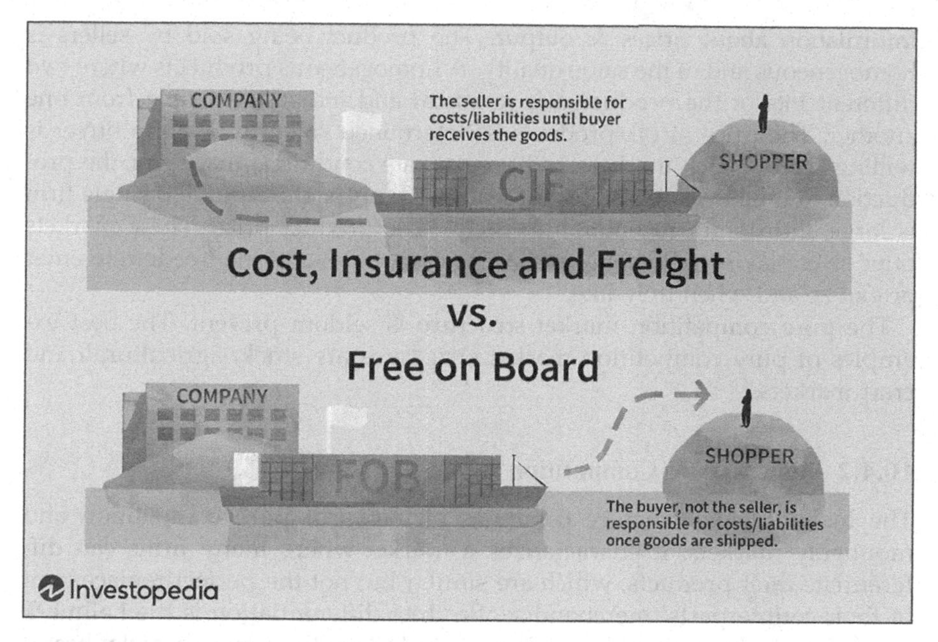

FIGURE 10.3
Shipping Agreement Basis-CIF versus FOB[5].

10.4 Market Structures

A basic understanding of market structure is necessary to appreciate how the oil price is controlled in the international market. Four basic types of market structures are defined based on the following questions:

- How easy or difficult is it for companies/investors to enter or exit the market?
- What is the distribution of the market share for the major firms?
- What is the number of trading companies in the market?
- What is the number of buyers, and how is their mutual relationship to strike a bargain with sellers?
- What is the number of sellers, and how is their mutual relationship?

10.4.1 Pure Competition

Pure or perfect competition is when the oil price is determined by the demand and supply position. It describes a market that has a large number and a broad range of buyers and sellers. Each buyer and seller has perfect information about prices & output. The product being sold by sellers is homogeneous and of the same quality. A homogeneous product is where two different lots of the products are identical and indistinguishable from one another. The price of oil/products is determined solely by what a buyer is willing to pay. Such market conditions create conditions that favor the production of the lowest possible cost leading to optimal output. No single firm is large and dominant enough to dictate the product price. These markets offer level playing field conditions to all companies and the freedom to enter or exit as and when they like.

The pure competition market structure is seldom present. The best examples of pure competition market structures are stock, agricultural, and craft markets.[6]

10.4.2 Monopolistic Competition

The monopolistic structure combines elements of pure competition and monopoly markets. It characterizes a market where many firms can differentiate their products, which are similar but not the perfect replacement of their counterparts' merchandise. Product differentiation is the hallmark of monopolistic competition. It can include the difference in style, brand name, location, packaging, advertisement, pricing strategies, and more. By developing a product line that appeals to a specific segment of buyers, companies carve out a niche for their products. This strategy can help companies maintain a steady demand for their goods.

Monopolistic competition imposes no barriers to entry or exit. It makes no difference to the market if one or two firms decide to join or leave the market. Depending on the mass appeal of their products, companies can set a high price for their product and earn supernormal profits for a short time, knowing fully well that their competitors will not be able to bring out matching products so quickly. However, in the long run, profits will normalize as several new companies will enter the market. To create demand for their product, the new firms will fix the price at a lower level to make inroads into the market share of old companies that earlier had a monopoly in the market. Fashion accessories, cosmetics, newspapers and magazines, TV channels, restaurants, and consumer goods manufacturing companies are examples of monopolistic competition.

10.4.3 Oligopoly Competition

An oligopoly may be described as limited competition and a partial monopoly. It is an imperfect market, with only a few sellers competing to sell their products. Goods or products sold by oligopoly may be homogenous or differentiated. Their products are similar but not perfect substitutes for each other. Their marketing strategy, policies, pricing, and output are based on their rivals' plans and performance. Oligopoly is dominated by a few firms/

companies who can form cartels to control and secure the market's supply side. These cartels use their clout to regulate production in a way that does not exceed the demand. Getting entry into an oligopoly is difficult because constituent members use their network to control feedstock, physical, and financial resources to create barriers for new entrants.

A good example of oligopoly competition is the OPEC (Organization of Petroleum Exporting Countries), whose primary goal is to maintain its stability. Comprising 14 member countries, OPEC is an intergovernmental organization that allocates oil production quotas of the member states and regulates oil production. It controls oil prices based on demand and volume of production or crude availability in the global market. In the case of excess supply of oil globally, OPEC applies cuts to the production quotas of its member states. In the event of short supply, it increases oil prices, which lowers the demand. Thus by regulating production, OPEC controls oil prices and maximizes its profit. It also uses tactics and strategies to stifle competition arising from factors such as rapidly rising shale oil production from the US.

Oligopoly can have different models based on the following characteristics:

10.4.3.1 Product Differentiation

If products are homogenous, viz., identical, and indistinguishable, it is a pure or perfect oligopoly. If products are heterogeneous or differentiated, it is termed as a differentiated or imperfect oligopoly.

10.4.3.2 Entry of Firms

If entry to the new firms is open, it is called an open oligopoly. If entry is restricted, it becomes a closed oligopoly.

10.4.3.3 Agreement

If the rival firms agree to follow a standard price policy instead of competing, it is known as a collusive oligopoly. Collusion in the form of an agreement is called "open collusion," but collaboration in the way of an "understanding" is known as tacit collusion. In case the rival firms act or compete independently, it is called a non-collusive oligopoly. The non-collusive model assumes no collusion between the selling companies, that is, there is no spoken or unspoken understanding between sellers regarding price fixation or market share. Selling companies compete with each other for sales and profits.

10.4.3.4 Price Leadership

If a firm leads to control and set the pricing mechanism/policy and other firms follow, it is called a partial oligopoly. If no firm is dominant enough to assume the role of price leader, it is called a full oligopoly.

10.4.4 Pure Monopoly

A monopoly is a market where there is a single provider of a product/goods/services with no close substitutes and alternatives. With no opportunities for entry or exit, the company wields extensive control over the price. As a result, monopolies can afford to reduce output in terms of quantity and quality, increases rates, and earn more profit.

Public utility services such as railways, road transport, electricity, water, and telephone in some countries are examples of "natural monopolies" due to their significance to the masses. These monopolies have the advantage of the economy of scale due to their size in the commercial landscape. The term economy of scale implies a proportionate saving in costs that can be achieved based on the sheer volume of business. Quite often, national governments or companies under their patronage execute such projects considering public welfare and government earnings enhancement. The government can also formulate policies to protect the operating company's business interests, denying entry to new players.

A comparison of the salient features of different market models is presented in Table 10.2.

10.5 Factors Affecting Oil Prices

Crude oil is a global commodity. It is at the core of the international energy and economic policies of the modern world. The crude oil market is very dynamic and volatile. Significant factors that have the potential to drive its cost up or down are discussed later.

10.5.1 Current Supply and Output

OPEC member countries produce about 40% of the world's crude oil demand, whereas their oil exports represent nearly 60% of the total petroleum traded internationally. OPEC exercises a quota system to control its market share and crude oil price. Crude oil prices go up in times when OPEC curtails its production. An increase in oil prices after 1970 forced the US and other countries to consider increasing their indigenous production and look for alternatives to meet their energy needs. High crude oil prices in 2008 created opportunities for exploitation of the high-cost shale oil.

Consequently, American shale oil production doubled between 2011 and 2014, driving down the oil price. OPEC's response to this has been to increase oil supply and push the cost per barrel down to a level where shale oil production is no longer economically viable. OPEC states can afford low oil prices due to the low cost of their production. On the contrary, shale

TABLE 10.2

Comparison of Markets Models

	Feature	Perfect Competition	Monopolistic Competition	Oligopoly	Monopoly
1	Number of Sellers	Many firms	Many firms	Few firms	Single firm
2	**Average size of the Firm**	Small	Small-medium	Large	Very large
3	**Competition**	High	Some	High	Low
4	Non-price Competition	None	Advertising and Product Differentiation	Huge Advertisement Expenditure	None
5	Barriers to Entry	Very Low	Low	High	High
6	Nature of Product	Homogenous (Indistinguishable)	Differentiated substitutes	Homogeneous or differentiated products	One type
7	**Consumer Demand**	Potentially unlimited relies on supply	Depends on what differentiation provides	Consumers in control of what and how much to buy	Firms produce if demand elastic. Demand low if the product not desirable
8	Market Knowledge	Complete	Incomplete	Incomplete	Incomplete
9	**Specific Market Trait**	Buyers can switch between sellers because of no difference in products and price	When prices are raised, "brand loyalty" helps maintain demand	Firms often form cooperation to stabilize the market and reduce risks	Control the market
10	**Output**	Unlimited	Varies with product differentiation and selling cost	Difficult to predict. It depends on the decision of rival companies.	Based on the seller's capability
11	Price Power	None, Price taker	Some, Price maker	Price rigidity, Price maker	High, Price maker

(*Continued*)

TABLE 10.2 (Continued)

Comparison of Markets Models

Feature		Perfect Competition	Monopolistic Competition	Oligopoly	Monopoly
12	**Profit**	Can make short-run economic profits but cannot make long-run economic profits	In the short-run economic benefits, in the long run, only average profits	Can make economic profits, in the short and long run	Can make economic profits in the short and long run
13	**Government Intervention**	Very little	Very little	Governments monitor that cartels and collusions do not manipulate the market	Governments can impose taxes or levies

producers need an oil price of US $40 to $50 a barrel to make its exploitation economic.

10.5.2 Future Supply and Reserves

The position of reserves and future supply is a significant factor that affects oil prices. Any uncertainty, risk, or threat to future oil supply can send shockwaves and make many countries nervous. That is why many countries have long-term agreements to secure their oil supplies. Due to volatility in oil prices, many large-oil-consuming countries considered building strategic reserves on what the US has. They tend to buy more than their requirements when the oil price is low and less when the oil price is high. This strategy can insulate their national economy from the effects of high oil prices to some extent and for some time, if not for very long.

Hydrocarbons are non-renewable sources of energy. Based on the present trend of oil consumption, the world's proven oil reserves are expected to last for 50 to 60 years. This estimate may arguably be contested on the grounds of mistaken assumptions. But there is no denying that with the steady consumption, the hydrocarbon reserves are continually diminishing. Scientists are working overtime to find a lasting solution to energy security before oil and gas become extinct. Recent successes and enthusiasm in harnessing renewable energy have caused worries among oil-rich nations. Sensing the future, they would like to price oil and gas prudently while the going is still good.

Oil prices also depend on oil futures contracts. Traders in the commodities market develop these contracts. Being the agreement to buy or sell oil at a specific date and price in the future, they consider current and future oil supply and demand of the major consumers.

10.5.3 Competition

The companies cannot ignore the competition. It results when two or more companies decide to produce the same or similar products to meet market demand. In such a situation, they are forced to price their product based on the competition and not necessarily on the cost or consumer demand. A competitive price may hurt a company's profit, but it allows it to hold onto its market share and avoid business loss. Pricing strategy plays a decisive role in a market-controlled environment that witnesses a higher level of competition.

Competing companies can sometimes use unique tactics to attract customers. The digital age has opened up new and numerous ways to reach out and educate customers. Blogs, SEO (Search Engine Optimization), e-mail marketing, social media, video channels, radio, TV, magazines, and newspapers are methods that companies use to highlight their products' strengths. Companies organize special events, promotions, and conferences to draw customers to their fold.

Competition pushes the organizations/companies to improve quality and innovate their products. Business compulsions can also bring the competing organizations/companies closer who may form cartels or consortia to buy or sell a product/service.

10.5.4 Demand for Oil

Crude oil supplies are crucial for developed as well as developing countries. Because of the importance of oil supplies, the fluctuation of oil prices can significantly affect the economy. In a free market, international oil prices are determined by the standard economic principle of demand and supply. The recent economic growth of China and India, despite global financial setbacks, is a success story that is partly fueled by lower oil and gas prices. Energy prices fell due to the diminishing demand for oil. The fiscal and monetary stimulus provided by the governments reversed the oil price trend later on. Any sudden or major changes in oil demand or production cuts affect the market and the oil price.

10.5.5 Political Events and Crises

Potential unrest or crises in oil-producing countries can have a massive impact on oil prices due to the fears that such a mess will limit supply. War, natural disasters, and new government leaders can also be the factors to influence crude oil pricing.

Figure 10.4 uses historical examples to illustrate how various political and economic events can dictate the international crude oil prices. Before 1970, crude oil prices generally stayed low and relatively stable. The first "oil-shock" was observed in 1973, which caused oil prices to jump threefold. When OPEC, dominated by Arab countries, declared an oil embargo against Canada, Japan, the Netherlands, the uk, and the US, these nations were seen as supporting Israel in the Arab–Israeli War. Iranian Revolution of 1978 and the Soviet invasion of Afghanistan in 1979 ignited fears over access to Middle Eastern oil. Iraq's invasion of Kuwait in 1990, the split of the Soviet Union in 1991, and the Asian financial crisis in 1997 that raised fears of a worldwide economic meltdown demonstrated the volatility of oil prices to international events. The global economic crisis engulfing the investment bank Lehman Brothers in 2008, the 9/11 terrorist attack on WTC, the impact of "Arab Spring" unrest in Egypt, Libya, and Tunisia in the same year are other examples that prove how such incidents can drive oil prices up and down.

Natural calamities such as hurricanes and earthquakes can damage offshore platforms, processing facilities, and infrastructure disrupting the supply chain and schedules. Such breakdowns can temporarily affect oil prices unless infrastructural gaps are bridged or alternate solutions are found to mitigate them.

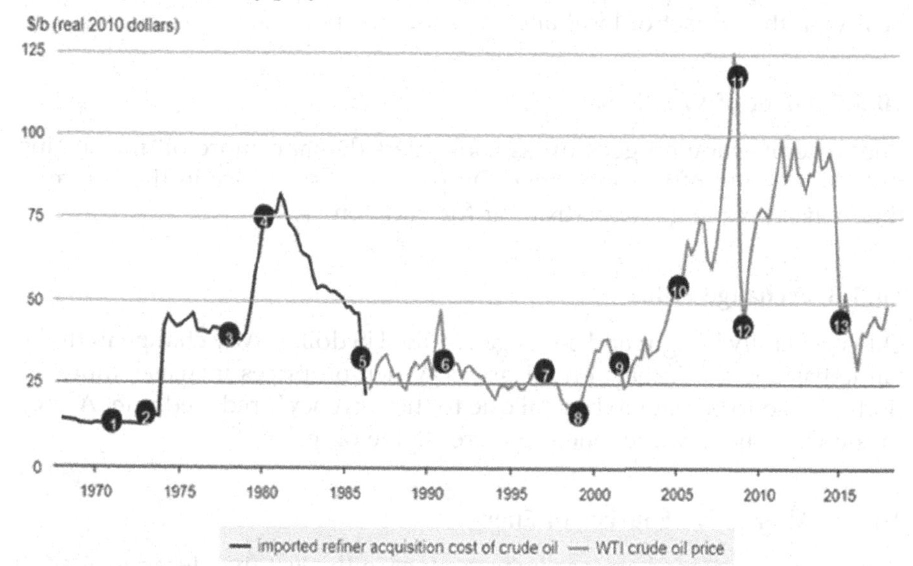

eia Source: U.S. Energy Information Administration, Thomson Reuters

1: US spare capacity exhausted
2: Arab Oil Embargo
3: Iranian Revolution
4: Iran-Iraq War
5: Saudis abandon swing producer role
6: Iraq invades Kuwait
7: Asian financial crisis
8: OPEC cuts production targets 1.7 mmbpd
9: 9-11 attacks
10: Low spare capacity
11: Global financial collapse
12: OPEC cuts production targets 4.2 mmbpd
13: OPEC production quota unchanged

FIGURE 10.4
History of Crude Oil Prices[7] *(Source- US Energy Information Administration, Refinitiv; June 2020).*

10.5.6 Political Environment

The political environment plays a crucial role in all business operations. It can introduce policies that provide a stimulus for growth or halt progress. Political stability ensures that the direction of the government's policies and actions are consistent over the term of its election. Lack of political stability causes uncertainties that can lead to risks on account of agitations, social unrest, sabotage, and any kind of economic threats. International political hazards for businesses include acts of terrorism, war, sanctions, and so forth.

Political decisions affect economic, legal, or social environments. Government policies and laws can influence tax rates, promote new technologies, and create new business opportunities and culture. National governments can play a decisive role in employment generation and develop a level playing field for the private and public sector enterprises. Political willpower is essential to curtail bureaucracy, reduce corruption, and eliminate red-tapism. On the other hand, companies should be ready to deal with the impact of local and international politics.

10.5.7 Effect of Weather

The price of crude oil goes up as consumers demand more oil for heating their homes and offices in winter. Oil prices can see a rise in the summertime when people go globe-trotting for vacation.

10.5.8 Exchange Rates

Oil is generally bought and sold against the US dollar. Any change in dollar value has a direct impact on oil prices. When oil prices increase, more US dollars need to be paid to buy oil due to the currency's reduced rate. A drop in the US dollar's value tends to increase the oil price.

10.5.9 Alternative Sources of Energy

Petroleum has gained extraordinary power in the last five decades or so. It is fundamental to sustain the development and growth of all developed and developing countries. But petroleum resources are finite and non-renewable, and their restoration is not possible once exhausted. Many petroleum-rich states, some very small, use the power of petroleum to script their destiny and influence critical international matters. The future landscape of petroleum geopolitics may get murkier by the increasing competition and conflict to gain access to petroleum resources. Not only this, but petroleum supplies also have to overcome the challenges of piracy and terrorism. The rising cost of petroleum exploration in the face of continually diminishing volume of easy oil and the environmental concerns have already forced the world to look for alternative options to ensure its energy security.

Hydroelectric energy, solar energy, wind energy, biomass energy, geothermal energy, and tidal power are commonly known as alternative energy sources. The most exciting and lucrative side of these energy sources is that they are green and do not harm the environment, renewable, and free. Being the gift of nature, they belong to anybody who captures them. Scientists around the world are engaged in advanced research on establishing the commercial viability of alternative energy sources. Out of all the other alternative energy sources, hydroelectric energy or hydroelectric power has been most commonly used in the present time.

A shift in the consumption of these energy patterns will significantly reduce modern civilization's dependence on petroleum and its products. It will prolong the life of the fast depleting traditional energy sources and protect the environment from carbon dioxide emission and air, soil, and water pollution.

10.6 Role of OPEC

Before the formation of OPEC (Organization of Petroleum Exporting Countries), the global petroleum industry was dominated by seven major transnational oil companies commonly referred to as "seven sisters" or the "Consortium for Iran" cartel. These companies were as follows:[8]

1. Anglo-Iranian (started as Anglo-Persian) Oil Company (now BP)
2. Gulf Oil (later part of Chevron)
3. Royal Dutch Shell
4. Standard Oil Company of California (SoCal, now Chevron)
5. Standard Oil Company of New Jersey (Esso, later Exxon, now part of ExxonMobil)
6. Standard Oil Company of New York (Socony-later Mobil, also now part of ExxonMobil)
7. Texaco (later merged into Chevron)

OPEC was established in Baghdad, Iraq, on 14 September 1960 with five founding member countries, namely, Iran, Iraq, Kuwait, Saudi Arabia, and Venezuela. Libya, UAE, Algeria, Nigeria, Ecuador, Gabon, Angola, Equatorial Guinea, and Congo are other OPEC member countries (Figure 10.5). Qatar, a member of OPEC since 1961, decided to pull out of it in January 2019 to focus on its gas business. Indonesia has been in and out of OPEC. It first joined OPEC in 1962 but left it in 2008 as it ceased to be a net oil-exporting country. The OPEC statute provides that any country with a "substantial net export of crude petroleum, which has fundamentally similar interests to those of member countries may become a full member of the Organization[9]." Indonesia returned to OPEC in 2016 but held on to its membership for barely a year. It decided to quit when a request for a 5% production cut was made.[10] It means that currently, OPEC has a total of 14 member countries. OPEC's role and business objectives are best spelled out in the OPEC Statute, which are as follows:[11]

> OPEC's principal aims are the coordination and unification of the petroleum policies of Member Countries and the determination of the best means for safeguarding their interest, individually and collectively. The Organization shall devise ways to stabilize prices in international oil markets to eliminate harmful and unnecessary fluctuations. Due regard shall be given at all times to the interest of the producing nations and to the necessity of securing a steady income to the producing countries, efficient, economic, and regular supply of petroleum to consuming nations, and a fair return on their capital to those investing in the petroleum industry.

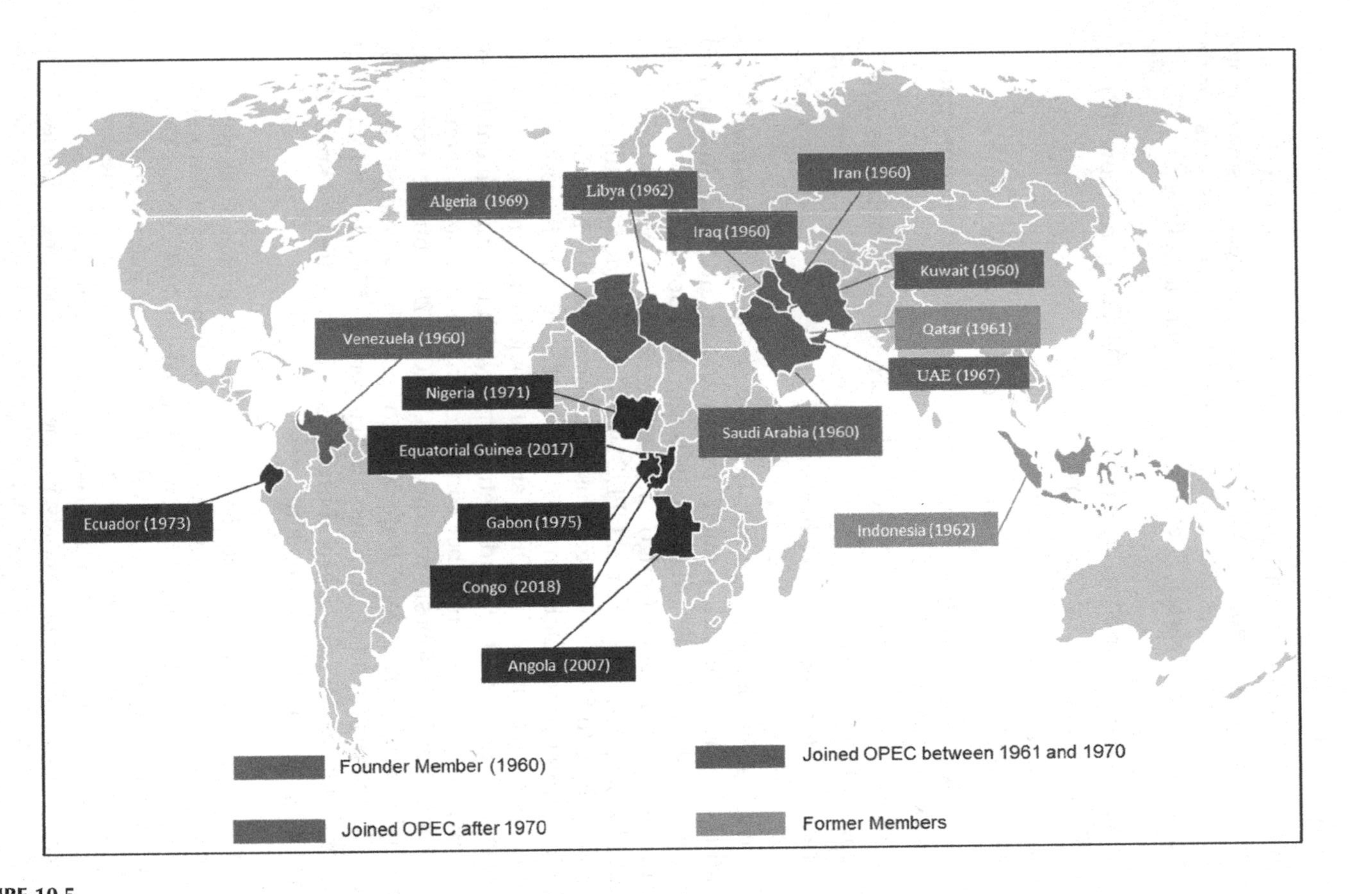

FIGURE 10.5
OPEC Member Countries.

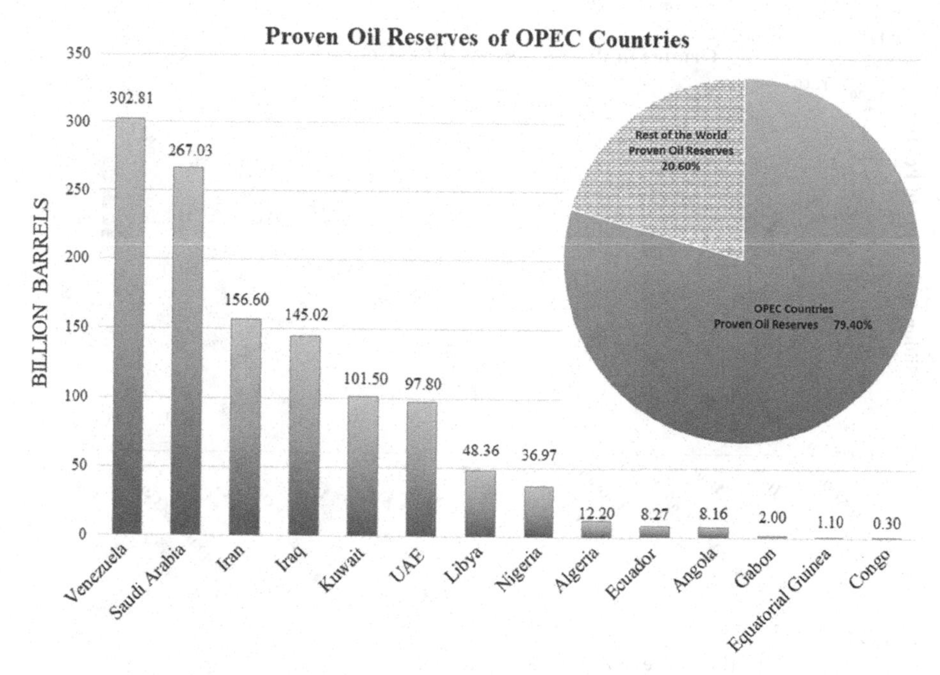

FIGURE 10.6
OPEC Share of World Crude Oil Reserves.

OPEC represents diverse cultures spread over three continents that speak different languages and practice different religions. Their unity lies in the shared business objectives of exporting crude and framing the petroleum policies and decisions in their best interest. OPEC is the largest source of future oil supply. The following factors give it enormous bargaining power to control the oil price in the international market.

- OPEC holds about 79% share[12] of the world proved crude oil reserves (Figure 10.6)
- The exponential increase in the demand for energy
- Scarcity of petroleum resources outside the domain of OPEC
- Surplus production capacity available with Middle East member countries of OPEC
- Low cost of oil production, particularly in the Middle East countries
- Research on alternative energy sources not providing a commercial solution yet

In the 1980s, the cartel took a new initiative to increase its influence by inviting other oil-exporting countries outside OPEC such as Mexico, Russia,

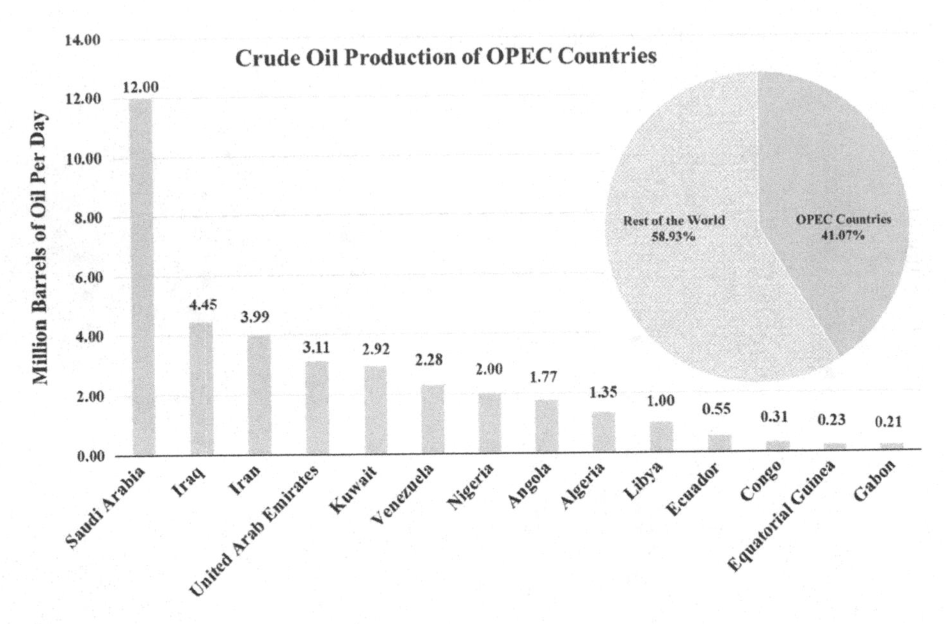

FIGURE 10.7
OPEC Share of World Crude Oil Production.

Norway, Egypt, and Oman as observers. Since then, this forum has served as an informal communication channel to discuss agreements and concerns.

According to 2019 estimates, OPEC is responsible for approximately 41% of the world crude oil production (Figure 10.7) of approximately 88 million BOPD.[13] Its member countries have an excess production capacity that can be turned on and off depending on the requirement. OPEC thrives on the opportunities and is tested on the threats to the supremacy of oil as stated below:

Need for oil: Emerging markets such as China and India need more oil to meet their energy needs. China's oil import dependence stands at about 55%, whereas India imports around 85% of its total crude oil requirement. Many countries, such as the US, have crude oil reserves to meet their demand, but they have to import crude oil to feed their refineries.

Scarcity of Oil Resource: The oil supply is finite as oil reserves are non-renewable and diminish every day. Many countries, some out of concern of running out of future oil supplies and some as a strategy to conserve their oil reserve, turn to imports impacting the production and distribution pattern of oil.

Political instability: Today, many oil-rich countries are politically unstable. Political instability can be a significant factor affecting oil production and supplies.

HSSE Concerns: Increasing expenditure on public health due to pollution caused by the oil used in vehicles/transport is a significant health concern. Environmental pollution caused by oil spills, toxic emissions wastewater disposal is on the increase. NGOs are becoming increasingly vocal against the use of fossil fuels. Many countries and regions are trying to reduce the effects of global warming. They are also promoting alternative sources of energy to reduce their dependence on oil.

The process of extraction of oil is complicated. Blowouts and well fires can cause massive scale damage to the environment, property, and human life. The cost of oil production and transportation has significantly increased and so has the risk to oil shipments due to international disputes, conflicts, or piracy incidents.

10.7 Crude Oil Classifications

It is believed that oil and natural gas were formed millions of years ago due to the thermal maturation of organic matter buried deep in the earth under extreme pressure and high temperature. Hydrocarbons occur underground in various forms: kerogen, asphalt, crude oil, natural gas, condensates, and coal in solid form.

Crude oils from different locations can vary in color from nearly colorless to tar-black, and quality as good as that of water to almost solid. Many criteria have been used in petroleum literature to classify crude oils. Some of those relevant to the pricing of oil will be discussed here.

10.7.1 Composition of Crude Oil

Crude oil is predominantly composed of carbon and hydrogen. Almost all crude oils range from 82 to 87% carbon by weight and 12 to 15% hydrogen[14] by weight, besides varying amounts of oxygen, Sulfur, nitrogen, and helium (Table 10.3). Crude oils are typically characterized by the type of hydrocarbon compound that is most prevalent in them and can be classified into four main categories:

Paraffins: Paraffins, also called alkanes, are major constituents of crude oil. They are straight or branched chain compounds of carbon and hydrogen, which have a general chemical formula C_nH_{2n+2}. At ambient temperature, paraffin containing less than five carbon atoms is in a gaseous state, with 5 to 15 carbon atoms is in a liquid state and carrying more than 15 carbon atoms is in the solid state. Paraffin is the desired content in crude and is primarily responsible for making the fuels that makeup 15% to 60% of crude oil and 5% or more wax. The shorter the chain of paraffin, the lighter the crude is. The octane number rating of branched-chain paraffin is

TABLE 10.3

Composition of Petroleum[15]

Hydrocarbons	Molecular Characteristics	Major Hydrocarbons	Remarks
Paraffins (Alkanes)	Straight carbon chain without any ring structure	Methane, ethane, propane, butane, pentane, hexane	General formula C_nH_{2n+2} Boiling point increases as the number of carbon atom increases. With number of carbon 25-40, paraffin becomes waxy.
Naphthenes	5 or 6 carbon atoms in ring. Saturated hydrocarbons containing one or more rings, each of which may have one or more paraffin side-chains	Cyclopentane, Methyl cyclopentane, Dimethyl cyclopentane, cyclohexane, 1,2 dimethyl cyclohexane.	General formula C_nH_{2n}. The average crude oil contains about 50% by weight naphthenes. Naphthenes are modestly good components of gasoline.
Aromatics	6 carbon atoms in ring with 3 around linkage	Benzene, Toluene, Xylene, Ethyl Benzene, Cumene, Naphthaline	General formula C_nH_{2n-6} Aromatics are not desirable in kerosene and lubricating oil. Benzene is carcinogenic and hence undesirable part of gasoline.
Asphaltics	Extremely complex chemical structure. 25-150 carbon atoms	The components of asphalt include four main classes of compounds: Naphthenes, Polar aromatics, Saturated hydrocarbons and asphaltenes	This crude oil is black in color, heavy and may occur in a semi-solid or liquid state. contains up to 1-6% of sulphur
Non-hydrocarbons	**Compounds**	**Remarks**	
Sulphur compounds	Hydrogen sulphide, Mercaptans	Undesirable due to foul odour 0.5% to 7%	
Nitrogen compounds	Quinotine, Pyradine, pyrrole, indole, carbazole	The presence of nitrogen compounds in gasoline and kerosene degrades the colour of product on exposure to sunlight. They may cause gum formation normally less than 0.2.	
Oxygen compounds	Naphthenic acids, phenols	Content traces to 2%. These acids cause corrosion problem at various stages of processing and pollution problem.	

higher, making it a more desirable constituent of gasoline than straight-chain paraffin. Paraffinic oils are virtually sulfur free.

Naphthenes: Hydrocarbon molecules with more than four carbon atoms tend to form a closed ring structure known as a cyclo compound. Saturated cyclo compounds are called naphthenes and have a general chemical formula C_nH_{2n}. Naphthenic crudes are inferior raw materials for making lubricants but can be processed to produce high-quality gasoline with relative ease. They can be refined to produce all liquid refinery products and leave behind dense asphalt-like residues. They can make up 30% to 60% of crude and are heavier and more viscous than their equivalent paraffin. Naphthenic base oils cannot be made free of sulfur.

Aromatics: Aromatics are unsaturated closed-ring compounds that constitute only a small percentage of most crudes. They have much less hydrogen in comparison to carbon than is found in paraffin. They are often solid or semi-solid compared to their equivalent paraffin that would generally be a viscous liquid under the same conditions. Aromatics in kerosene and lubricating oils result in soot when burnt and are therefore not desirable. The most common aromatic in crude oil is benzene. Benzene is a common aromatic in crude oil. It is an undesirable part of gasoline, which is carcinogenic and pollutes the environment. In real life, many aromatic compounds have a distinctive odor.

Asphaltic: The crude oils where asphalt content is high are termed as asphaltic crudes. Asphaltic crude is also known as pitch, bitumen, or asphalt. This crude oil is black in color, dense, and may occur in a semi-solid or liquid state. Asphalt is a byproduct of a residue of the crude oil refining process. Its high viscosity and "stickiness" help in the construction of roads. It typically contains up to 1% to 6% of sulfur, which is undesirable in crude oils.

Sulfur, nitrogen, oxygen, and helium are some non-hydrocarbon contents that may also be present in small but often essential quantities. Generally, the heavier the crude oil, the higher it's sulfur content. The total sulfur content in crude oil varies from less than 0.05% in medium-heavy oils to up to 7% or more in heavy oils. The excess sulfur needs to be removed before this lower-quality crude can be refined into diesel fuel or gasoline. Sulfur generates SO_2, which is a significant pollutant. Many countries have introduced regulations to limit the sulfur content in refined fuels such as diesel and petrol used in transport vehicles.

The heavier oils contain the most oxygen. The oxygen content of crude oils is usually less than 2% by weight. Nitrogen and helium are also is present in small quantities in crude oils. Nitrogen is generally present in amounts of less than 0.1% by weight. A helium content of 0.3% or more is considered necessary for commercial helium extraction in the US.[16] Sodium chloride also occurs in most crudes and is usually removed like sulfur.

10.7.2 Sweet and Sour Crude Oils

Crude oil, produced from different oil fields across the world, has different chemical properties. An important classification of crude oils is based on the sulfur content based on which crude oil is termed sweet or sour. Crude oil with no or low sulfur content is classified as "sweet," crude oil with a higher sulfur content is classified as "sour." The terminology dates back to the old practice of tasting crude oil to determine its quality. Sweet crude has a mild sweet taste and a pleasant odor due to low sulfur content, generally lower than 0.5% by weight.

Sulfur is considered an impurity and an undesirable fraction in crude. The level of sulfur in sour crudes is more than 0.5% by weight. The sour crude derives its name from bitter taste and pungent smell due to higher Sulfur content. Generally, the heavier the crude oil, the higher it's sulfur content. Sulfur may be present in the form of elemental sulfur or hydrogen sulfide. Most Sulfur in crude is bonded to carbon atoms. Sour oil may also contain carbon dioxide making it more corrosive. It is more challenging to produce, process, transport, and refine sour crude than sweet crude. Special care needs to be taken as sour crudes with high hydrogen sulfide quantities can cause serious health problems.

The sulfur content may vary from one geographical location to another or sometimes even between two wells in the same field/reservoir. The sulfur content in crude oil can harm the catalysts and inhibit their activity. Catalysts are costly materials that are used to promote chemical reactions in crude oil refining. Sulfur can interfere with the functioning of catalysts and inhibit their activity. It affects the cost of refining and end-product quality.

Light sweet crude is easier to extract, refine, and transport than sour crude. It also produces a higher percentage of value-added products such as gasoline, diesel, and aviation fuel. On the other hand, heavy sour oil provides a higher percentage of lower value-added products with distillation and requires additional processing to give lighter products. Therefore, sweet crudes command a premium over sour oil.

10.7.3 Light and Heavy Crude Oils

It is common to classify crude oils based on the API gravity named after American Petroleum Institute. API gravity is a commonly used index of the density or gravity of crude oil or refined products and is mathematically expressed as follows:

$$^{\circ}API = \frac{141.5}{SG} - 131.5 \qquad (10.1)$$

where °API = Degrees API Gravity and SG = Specific Gravity of crude oil or product at 60°F

It is clear that API gravity is indirectly proportional to the specific gravity and is expressed in degrees. API gravity is scalable to the specific gravity of pure water, which has a specific gravity of 1. An oil with an API gravity of more than 10 is lighter than water and will float on the water surface. An oil with an API less than 10 is heavier and will sink in water.

The use of numbers 141.5 and 131.5 and the unit in this equation appears to have originated after the French scientist Antoine Baumé in the 18th century developed a hydrometer to measure the liquid densities. The Baumé hydrometer has two scales that are calibrated to measure the specific gravity of liquids. The first scale measures the specific gravity of liquids that are heavier than water, and the second the specific gravity of the liquids lighter than water. The units of the density measurements used different formats such as degrees Baumé, B°, or Bé°. For fluids lighter than water, Baume's gravity was determined from the following expression:[17]

$$degrees\ Baumé = \frac{141}{SG} - 131 \tag{10.2}$$

The above equation is close to that of the °API. The only difference is that the API equation uses numbers 141.5 and 131.5 instead of 140 and 130 used by *Baumé*. It appears that initial specific gravity measurements came from hydrometers with a modulus of 141.5 instead of the Baumé 140. The number 131.5 in the API equation represents the cut off for light crude. Anything less than this number refers to medium crude oil, and less than 122.3 is heavy crude. For calibration purposes, the Baumé degree scale for liquids lighter than water was set with 10° Baumé to the density of water.[17] API seems to have adopted *Baumé* formulation long ago, which is still in use.

The petroleum industry uses API Gravity to classify various crude oils (Figure 10.8). A generally acceptable classification is:

- **Light**: higher than 31.1° API
- **Medium**: between 22.3° and 31.1° API
- **Heavy**: between 10.0° and 22.3° API
- **Extra Heavy**: less than 10.0° API

This classification is not very rigid, as all parties do not use the same grading.

Crude oils are generally classified as bitumen, heavy oils, medium, and light oils based on API Gravity or specific gravity. Bitumen is a very heavy,

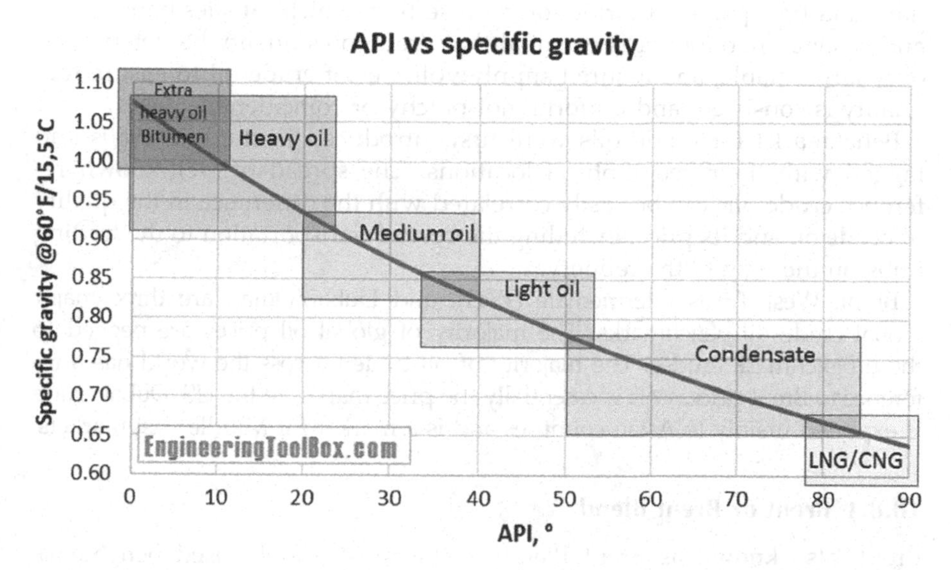

FIGURE 10.8
Grades of Various Crude Oils and Products on API Scale[18].

immobile, and degraded petroleum residue that may occur naturally in oil sands. It does not flow into the wellbore.

Heavy crude has no mobility. It cannot be produced, transported, and refined by conventional methods because it has high concentrations of Sulfur and several metals. However, heavy crudes can be recovered using thermal methods and chemical methods. Thermal methods improve crudes' mobility, and chemical methods use high molecular weight polymer solutions to increase oil recovery by decreasing the water-oil mobility ratio. Refineries have to build separate units and use exclusive technologies to process the recovered heavy crudes at higher costs.

Medium and light oils have higher mobility through porous rocks and can be recovered quickly through production wells at a lower cost than heavy oils. Light crude has a higher percentage of light hydrocarbons extracted with simple distillation at a refinery and fetch a higher value.

10.8 Benchmark Crude Oils

Benchmark refers to a particular standard or quality against which products, services, and work processes may be evaluated and compared. A benchmark or marker crude represents a crude oil with specific quality in API gravity and sulfur content. One of the critical things that benchmarking does is link the quality of crude oil with its price. Benchmarking is a standard and transparent criterion for price-setting, which satisfies both buyers and sellers. Therefore, it is essential that crude oil's quality be determined over large, stable, and assured supply volumes of crude oil to ensure that quality is consistent and uniform, not patchy or coincidental.

Benchmarks for crude oils were first introduced in the mid-1980s and tagged with their geographical locations. The spread of well-known reference crude oils can be easily correlated with the difference in the quality of crude oil and its price, including the cost of transportation to the trading hubs on the way to the refinery.

Brent, West Texas Intermediate (WTI), and Dubai/Oman are three major global crude oil benchmarks. The majority of global oil prices are pegged to these benchmark crudes. The majority of oil traded across the world has a reference to Brent price. WTI is essentially the price marker in the US. Dubai crude is exported mainly to Asian countries and is a marker for Middle East markets.

10.8.1 Brent or Brent Blend

Brent, also known as Brent Blend, is the most widely used benchmark which derives its name from the Brent oil and gas field located in the North Sea. It includes four separate light, sweet crude streams produced from

Brent and Forties oilfields offshore the UK and Ekofisk and Oseberg oilfields offshore Norway.

Brent is a light and sweet crude with API gravity of 38.06 (specific gravity of 0.835) and a sulfur content of 0.37%. Brent Blend is rated as a premium quality oil used to extract gasoline and middle distillates in refineries.

10.8.2 West Texas Intermediate

WTI is a high-quality, light, and sweet crude oil produced from US oilfields in Texas, Louisiana, and North Dakota. This oil with a specific gravity of 0.827 (API gravity of 39.6°) and a sulfur content of 0.24% is lighter and sweeter than Brent oil.

WTI is in high demand due to its high yield of gasoline in refineries than any other crude type. It is also used as a benchmark for crude oil imported from Mexico, South America, and Canada.

10.8.3 Dubai and Oman

Dubai and Oman are the third major benchmark crude. The prices of Dubai and Oman crude oils are averaged together to create a standard benchmark for the oil that comes from the Persian Gulf and is sold in Asian markets.

Dubai crude is also light, though not as much as Brent and WTI. It has an API gravity of approximately 31° and a specific gravity of 0.871. However, its sulfur content is much higher at 2%, making it several times sourer than Brent and WTI Crudes.

Saudi Arabia's state-owned oil company, Saudi Aramco, uses the Dubai/Oman benchmark for oil exports to Asian countries.

10.8.4 OPEC Reference Basket

The OPEC Basket or OPEC Reference Basket (ORB) is a mix of light and heavy crudes produced by OPEC member countries. Initially, it included a weighted average price for petroleum blends produced by seven members, namely, Algeria, Saudi Arabia, Indonesia, Nigeria, Dubai, Venezuela, and the Mexican Isthmus. However, the new ORB has been expanded to include the weighted average of 11 crude oil blends from many member countries. These countries are Algeria, Indonesia, Iran, Iraq, Kuwait, Libya, Nigeria, Qatar, Saudi Arabia, the UAE, and Venezuela.[19]

The ORB is a mix of light and heavy crudes with an average degree of API of 32.7 and a much higher percentage of sulfur (1.7%). Consequently, it is heavier than both Brent and WTI crude oils and is accordingly discounted in price.

10.8.5 Other Blends

Apart from the above, many other benchmark crudes are used for price referencing around the world. Some of these benchmarks are -

10.8.5.1 Tapis Crude

Tapis crude is produced from one of Malaysia's most extensive offshore oilfields known as Tapis. It is a very light and sweet crude with API gravity of 43° to 45° with about 0.04% sulfur content. It is far more superior in quality than WTI and Brent and is used as a pricing benchmark in Singapore, Asia, and Australia. The price of Tapis crude in Singapore is often considerably higher than the cost of benchmark crude oils such as Brent or WTI.

10.8.5.2 Bonny Light

Bonny Light is a light and sweet benchmark crude oil produced from the Niger Delta basin in Nigeria. It is named after the prolific hydrocarbon region near the city of Bonny. It has an API gravity of 35.2° to 37.5° (specific gravity 0.85) and low sulfur content of 0.12% to 0.16%. American and European refineries prefer this grade of crude due to its proximity, good quality, and moderate sulfur content than those of WTI and Brent.

10.8.5.3 Mexico Isthmus

Mexico's Isthmus is a light and sour crude that is produced in Mexico. It has an API gravity of 33.74° and sulfur content of 1.45%. Isthmus crude oil is extracted in the Campeche zone in Mexico and the Gulf of Mexico.

10.8.5.4 Maya

Maya is a heavy sour crude that is a blend of crude oils from two giant oilfields Cantarell and Ku Maloob Zaap in Mexico. It has an API gravity between 21° and 22° and sulfur content of 3.4%. Maya has an ideal customer base of about 25 refiners in the Americas, Europe, and the Far East.[21] They are capable of refining the high sulfur content crude in their sophisticated refineries.

Figure 10.9 presents a distribution of various benchmark crudes discussed above and many more used for pricing the crudes based on their API gravity and sulfur content.

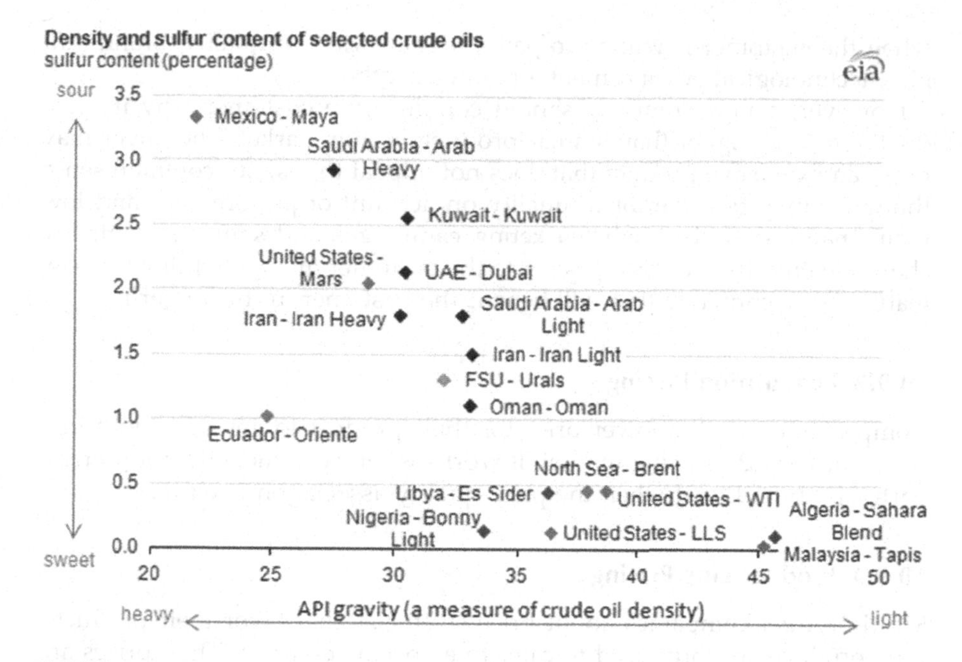

FIGURE 10.9
Distribution of Benchmark Crudes Based on Quality[20] (*Source: US.Energy Information Administration, July 2012*).

10.9 Pricing Strategies

It is widely believed that price is set with the motive of profit maximization. In reality, producers may have multiple objectives and can use various pricing methods to achieve these objectives. The main features of some of the key pricing strategies are given below:

10.9.1 Full Cost Pricing

It is the most common method based on the assumption that businesses will earn a profit. It is also a straightforward method where the producer decides a fixed mark-up over its production cost. Mark-up is the amount of expenditure on overheads and profit added to the cost price of the product.

10.9.2 Skimming Pricing

Producers or marketers use the skimming price strategy to recover the sunk capital quickly before the onset of competition. This strategy delivers results when the customer is willing to pay a higher price for quality, brand affinity, technological advancement, or status symbol.

However, a new company should convince the customer why its product's price is higher than similar products in the market. The buyer may reject an expensive product that does not appeal to his/her common sense though it may be superior in quality on account of performance and low maintenance cost. Through marketing campaigns and staff, the company should justify the premium pricing to the customer by highlighting the new features or technology that will benefit the customer in the longer run.

10.9.3 Penetration Pricing

Companies may set a lower price for their product to get a foothold in a competitive and growing market. It works when customers do not prefer a particular brand, and when the price-quality association is weak.

10.9.4 Product Line Pricing

Multi-product companies adopt price strategies that favor their products. Line products are produced to cater to a specific demand. Their prices are bundled together and may be offered at a discounted rate to the customer. In other words, the sum of the product line's cost is less than the amount of the price of the individual line products.

Companies may introduce premium pricing for a particular product or optional features if their product line has optional accessories.

10.9.5 Dual Pricing

It refers to one product with two prices, as in the case of petrol, diesel, or kerosene. These products may have different grades, and rates may be linked to a specific category. In countries that offer fuel subsidies, the customer can choose between subsidized lower quality fuel and unsubsidized higher quality fuel.

Dual pricing can lead to malpractices such as adulteration, smuggling. Many countries have experimented with dual pricing. However, its success and viability require strong enforcement measures.

10.10 Pricing Criteria

A common-sense approach to pricing rests on the premise that a product's price should be higher than its production cost. It is, therefore, crucial for the producing company to determine the value of the product carefully.

Following are the primary criteria that decide the price of crude oil:

10.10.1 Production Cost

Production cost in oil and gas is a generic term used to combine the expenditure incurred on exploration, development, and extraction of oil and gas from the reservoir. In the exploration stage, oil and gas are discovered to create an asset. The development stage allows oil and gas to be accessed and recovered by drilling a definite number of wells. The extraction or production stage includes programs or activities to lift crude oil and maintain oil wells, equipment, and facilities. This stage helps to convert the oil and gas asset into a product that can earn direct revenue.

Production cost typically includes expenditure incurred on wages, materials, supplies, services, lease rentals, and general overhead. Government taxes or royalties are also part of the production cost.

The production cost per unit barrel of oil is a more meaningful parameter obtained by dividing the production cost (sum of exploration, development, and extraction costs) by the number of produced oil barrels. The unit cost of production is a direct measure to compare costs across various fields and companies. It provides a rationale to set an appropriate sales price. Prices higher than the unit cost of production result in profits. Prices lower than the unit cost of production result in losses.

Figure 10.10 shows the unit cost of producing one barrel of crude oil in different oil-producing countries. Based on this chart, the unit cost of oil production is the highest in the UK and the lowest in Kuwait.

It must be acknowledged that the unit cost of oil production is going up as the petroleum industry transitions from "easy" to "difficult" oil. Future oil production will include the extraction of unconventional resources such as deep offshore waters, tight sands, and shale oil to apply new technologies and higher capital outlays.

10.10.2 Crude Oil Quality: API and Sulfur Content

The quality of crude oil produced around the world is variable. The two main quality characteristics that have direct relevance and association with its market value and price are density or API gravity and sulfur content. As discussed above in sections 10.7.2 and 10.7.3, the crude oils may be characterized as sweet or sour based on their sulfur content and light or heavy, based on their density or API gravity.

Sweet crude oils with low sulfur content and higher degrees of API gravity or lower density are priced higher. The heavy sour grades of crudes are inferior in quality and require far more sophisticated and expensive refinery processes. The light and sweet crude oils yield much more refined high-value products than heavy sour crudes (Figure 10.11). As a result, light and sweet crudes are priced higher than heavy sour crudes.

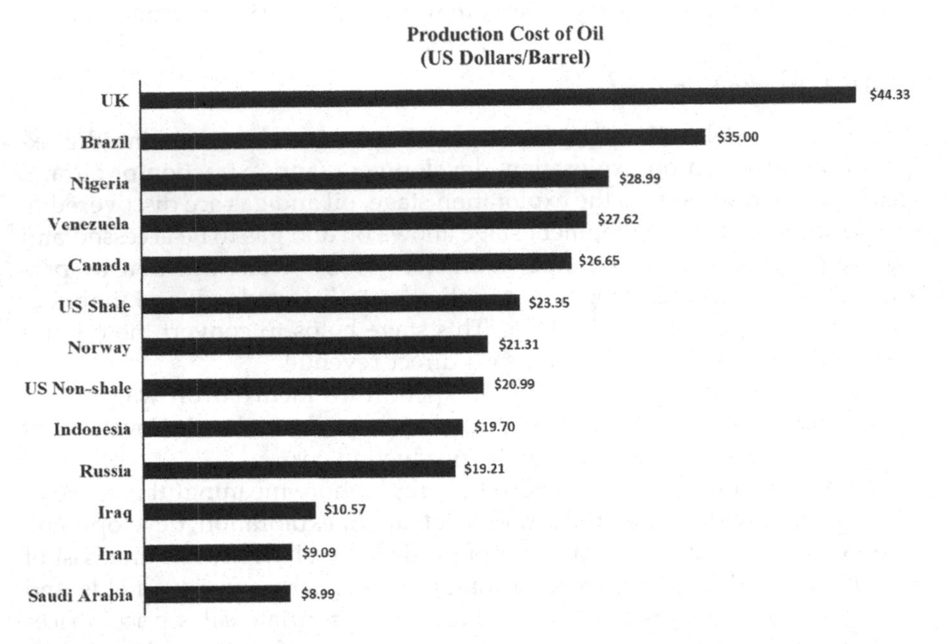

FIGURE 10.10
Unit cost of Production of Crude Oil[22].

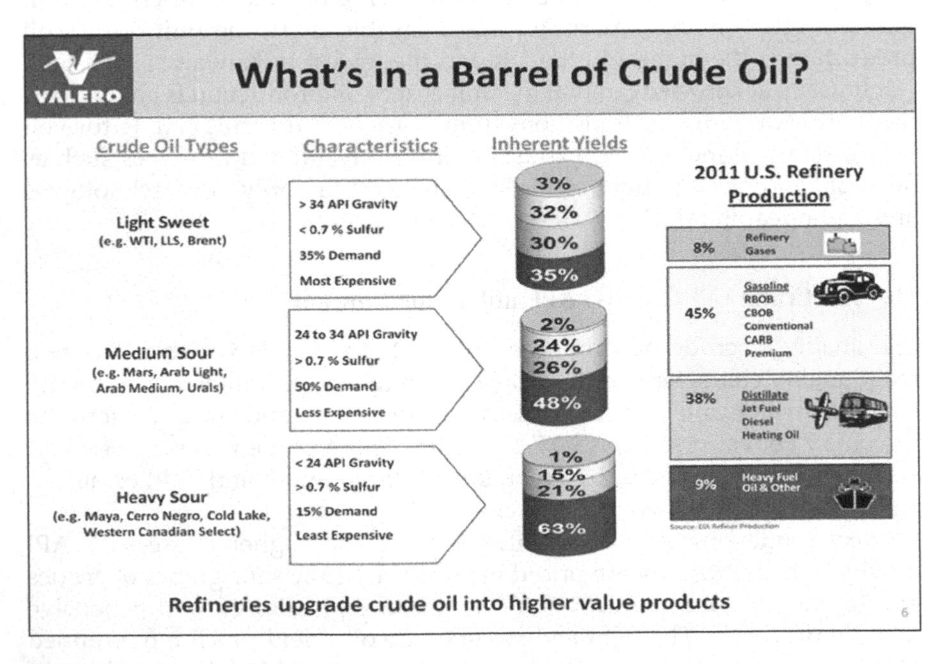

FIGURE 10.11
Typical Crude Oil Types and Their Inherent Yields[23].

10.10.3 Geographical Location

Geographical location is a factor that impacts crude oil prices both on the buyer or seller side. It is, therefore, an essential and integral part of shipping agreements. Crude oil may come from onshore or offshore field locations. It is processed at gathering centers and finally dispatched to the tank farms. It is then transported from tank farms to the export terminals from where it is shipped to the buyer country's port. In case the refinery identified to refine it is not located nearby, the oil cargo has to further travel to the refinery by road, rail, or pipeline. All this movement of oil by different modes adds to the cost of oil.

A significant amount of expenditure and effort on transportation can be saved by setting up the oil refineries near the oilfields, ports, or the market that uses refined products. However, an oil refinery located close to the oilfield faces the risk of redundancy when the oilfield is depleted.

In the USA, the world's most massive refinery clusters are concentrated close to the oil-rich areas along the Gulf Coast of Texas and Louisiana. There are some large and small refineries located on the east coast and mid-west region of the USA to process crude oil from the Middle East countries and Venezuela. The US has built extensive infrastructure to transport refined products by pipelines, railroads, or barges to the bulk terminals markets from where it is distributed to retail outlets.

European countries have located their refineries near manufacturing industries because they use petroleum products as their feedstock. Lindsey oil refinery at Immingham, Nynas oil refineries at Dundee and Ellesmere Port in the UK, Nynas oil refinery at Nynashamn in Sweden, BP refinery at Europoort Rotterdam in the Netherlands, Petrobrazzi refinery at Brazii de Sus, and Petromidia refinery at Navodari in Romania have the benefit of the coastal location. These refineries have the added advantage of proximity to the vast European market with a significant manufacturing industry that uses petroleum products. Most of the Russian refining capacity is also close to the manufacturing or consumption centers.

Middle-East countries are some of the largest oil exporters of crude oil. These countries have set up their refineries/petrochemical complexes on the coast to simplify exports. Saudi Aramco's refinery with Shell (SASREF), Yanbu Refinery, and their Refining and Petrochemical complex with Total in Saudi Arabia; and KNPC's Mina Al-Ahmadi and Shuaiba refineries in Kuwait are located on the coastline.

10.11 Crude Oil Blending Opportunities

Blending involves mixing two or more substances to form a homogeneous mixture of uniform properties. Crude oil blending is often undertaken to upgrade or downgrade crude oils to meet the pipeline/refinery/export target spec range for crude at the lowest cost.

The crude oil supply chain consists of production wells, GCs, and oil terminals with storage facilities connected with export terminals or refineries by pipelines. Oil wells produce crude oil of different varieties from different reservoirs into a GC, where this oil is collected and separated from associated water and gas. Crude blending, not necessarily deliberate, starts at the GC level as crude oil properties from various wells in the same field may not be consistent and uniform.

Oil is then transported to the Process Terminals, where different varieties of crude oils are received and processed to adjust their quality parameters (BS&W, API, sulfur contents, etc.). If necessary available crude oils are blended in the right proportions to achieve the target quality specifications set for the local refineries or export to foreign countries. Process terminals can facilitate the blending of varying grades of crude oils "In Tanks" or "In-Line" depending on the available facilities and requirements. In-tank mixing takes time, requires large storage capacities, and is not flexible in batch sizing. In-line blending does not have these disadvantages. High oil density, viscosity, and pour point of some oils can cause transportation problems through the pipeline and increase horsepower requirements. The problem can also be solved by inline blending the problematic crude with light oil or an equivalent hydrocarbon.

Figure 10.12 presents a real-life example of different hydrocarbons produced by four hypothetical oilfields named Alpha, Beta, Gamma, and Delta. Alpha and Beta are two light and sweet crude oil-bearing fields. Alpha is the largest oilfield that produces 250,000 BOPD of light and sweet crude oil of 34° API and sulfur content of 0.28%. Beta provides 100,000 BOPD of 50° API condensate with 0.05% sulfur. Gamma and delta are medium sour and heavy sour crude oilfields, respectively. Gamma produces 70,000 BOPD of 28° degrees API crude, which has a sulfur content of 0.6%. Delta provides a heavy oil of 21° API and sulfur content of 1.4%.

The refinery is built to process light and sweet crude oil. The producing company has committed to deliver the crude oil 35° API and sulfur content of 0.4% to the refinery or better. The producing company has several options. Only two options that demonstrate the benefit of crude oil blending on its price before the refinery are discussed below.

Option A: Transfer the "over-spec" (specifications are much better than required) crude/hydrocarbons of oilfield Alpha and Beta to the refinery separately. Each of these streams is light and sweet, so no problem for the refinery. This option is commercially not attractive to the producing company because it misses utilizing the crude oil streams of Gamma and delta oilfields.

Option B: Blend the streams from three or four oilfields to check if the producing company can achieve the target quality specifications. This exercise demonstrates that medium-sour and heavy-sour crude oils can be

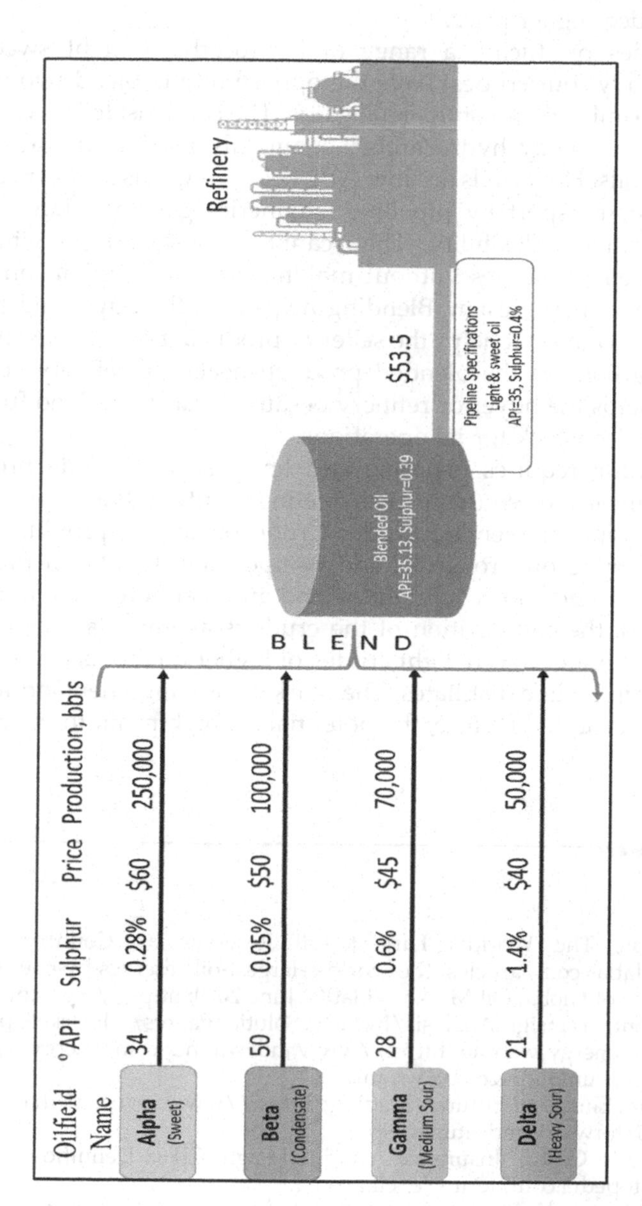

FIGURE 10.12
Benefit of the Crude Oil Blending: An Example.

blended with the light and sweet hydrocarbons from Alpha and Beta oilfields and still be within the refinery's quality requirements. By adopting this approach, the producing company can sell its medium and heavy sour crudes at a much higher price.

Oil companies producing a range of hydrocarbons (light sweet condensates to heavy sour crudes) have the opportunity to blend two or more streams to maximize their commercial gains. The condensate is a very light, high API, low viscosity hydrocarbon, commonly used to dilute heavier, high viscosity oils. Heavy oils are low API, heavy, high viscosity crudes that are difficult to transport by pipelines. Refineries generally have a fixed range of operational flexibility. They cannot process crudes with wide variation in their properties without making time and capital-consuming changes in the refinery design. Blending helps both the buyer and seller to achieve their objectives. It helps the seller or producing company sell lower-grade hydrocarbons as the blended product meets the refinery specifications. It also helps the buyer or refinery because it has to look no further to find a suitable feedstock for its operations.

Crude blending requires planning and knowledge of crude properties and their volumes and would be available in the future. It also requires the target specification of blended crude. Crude oil assays provide this information on crude oil properties and composition. Crude oil properties manifest a linear behavior when mixed and are, therefore, easy to predict.

For refineries, the composition of the crude is as important as its price. Some refineries prefer to buy light crudes of higher quality because of their high yield of high-value distillates. The market has a high demand for these refined products and, therefore, the potential for high profit to the refinery.

Notes

1 World Atlas, The World's Largest Oil Reserves by Country; https://www.worldatlas.com/articles/the-world-s-largest-oil-reserves-by-country.html
2 The Structure of Global Oil Markets, Platt's, June 2010; https://www.platts.com/IM.Platts.Content/InsightAnalysis/IndustrySolutionPapers/oilmarkets.pdf
3 index mundi energy website; https://www.indexmundi.com/energy/?product=oil&graph=consumption&display=rank
4 Investopedia, Guide to Futures Trading; https://www.investopedia.com/ask/answers/06/forwardsandfutures.asp
5 Investopedia, Cost, Insurance and Freight-CIF Definition https://www.investopedia.com/terms/c/cif.asp
6 AU Online, Aurora University; "A guide to types of Market structures?" https://online.aurora.edu/types-of-market-structures/
7 US Energy Information Administration, Refinitiv https://www.eia.gov/finance/markets/crudeoil/spot_prices.php

8 Wikipedia The Free Encyclopedia, Anglo-Persian Oil Company; https://en.wikipedia.org/wiki/Anglo-Persian_Oil_Company#Consortium
9 Organization of Petroleum Exporting Countries, OPEC Statute; https://www.opec.org/opec_web/en/publications/345.htm
10 Wikipedia The Free Encyclopedia, OPEC; https://en.wikipedia.org/wiki/OPEC
11 Organization of Petroleum Exporting Countries, OPEC Statute; https://www.opec.org/opec_web/en/publications/345.htm
12 OPEC Share of Crude Oil reserves; https://www.opec.org/opec_web/en/data_graphs/330.htm
13 Wikipedia The Free Encyclopedia, List of countries by oil production; https://en.wikipedia.org/wiki/List_of_countries_by_oil_production
14 Encyclopedia Britannica, Crude Oil; https://www.britannica.com/science/crude-oil
15 National Programme on Technology Enhanced Learning (NPTEL) website; https://nptel.ac.in/courses/103107082/module6/lecture1/lecture1.pdf
16 Wikipedia The Free Encyclopedia, "Helium production in the United States"; https://en.wikipedia.org/wiki/Helium_production_in_the_United_States
17 Wikipedia The Free Encyclopedia, "Baumé scale"; https://en.wikipedia.org/wiki/Baumé_scale
18 Engineering ToolBox, (2007). API Gravity. [online] Available at: https://www.engineeringtoolbox.com/api-gravity-d_1212.html [Accessed 20/08/2020]
19 Organization of Petroleum Exporting Countries, Composition of the new OPEC Reference Basket; https://www.opec.org/opec_web/en/press_room/1026.htm
20 US Energy Information Administration, Crude oils have different quality characteristics; https://www.eia.gov/todayinenergy/detail.php?id=7110
21 RBN Energy website; https://rbnenergy.com/it-aint-heavy-its-my-maya-impact-of-changes-to-the-mexican-heavy-crude-benchmark
22 Wikipedia The Free Encyclopedia; https://en.wikipedia.org/wiki/Price_of_oil
23 US Securities and Exchange Commission website, What's in a barrel of crude oil?; https://www.sec.gov/Archives/edgar/data/1035002/000119312513015343/g467558467558z0006.jpg

Acronyms and Abbreviations

ADB	Asian Development Bank
AfDB	African Development Bank
AHM	Assisted History Matching
AL	Artificial Lift
API	American Petroleum Institute
APOC	Anglo-Persian Oil Company
BAPSCO	Bahrain Petroleum Company
BHP	Bottom Hole Pressure
BOD	Board of Directors
BP	British Petroleum
BPP	Bubble Point Pressure
BSC	Balanced Score Card
BS&W	Bottom Sediment and Water
BTEX	Benzene, Toluene, Ethylbenzene, and Xylene
BTU	British Thermal Units
CAPEX	Capital Expenditure
CCE	Constant Composition Expansion
CDP	Conceptual Development Plan
CEO	Chief Executive Officer
CIF	Cost, Insurance, and Freight
CNOOC	China National Offshore Oil Corporation
C-O	Carbon-Oxygen Log
CPR	Cardio Pulmonary Resuscitation
CSR	Corporate Social Responsibility
CSS	Cyclic Steam Stimulation
CWE	Collaborative Working Environment
DCF	Discounted Cash Flow
DART	Downfall Air Receiver Technology
DNA	Deoxyribose Nucleic Acid
DP	Dynamic Positioning
DP-DP	Dual Porosity Dual Permeability
DCA	Decline Curve Analysis
DE	Drilling Efficiency
EBRD	European Bank for Reconstruction and Development
EEZ	Exclusive Economic Zone

EI	Exploitation Index
EOR	Enhanced Oil Recovery
EPF	Early Production Facility
ERM	Enterprise Risk Management
ESP	Electrical Submersible Pump
EUR	Estimated Ultimate Recovery
EURF	Estimated Ultimate Recovery Factor
EW	Effluent Water
EWD	Effluent Water Disposal
E&P	Exploration and Production
FDP	Field Development Plan
FDI	Foreign Direct Investments
FEED	Front End Engineering Design
FEL	Front End Loading
FOB	Free on Board
FPSO	Floating Production, Storage, & Offloading
FVF	Formation Volume Factor
GIS	Geographic Information System
GC	Gathering Center
GDP	Gross Domestic Product
GL	Gas Lift
GOC	Gas Oil Contact
GOR	Producing Gas Oil Ratio
GRC	Governance, Risk, and Compliance
GTL	Gas to Liquid
GU	Gas Utilization
G&G	Geology and Geophysics
HAP	Hazardous Air Pollutant
HP	High Pressure
HPAM	Hydrolyzed Polyacrylamide
HRM	Human Resource Management
HSE	Health, Safety, and Environment
HSSE	Health, Safety, Security, and Environment
IADB	Inter-American Development Bank
IAM	Integrated Asset Models
ICD	Inflow Control Device
ICV	Inflow Control Valve
IDP	Initial Development Plan
IEO	International Energy Outlook
IEEE	Institute of Electrical and Electronics Engineers
II	Injectivity Index
IIoT	Industrial Internet of Things
IMF	International Monetary Fund
INOC	Iraq National Oil Company
IOC	International Oil Company

IPDS	International Petroleum Data Standards
IRR	Internal Rate of Return
ISO	International Standards Organization
IT	Information Technology
JAPEX	Japan Exploration Petroleum Company
JV	Joint Venture
KMG	KazMunayGas (National Operator of Kazakhstan)
KNPC	Kuwait National Petroleum Company
KOC	Kuwait Oil Company
KPI	Key Performance Indicator
LP	Low Pressure
LNG	Liquefied Natural Gas
LPG	Liquefied Petroleum Gas
MBE	Material Balance Equation
MCF	Thousand Cubic Feet
MDT	Modular Dynamic Test
MIC	Microbially-Influenced Corrosion
MMP	Minimum Miscible Pressure
MP	Middle Pressure
MWD	Measurement While Drilling
NCW	Non-Conventional Wells
NDT	Non-destructive Testing
NIOC	National Iranian Oil Company
NIS	Naftna industrija Srbije (Petroleum Industry of Serbia)
NNPC	Nigerian National Petroleum Corporation
NOC	National Oil Company
NORM	Naturally Occurring Radioactive Material
NPV	Net Present Value
NTG	Net to Gross
OECD	Organization for Economic Co-operation and Development
ONGC	Oil & Natural Gas Corporation Ltd.
OOIP	Original Oil in Place
OPEC	Organization of Petroleum Exporting Countries
OPEX	Operating Expenditure
ORB	OPEC Reference Basket
OWC	Oil Water Contact
OWIU	Oil Well Inventory Utilization
PDO	Petroleum Development Oman
PDVSA	Petronas Petróleos de Venezuela, S.A.
Pemex	Petróleos Mexicanos
Pertamina	Perusahaan Pertambangan Minyak dan Gas Bumi Negara
PEST	Political, Environmental, Social, and Technological
Petrobras	Petróleo Brasileiro S.A.

PetroVietnam Vietnam Oil and Gas Group

PI	Productivity Index
PI	Profitability Index
PIR	Profit-to-Investment Ratio
PLC	Programmable Logic Controller
PLT	Production Logging Tool
PNL	Pulsed Neutron Lifetime Log
POC	Private Oil Company
PPDM	Professional Petroleum Data Management
PRODML	Production Markup Language
PSA	Production Sharing Agreements
PSC	Production Sharing Contracts
PVT	Pressure Volume Temperature
PWD	Produced Water Disposal
PWU	Produced Water Utilization
P&A	Plug and Abandon
QC	Quality Check
QGPC	Qatar General Petroleum Corporation
RCA	Routine Core Analysis
RD	Recovery Discriminator
RESQML	Reservoir (Characterization, Earth, and Reservoir) Models Markup Language
RFT	Repeat Formation Test
RHI	Reservoir Health Indicator
RNA	Ribo Nucleic Acid
RPM	Relative Permeability Modification
R-P Ratio	Reserves to Production Ratio
RRR	Reserve Replacement Ratio
RST	Reservoir Saturation Tool
RTU	Remote Terminal Unit
R&D	Research and Development
R&T	Research and Technology
SAGD	Steam Assisted Gravity Drainage
SASREF	Saudi Aramco's refinery with Shell
SBHP	Static Bottom Hole Pressure
SCADA	Supervisory Control and Data Acquisition
SCAL	Special Core Analysis
SEO	Search Engine Optimization
SIG	Special Interest Groups
SMART	Specific, Measurable, Attainable, Relevant, and Time-bound
SME	Subject Matter Expert
SOCAR	State Oil Company of Azerbaijan Republic
SPE	Society of Petroleum Engineers

SRA	Sulfate Reducing Archaea
SRB	Sulfate Reducing Bacteria
SRM	Sulfate Reducing Microbes
Sudapet	Sudan National Petroleum Corporation
Sonatrach	Société Nationale pour la Recherche, la Production, le Transport, la Transformation, et la Commercialisation des Hydrocarbures
SWOT	Strengths, Weaknesses, Opportunities, and Threats
TDS	Total Dissolved Solids
TDT	Thermal Decay Time
TEG	Tri Ethylene Glycol
THAI	Toe to Heel Air Injection
TSS	Total Suspended Solids
TVM	Time Value of Money
UAVs	Unmanned Aerial Vehicles
UK	United Kingdom
UNCLOS	United Nations Convention on the Law of the Sea
US	United States
UTF 8	Unicode Transformation Format (8-bit)
UTM	Universal Transverse Mercator System of Coordinates
VOC	Volatile Organic Compounds
VRR	Voidage Replacement Ratio
VSE	Volumetric Sweep Efficiency
WC	Water Cut
WHI	Well Health Indicator
WITSML	Well-site Information Transfer Standard Markup Language
WOR Qo	Oil Rate from Rigless and Rig Work Over Activities
WPC	World Petroleum Council
WQ	Well Quality
WTC	World Trade Center
WTI	West Texas Intermediate
YPF	Yacimientos Petrolíferos Fiscales

Notations

Ø	Porosity
K	Permeability
S_o	Oil Saturation
S_g	Gas Saturation
S_w	Water Saturation
N_c	Capillary Number
σ	Inter Facial Tension
μ	Fluid Viscosity
S_{or}	Residual Oil Saturation
S_{org}	Residual Oil Saturation to Gas
R_{si}	Initial Solution Gas Oil Ratio
S_{gc}	Critical Gas Saturation
S_{orw}	Residual Oil Saturation to Water
S_{wirr}	Irreducible Water Saturation
(k/μ)	Mobility of a fluid
(k_g/μ_g)	Mobility of gas
P90	Cumulative/Exceedance probability
P50	Cumulative/Exceedance probability
P10	Cumulative/Exceedance probability
Pc	Capillary Pressure
E_v	Volumetric Sweep Efficiency
E_A	Areal Sweep Efficiency
E_I	Vertical Sweep Efficiency
M	Mobility Ratio

Index

E

F

G

H

I

K

S

T

U

Printed in the United States
by Baker & Taylor Publisher Services